The IMA Volumes in Mathematics and its Applications

Volume 142

Series Editors
Douglas N. Arnold Arnd Scheel

Institute for Mathematics and its Applications (IMA)

The **Institute for Mathematics and its Applications** was established by a grant from the National Science Foundation to the University of Minnesota in 1982. The primary mission of the IMA is to foster research of a truly interdisciplinary nature, establishing links between mathematics of the highest caliber and important scientific and technological problems from other disciplines and industries. To this end, the IMA organizes a wide variety of programs, ranging from short intense workshops in areas of exceptional interest and opportunity to extensive thematic programs lasting a year. IMA Volumes are used to communicate results of these programs that we believe are of particular value to the broader scientific community.

The full list of IMA books can be found at the Web site of the Institute for Mathematics and its Applications:

http://www.ima.umn.edu/springer/volumes.html

Douglas N. Arnold, Director of the IMA

* * * * * * * * * *

IMA ANNUAL PROGRAMS

1982–1983	Statistical and Continuum Approaches to Phase Transition
1983–1984	Mathematical Models for the Economics of Decentralized Resource Allocation
1984–1985	Continuum Physics and Partial Differential Equations
1985–1986	Stochastic Differential Equations and Their Applications
1986–1987	Scientific Computation
1987–1988	Applied Combinatorics
1988–1989	Nonlinear Waves
1989–1990	Dynamical Systems and Their Applications
1990–1991	Phase Transitions and Free Boundaries
1991–1992	Applied Linear Algebra
1992–1993	Control Theory and its Applications
1993–1994	Emerging Applications of Probability
1994–1995	Waves and Scattering
1995–1996	Mathematical Methods in Material Science
1996–1997	Mathematics of High Performance Computing
1997–1998	Emerging Applications of Dynamical Systems
1998–1999	Mathematics in Biology

Continued at the back

Douglas N. Arnold Pavel B. Bochev
Richard B. Lehoucq Roy A. Nicolaides
Mikhail Shashkov
Editors

Compatible Spatial Discretizations

Springer

Douglas N. Arnold
Institute for Mathematics and its Applications
University of Minnesota
Minneapolis, MN 55455
USA
http://www.ima.umn.edu/~arnold/

Pavel B. Bochev
Sandia National Laboratories
Computational Mathematics and Algorithms Department
Alburquerque, NM 87185-1110
USA
http://math.uta.edu/~bochev/

Richard B. Lehoucq
Sandia National Laboratories
Computational Mathematics and Algorithms Department
Alburquerque, NM 87185-1110
USA
http://www.cs.sandia.gov/~rlehoucq

Roy A. Nicolaides
Department of Mathematical Sciences
Carnegie Mellon University
Pittsburgh, PA 15213-3890
USA
http://www.math.cmu.edu/people/fac/nicolaides.html

Mikhail Shashkov
Theoretical Division
Los Alamos National Laboratory
Los Alamos, NM 87545
USA
http://cnls.lanl.gov/~shashkov

Mathematics Subject Classification (2000): 65N06, 65N12, 65N30, 65N35, 65D25, 65D30, 58A10, 58A12, 14F40, 58A15, 58A14, 55U10, 55U15, 53A45

Library of Congress Control Number: 2006921649

ISBN-10: 0-387-30916-0
ISBN-13: 978-0387-30916-3

Printed on acid-free paper.

Printed in the United States of America. (MVY)

Camera-ready copy provided by the IMA.

9 8 7 6 5 4 3 2 1

springer.com

FOREWORD

This IMA Volume in Mathematics and its Applications

COMPATIBLE SPATIAL DISCRETIZATIONS

contains papers presented at a highly successful IMA Hot Topics Workshop: Compatible Spatial Discretizations for Partial Differential Equations. The event which was held on May 11-15, 2004 was organized by Douglas N. Arnold (IMA, University of Minnesota), Pavel B. Bochev (Computational Mathematics and Algorithms Department, Sandia National Laboratories), Richard B. Lehoucq (Computational Mathematics and Algorithms Department, Sandia National Laboratories), Roy A. Nicolaides (Department of Mathematical Sciences, Carnegie-Mellon University), and Mikhail Shashkov (MS-B284, Group T-7, Theoretical Division, Los Alamos National Laboratory). We are grateful to all participants and organizers for making this a very productive and stimulating meeting, and we would like to thank the organizers for their role in editing this proceeding.

We take this opportunity to thank the National Science Foundation for its support of the IMA and the Department of Energy for providing additional funds to support this workshop.

Series Editors

Douglas N. Arnold, Director of the IMA

Arnd Scheel, Deputy Director of the IMA

PREFACE

In May 2004 over 80 mathematicians and engineers gathered in Minneapolis for a "hot topics" IMA workshop to talk, argue and conjecture about compatibility of spatial discretizations for Partial Differential Equations. We define *compatible*, or *mimetic*, spatial discretizations as those that inherit or mimic fundamental properties of the PDE such as topology, conservation, symmetries, and positivity structures and maximum principles.

The timing and place for this workshop were not incidental. PDEs are one of the principal modeling tools in science and engineering and their numerical solution is the workhorse of computational science. However, historically, numerical methods for PDEs such as finite differences (FD), finite volumes (FV) and finite elements (FE) evolved separately and until recently, in relative isolation from each other. This situation started to change about two decades ago when researchers working in these areas began to realize that robust and accurate discrete models share more than just a passing resemblance to each other. While FD, FV and FE methods have all developed specific approaches to compatibility, their successful discrete models were found to operate in what essentially came down to a discrete vector calculus structure replete with algebraic versions of the vector calculus identities and theorems.

Because of their more explicit reliance on grid topology, FD and FV methods recognized the role of geometry earlier than FE methods. For FEM compatibility criteria evolved from variational theories and assumed the form of powerful, but non-constructive inf-sup conditions. This changed in the 80s with the pioneering work of Bossavit who brought to light fundamental connections between the DeRham complex and compatible FEs for the Maxwell's equations. Consequently, research in applications of differential geometry, exterior calculus and algebraic topology to numerical PDEs intensified. This research led to important advances in understanding of spatial compatibility and connections between different compatible discrete models. Among the payoffs from this work were development of new stable FE models for linear elasticity and rigorous convergence analysis of mimetic FD by variational tools.

Thus, the organizers felt that the time was ripe for the researchers working in this field to get together and compare notes. The relevance of the topic and its impact on computational sciences helped to attract attendees from a broad cross-section of the community. The stature of IMA, its tradition and experience in organizing small focused workshops and its dedicated staff made the Institute a natural venue for this gathering.

This volume, co-edited by the workshop organizers, is representative of the topics discussed during the meeting. The papers, based on a subset of the plenary talks, offer the reader a snapshot of the current trends and developments in compatible and mimetic spatial discretizations. Abstracts and presentation slides from the workshop can be accessed at `http://www.ima.umn.edu/talks/workshops/5-11-15.2004/`.

While many of the contributions in this volume address questions regarding spatial compatibility, each paper offers a unique perspective and insight into specific techniques and approaches. Arnold *et al* focus on a *homological approach* to stability of mixed FE which, in the recent years, has greatly contributed to the understanding of mixed methods and the development of stable methods for previously intractable problems. The first part of their contribution deals with two polynomial versions of the DeRham complex. One complex involves homogeneous polynomial spaces of decreasing degree and the second is obtained with the help of the Kozsul differential. The two polynomial complexes contain generalizations of well-known finite element pairs such as Raviart-Thomas, BDM and Nedelec elements of first and second kinds. Then, they proceed to show how to use polynomial sub-complexes and commuting diagrams to obtain stability of mixed methods. The second part of Arnold *et al* deals with application of the homological approach to *mixed linear elasticity*. They show that a differential complex relevant to mixed linear elasticity can be obtained from the DeRham complex. An analogous construction is used to develop a discrete elasticity complex from a polynomial DeRham complex and results in new stable finite element spaces for mixed linear elasticity.

The paper by Boffi examines compatibility issues that arise in mixed finite element approximations to *eigenvalue problems*. A surprising counterexample shows that the classical Brezzi theory, which provides sufficient compatibility conditions for mixed methods, is not enough to guarantee the absence of spurious modes in mixed approximations of eigenvalue problems. After theoretical explanation and practical demonstration of this behavior, Boffi proceeds to develop sufficient and necessary conditions for correct mixed eigenmode discretizations and then gives several examples for possible application of the eigenvalue compatibility theory. Among other things, Boffi shows that good approximation of evolution problems in mixed form is contingent upon spectral convergence of the related eigenvalue problem, that is, it is also a subject to compatibility conditions beyond that of the classical Brezzi theory.

Application of algebraic topology to compatible discretizations is the central topic of Bochev and Hyman. They use two basic mappings between differential forms and cochains to define a framework that supports mutually consistent operations of differentiation and integration. This is accomplished by a set of *natural* operations that induce a set of *derived* discrete operations. The resulting framework has a combinatorial Stokes theorem and preserves the invariants of the De Rham cohomology groups.

The key concept of their approach is the natural inner product on cochains. This inner product is sufficient to generate a combinatorial Hodge theory on cochains but avoids complications attendant in the construction of efficient discrete Hodge-star operators. The framework provides an abstraction that includes examples of mixed FE, mimetic FD and FV methods. The paper also describes how these methods result from a choice of a reconstruction operator and explains when they are equivalent.

An interesting perspective on compatibility and how it affects *Discontinuous Galerkin* (DG) methods is presented in the paper by Barth. Because of a number of valuable computational properties, DG methods are attracting significant attention. Their origins for elliptic problems can be traced to interior penalty methods and so they are not compatible in the sense of mixed finite element methods. Using the Maxwell's equations and ideal MHD, Barth draws attention to the different roles played by their *involutions* for the formulation of energy-stable DG methods. The Maxwell's equations are naturally expressed in symmetric form, while symmetrization of MHD utilizes the involution as a necessary ingredient. This leads to fundamental differences in energy stability of the associated DG methods. Barth shows that imposing continuity of the magnetic flux at interelement boundaries is beneficial for energy stability of DG for MHD, while, somewhat counterintuitively, this condition is not required for DG discretizations of the Maxwells equations.

A *co-volume* approach to compatible discretizations is discussed by Trapp and Nicolaides. Building upon a solid body of work in classical FV methods, they use Voronoi-Delaunay grids to discretize differential forms. Their approach exploits the Voronoi-Delaunay grid complex to obtain a primal and a dual set of discrete forms connected by a local discrete Hodge operator. This leads to algebraic PDE models with particularly simple and attractive structure and a discrete setting where both the primal and the dual discrete differential operators have *local* stencils. In addition, the primal and dual operators are adjoint with respect to a co-volume inner product, which immediately gives rise to a discrete Hodge decomposition. To illustrate the co-volume approach, Trapp and Nicolaides develop compatible discretizations for two instances of the Hodge Laplacian in three-dimensions.

The two contributions by Wheeler and Yotov, and by Aavatsmark *et al* examine compatible methods for problems arising in *reservoir simulation* and *porous media* flows. The task of devising compatible methods for these applications is greatly complicated by the need to reconcile mathematical compatibility conditions with grid structure imposed by *geological features* such as layering, faults and crossbeddings. As a result, methods for geophysical applications have traditionally favored quadrilateral and hexahedral grids, which can cause some problems in the reconstruction of vector fields from normal components. In addition, permeability tensor in reservoir models often has strong anisotropy and/or discontinuities along geological features. The two papers offer two alternative approaches that lead to cell-centered, locally conservative schemes. Aavatsmark *et al* adopt a Finite Volume approach based on the concept of multipoint flux approximation (MPFA). In this approach, fluxes are defined by using linear reconstruction of the potential subject to specific flux and potential continuity conditions. In contrast, Wheeler and Yotov start from a mixed variational formulation and then design a quadrature rule that allows for a local elimination of the velocities and results in a symmetric and positive definite cell-centered potential matrix. The result is a method that is related to MPFA and has a variational formulation. This allows them to leverage approximation theory from mixed methods and prove second order convergence of the scalar at the cell-centers.

A hallmark of many compatible discretizations, such as Raviart-Thomas elements, Nedelec elements or mimetic Finite Differences, is the use of normal or tangential vector components. This enables discrete versions of the divergence and the Stokes theorems but poses problems when vector fields are needed to compute *vector derived quantities* such as kinetic energy or advective terms. The reconstructed fields may fail to provide local conservation of the kinetic energy and the momentum. Reconstruction of vector fields from dispersed data is the subject of the contribution by Perot *et al.* Their paper discusses relationship between three low order reconstruction operators. Two of these operators are related to mimetic finite difference and finite element methods, respectively. The third one is a new reconstruction approach proposed by the authors. Perot *et al* discuss how explicit reconstruction can be used to define discrete Hodge star operators. The paper then focuses on reconstruction approaches that can provide *local conservation* for vector derived quantities such as momentum and kinetic energy.

Software frameworks and computational experiments for compatible methods are communicated in the papers by Demkowicz and Kurtz, and by White *et al.* Both papers consider compatible methods for the Maxwell's equations. White *et al* describe an extensible, object-oriented C++ framework that closely mimics the structure of differential form calculus. The emphasis is on *high-order* finite element basis functions that form a discrete De Rham complex and have the relevant commuting diagram properties. As a result, any electromagnetics problem that can be cast in the language of differential forms can be easily modeled by their framework. The flexibility of the framework is illustrated by solving resonant cavity, wave propagation and eddy current problems. Demkowicz and Kurtz develop an *hp*-adaptive implementation of a coupled finite element/infinite element approximation for *exterior wave propagation* problems. The novel aspect of the paper is a family of infinite elements that satisfies an exact sequence property. The elements in the new sequence are obtained by multiplying basis functions from a standard polynomial De Rham complex by an exponential factor that comes from the far-field pattern. The exactness is with respect to similarly modified differential operators. A series of experiments confirms stability of the coupling and exponential rate of convergence obtained by automatic *hp*-adaptivity.

In closing, the editors want to thank the authors for contributing to this volume and their cooperation in the editorial process. Special thanks are also due to Patricia V. Brick and Dzung N. Nguyen for the excellent coordination of the production schedule and assistance in the final preparation of the papers for the publisher. Dr. C. Romine, formerly of the DOE'S MICS Applied Mathematics Research program, offered enthusiastic support and encouragement during the preparation of the workshop. His help is greatly appreciated. Funding for the workshop was provided by the DOE Office of Science's Advanced Scientific Computing Research (ASCR) Applied Mathematics Research Program.

Douglas N. Arnold
Institute for Mathematics and its Applications
University of Minnesota

Pavel B. Bochev
Computational Mathematics and Algorithms Department
Sandia National Laboratories

Richard B. Lehoucq
Computational Mathematics and Algorithms Department
Sandia National Laboratories

Roy A. Nicolaides
Department of Mathematical Sciences
Carnegie Mellon University

Mikhail Shashkov
Theoretical Division
Los Alamos National Laboratory

CONTENTS

NUMERICAL CONVERGENCE OF THE MPFA O-METHOD FOR GENERAL QUADRILATERAL GRIDS IN TWO AND THREE DIMENSIONS

IVAR AAVATSMARK*, GEIR TERJE EIGESTAD†, AND
RUNHILD AAE KLAUSEN‡

Abstract. This paper presents the MPFA O-method for quadrilateral grids, and gives convergence rates for the potential and the normal velocities. The convergence rates are estimated from numerical experiments. If the potential is in $H^{1+\alpha}$, $\alpha > 0$, the found L^2 convergence order on rough grids in physical space is $\min\{2, 2\alpha\}$ for the potential and $\min\{1, \alpha\}$ for the normal velocities. For smooth grids the convergence order for the normal velocities increases to $\min\{2, \alpha\}$. The O-method is exact for uniform flow on rough grids. This also holds in three dimensions, where the cells may have nonplanar surfaces.

Key words. Control-volume discretization, anisotropy, inhomogeneity, convergence.

AMS(MOS) subject classifications. 65M06, 76S05, 35R05.

1. Introduction. We consider a control-volume discretization of the model equation

$$\operatorname{div} \boldsymbol{q} = Q, \qquad \boldsymbol{q} = -\boldsymbol{K} \operatorname{grad} u \tag{1.1}$$

on a quadrilateral grid. The conductivity $\boldsymbol{K}$ is required to be symmetric and positive definite.

Our applications are solutions of multiphase flow equations in reservoir simulation. These equations contain an elliptic operator similar to the left-hand side of (1.1), and this motivates our study. The multiphase flow equations in reservoir simulation have properties which constrain the choice of grid and discretization technique used for the elliptic operator. By reformulation of the flow equations, a coupled set of parabolic equations appear. However, one of these equations (the pressure equation) has an elliptic character, while the other equations (the saturation equations) have hyperbolic character with a strongly nonlinear convective term. Phase transitions which are strongly pressure dependent, may occur.

Due to the hyperbolicity and the strong nonlinearity of the saturation equations, we require that the discretization scheme should be locally conservative. Also, since the phase transitions are pressure dependent, we require that the pressure is evaluated at the same point as the saturations.

*Center for Integrated Petroleum Research, University of Bergen, NO-5020 Bergen, Norway (ivar.aavatsmark@cipr.uib.no).

†(geirte@mi.uib.no).

‡(runhildk@ifi.uio.no).

This motivates the use of a control-volume scheme for (1.1), with evaluation of the dependent variable u at the center of the cells.

Stability for nonlinear hyperbolic equations is normally achieved by requiring that the chosen scheme is monotone. In reservoir simulation, stability is accomplished by upstream weighting of the phase flow. In a fully implicit scheme for the flow equations, a simple upstream weighting can only be done if the method for the elliptic operator in (1.1) yields the flux at the edges as an explicit function of the potential u at some neighboring cell centers.

The grids used in reservoir simulation are normally quadrilateral grids with an aspect ratio which strongly deviates from unity. To avoid the difficulties of upscaling, the grid layering is normally determined by the geological layering. This often yields almost rectangular grids with homogeneous cell properties. At faults or in near-well regions, grids with a more complex geometry may be preferred.

In reservoir simulation the conductivity $\boldsymbol{K}$ of (1.1) is given by the absolute permeability. It is a tensor which often has a strong anisotropy. Because of the symmetry of the tensor, the principal directions are orthogonal. The principal directions are often aligned with, and normal to, the grid layering. For layers with varying thickness, this is only approximately fulfilled. If the layers contain crossbeddings, the principal directions of the tensor may be arbitrary.

The absolute permeability may vary strongly in reservoir simulation. Since the potential node should be located at the cell centers, it is therefore important that the discrete resistance between two nodes honors the strong heterogeneity. This means that for one-dimensional flow, the method should give a conductance equal to a harmonic average of the cell conductances.

In summary, we will describe a control-volume method for equation (1.1) which yields the flux at the edges as an explicit function of the potential at the cell centers. The conductivity should be symmetric and positive definite, but its principal directions may be arbitrary compared to the grid directions. The discrete resistance between cell nodes must honor the heterogeneity. We will confine ourselves to quadrilateral grids.

One method with the above properties is the MPFA (Multipoint Flux Approximation) method. It can be applied to quadrilateral grids [1, 2, 4, 8, 18] and to unstructured grids [3, 5, 6, 7, 17], see [1] for a more complete bibliography.

In this paper we introduce the method in a new way which emphasizes the connection between anisotropy and grid skewness. Then we present convergence results for the method. These supplement the results of [9].

There are many variants of the MPFA method; in this paper we only discuss the method known as the *O-method*.

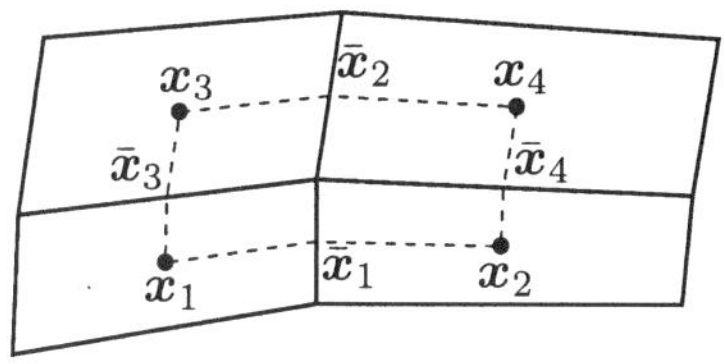

FIG. 1. *Interaction volume (bounded by the dashed lines).*

2. The MPFA O-method. In this section we derive the equations for the MPFA O-method in two dimensions. Consider the four quadrilateral cells with a common vertex in Fig. 1. The cells have cell centers $\boldsymbol{x}_k$, and the edges have midpoints $\bar{\boldsymbol{x}}_i$. The points are enumerated locally as shown in the figure. Between the cell centers and the midpoints of the edges we draw lines (shown as dashed lines in the figure). These lines bound an area around each vertex which is called an *interaction volume* (also referred to as an interaction region in previous papers). Hence, the interaction volume in the figure is the polygon with corners $\boldsymbol{x}_1\bar{\boldsymbol{x}}_1\boldsymbol{x}_2\bar{\boldsymbol{x}}_4\boldsymbol{x}_4\bar{\boldsymbol{x}}_2\boldsymbol{x}_3\bar{\boldsymbol{x}}_3$.

Within the interaction volume there are four half edges. Below, we will show how to determine the flux through these half edges from the interaction between the four cells. When the fluxes through the four half edges in an interaction volume around a vertex are determined, we may repeat the procedure for the interaction volumes of the other vertices. In this way, the flux through all the half edges in a grid will be determined. When the fluxes through the two half edges of an entire edge are known, we may add them to get an expression for the flux through the entire edge. An assembly procedure may then be performed to construct a system of difference equations corresponding to Eq. (1.1).

This procedure also holds for the half edges at the boundary of a domain, if the boundary conditions are given as homogeneous Neumann conditions. Outside the real cells we can put a strip of artificial cells with vanishing conductivity. The same procedure as described above for the interaction volumes around the vertices at the boundary then gives the flux through the half edges separating the real boundary cells. More general boundary conditions are discussed in [9].

We now show how the fluxes through the four half edges in an interaction volume may be determined. In each of the four cells of the interaction volume, the potential u is expressed as a linear function. The value of the potential in each cell center determines one of the coefficients in each cell for these linear functions. The linear function determines the flux through the half edges of the cell and the potential at the half edges. We require that the fluxes through the half edges in an interaction volume are continuous, and that the potentials at the midpoints of the edges are continuous. This yields eight equations for the determination of the unknown coefficients of the linear functions in the cells.

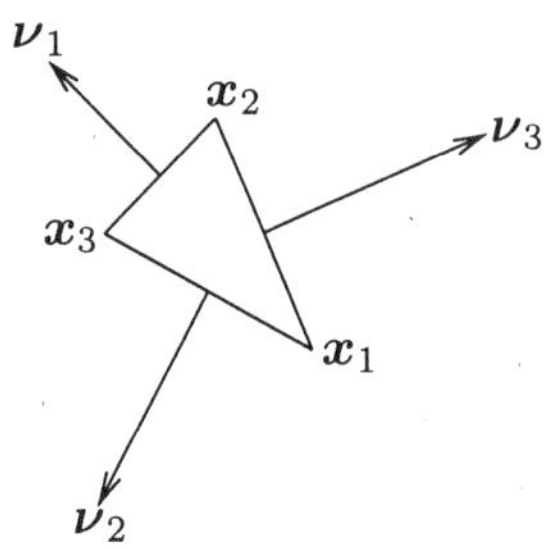

FIG. 2. *Triangle with edge normals* $\boldsymbol{\nu}_i$.

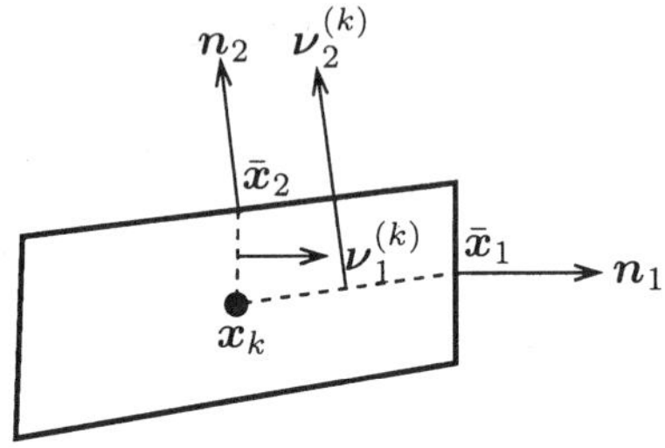

FIG. 3. *Normal vectors in cell k.*

Every linear function is decribed by three coefficients, but one of them is already determined throught the potential value at the cell center. All together there are therefore eight unknown coefficients for the linear functions. They are determined through the eight continuity equations. Note that the continuity principles used here, are exactly the same as the principles used to derive the classical two-point flux formula [1].

Every cell is shared among four interaction volumes. The linear functions for the potential in a cell, may vary from interaction volume to interaction volume. This does not cause any difficulties, since the linear functions are only used to determine an expression for the flux. In the resulting difference equations, only the potential at the cell centers appears.

For each interaction volume, the linear functions in each cell may be determined in the following way. On a triangle with corners $\boldsymbol{x}_i$, $i = 1, 2, 3$, any linear function may be described by

$$u(\boldsymbol{x}) = \sum_{i=1}^{3} u_i \phi_i(\boldsymbol{x}). \tag{2.1}$$

Here, u_i is the value of $u(\boldsymbol{x})$ at vertex i, and $\phi_i(\boldsymbol{x})$ is the linear basis function defined by $\phi_i(\boldsymbol{x}_j) = \delta_{i,j}$. The gradient is easily calculated to be

$$\operatorname{grad} \phi_i = -\frac{1}{2F} \boldsymbol{\nu}_i, \tag{2.2}$$

where F is the area of the triangle. Here, $\boldsymbol{\nu}_i$ is the outer normal vector of the edge lying opposite to vertex i, see Fig. 2. The length of $\boldsymbol{\nu}_i$ equals the length of the edge to which it is normal. For these normal vectors the following relation holds

$$\sum_{i=1}^{3} \boldsymbol{\nu}_i = \mathbf{0}. \tag{2.3}$$

Thus, the gradient expression of the potential in the triangle may be written

$$\operatorname{grad} u = -\frac{1}{2F} \sum_{i=1}^{3} u_i \boldsymbol{\nu}_i = -\frac{1}{2F} \big[(u_2 - u_1) \boldsymbol{\nu}_2 + (u_3 - u_1) \boldsymbol{\nu}_3 \big]. \tag{2.4}$$

Now consider the grid cell in Fig. 3. The grid cell has index k and cell center $\boldsymbol{x}_k$. Using local indices, the midpoints on the edges are denoted $\bar{\boldsymbol{x}}_1$ and $\bar{\boldsymbol{x}}_2$, and the associated normals on the connection lines between the cell center and the midpoints of the edges are denoted $\boldsymbol{\nu}_2^{(k)}$ and $\boldsymbol{\nu}_1^{(k)}$, see Fig. 3. Later, it will appear suitable to let the vectors $\boldsymbol{\nu}_i^{(k)}$ point in the direction of increasing global cell indices. In this cell we therefore reverse the direction of these vectors. Other locations of the points $\bar{\boldsymbol{x}}_1$ and $\bar{\boldsymbol{x}}_2$ on the edges are also allowed [8], but that will not be considered in this paper. Using the formula (2.4) on the triangle $\boldsymbol{x}_k\bar{\boldsymbol{x}}_1\bar{\boldsymbol{x}}_2$ yields

$$\operatorname{grad} u = \frac{1}{2F_k}\left[\boldsymbol{\nu}_1^{(k)}(\bar{u}_1 - u_k) + \boldsymbol{\nu}_2^{(k)}(\bar{u}_2 - u_k)\right], \tag{2.5}$$

where $\bar{u}_i = u(\bar{\boldsymbol{x}}_i)$, $i = 1, 2$, and $u_k = u(\boldsymbol{x}_k)$. Obviously, for Eq. (2.5) to be valid, the vectors $\boldsymbol{\nu}_1^{(k)}$ and $\boldsymbol{\nu}_2^{(k)}$ have to be linearly independent. Each of the edges can be associated with a global direction, defined through the unit normal $\boldsymbol{n}_i$. We will also let $\boldsymbol{n}_i$ point in the direction of increasing global cell indices. The flux through half edge i as seen form cell k is denoted $f_i^{(k)}$. The flux may now be determined from the gradient of the potential in the cell. For the fluxes in the cell in Fig. 3, the following expression appears

$$\begin{aligned}\begin{bmatrix} f_1^{(k)} \\ f_2^{(k)} \end{bmatrix} &= -\begin{bmatrix} \Gamma_1 \boldsymbol{n}_1^{\mathrm{T}} \\ \Gamma_2 \boldsymbol{n}_2^{\mathrm{T}} \end{bmatrix} \boldsymbol{K}_k \operatorname{grad} u \\ &= -\frac{1}{2F_k}\begin{bmatrix} \Gamma_1 \boldsymbol{n}_1^{\mathrm{T}} \\ \Gamma_2 \boldsymbol{n}_2^{\mathrm{T}} \end{bmatrix} \boldsymbol{K}_k \begin{bmatrix} \boldsymbol{\nu}_1^{(k)} & \boldsymbol{\nu}_2^{(k)} \end{bmatrix} \begin{bmatrix} \bar{u}_1 - u_k \\ \bar{u}_2 - u_k \end{bmatrix},\end{aligned} \tag{2.6}$$

where Γ_i is the length of half edge i. By defining the matrix

$$\begin{aligned}\boldsymbol{G}_k &= \frac{1}{2F_k}\begin{bmatrix} \Gamma_1 \boldsymbol{n}_1^{\mathrm{T}} \\ \Gamma_2 \boldsymbol{n}_2^{\mathrm{T}} \end{bmatrix} \boldsymbol{K}_k \begin{bmatrix} \boldsymbol{\nu}_1^{(k)} & \boldsymbol{\nu}_2^{(k)} \end{bmatrix} \\ &= \frac{1}{2F_k}\begin{bmatrix} \Gamma_1 \boldsymbol{n}_1^{\mathrm{T}} \boldsymbol{K}_k \boldsymbol{\nu}_1^{(k)} & \Gamma_1 \boldsymbol{n}_1^{\mathrm{T}} \boldsymbol{K}_k \boldsymbol{\nu}_2^{(k)} \\ \Gamma_2 \boldsymbol{n}_2^{\mathrm{T}} \boldsymbol{K}_k \boldsymbol{\nu}_1^{(k)} & \Gamma_2 \boldsymbol{n}_2^{\mathrm{T}} \boldsymbol{K}_k \boldsymbol{\nu}_2^{(k)} \end{bmatrix},\end{aligned} \tag{2.7}$$

Eq. (2.6) may be written in the form

$$\begin{bmatrix} f_1^{(k)} \\ f_2^{(k)} \end{bmatrix} = -\boldsymbol{G}_k \begin{bmatrix} \bar{u}_1 - u_k \\ \bar{u}_2 - u_k \end{bmatrix}. \tag{2.8}$$

Now consider the interaction volume in Fig. 4. Through the normal vectors introduced here, the matrix $\boldsymbol{G}_k$ is defined for all the four cells. Thus,

$$\begin{bmatrix} f_1^{(1)} \\ f_3^{(1)} \end{bmatrix} = -\boldsymbol{G}_1 \begin{bmatrix} \bar{u}_1 - u_1 \\ \bar{u}_3 - u_1 \end{bmatrix}, \qquad \begin{bmatrix} f_1^{(2)} \\ f_4^{(2)} \end{bmatrix} = -\boldsymbol{G}_2 \begin{bmatrix} u_2 - \bar{u}_1 \\ \bar{u}_4 - u_2 \end{bmatrix}, \tag{2.9}$$

$$\begin{bmatrix} f_2^{(3)} \\ f_3^{(3)} \end{bmatrix} = -\boldsymbol{G}_3 \begin{bmatrix} \bar{u}_2 - u_3 \\ u_3 - \bar{u}_3 \end{bmatrix}, \qquad \begin{bmatrix} f_2^{(4)} \\ f_4^{(4)} \end{bmatrix} = -\boldsymbol{G}_4 \begin{bmatrix} u_4 - \bar{u}_2 \\ u_4 - \bar{u}_4 \end{bmatrix}. \tag{2.10}$$

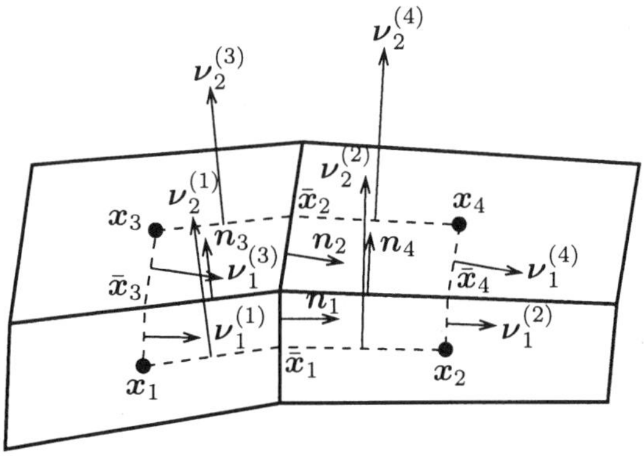

FIG. 4. *Normal vectors with local numbering in an interaction volume.*

Here, as before, $u_k = u(\boldsymbol{x}_k)$ and $\bar{u}_i = u(\bar{\boldsymbol{x}}_i)$, see Fig. 4. Compared to cell 1, we have reversed the directions of $\boldsymbol{\nu}_1^{(2)}$, $\boldsymbol{\nu}_2^{(3)}$, $\boldsymbol{\nu}_1^{(4)}$, and $\boldsymbol{\nu}_2^{(4)}$ (see Fig. 4). The differences $\bar{u}_1 - u_2$, $\bar{u}_3 - u_3$, $\bar{u}_2 - u_4$, and $\bar{u}_4 - u_4$ therefore appear in the expressions (2.9) and (2.10) with opposite sign.

The continuity conditions for the fluxes now yield

$$
\begin{aligned}
f_1 &= f_1^{(1)} = f_1^{(2)}, \\
f_2 &= f_2^{(4)} = f_2^{(3)}, \\
f_3 &= f_3^{(3)} = f_3^{(1)}, \\
f_4 &= f_4^{(2)} = f_4^{(4)}.
\end{aligned}
\tag{2.11}
$$

Using the expressions (2.9) and (2.10), these equations become

$$
\begin{aligned}
f_1 &= -g_{1,1}^{(1)}(\bar{u}_1-u_1)-g_{1,2}^{(1)}(\bar{u}_3-u_1) = g_{1,1}^{(2)}(\bar{u}_1-u_2)-g_{1,2}^{(2)}(\bar{u}_4-u_2), \\
f_2 &= g_{1,1}^{(4)}(\bar{u}_2-u_4) + g_{1,2}^{(4)}(\bar{u}_4-u_4) = -g_{1,1}^{(3)}(\bar{u}_2-u_3) + g_{1,2}^{(3)}(\bar{u}_3-u_3), \\
f_3 &= -g_{2,1}^{(3)}(\bar{u}_2-u_3) + g_{2,2}^{(3)}(\bar{u}_3-u_3) = -g_{2,1}^{(1)}(\bar{u}_1-u_1)-g_{2,2}^{(1)}(\bar{u}_3-u_1), \\
f_4 &= g_{2,1}^{(2)}(\bar{u}_1-u_2)-g_{2,2}^{(2)}(\bar{u}_4-u_2) = g_{2,1}^{(4)}(\bar{u}_2-u_4) + g_{2,2}^{(4)}(\bar{u}_4-u_4).
\end{aligned}
\tag{2.12}
$$

The Eqs. (2.12) contain the edge values $\bar{u}_1$, $\bar{u}_2$, $\bar{u}_3$, and $\bar{u}_4$. Tacitly we have here used the same expression for the edge value of the cells at each side of an edge, and thereby implicitly demanded continuity of the potential at the points $\bar{\boldsymbol{x}}_1$, $\bar{\boldsymbol{x}}_2$, $\bar{\boldsymbol{x}}_3$, and $\bar{\boldsymbol{x}}_4$.

If the matrix $\boldsymbol{G}_k$ is diagonal for all cell indices k, the grid is called ***K**-orthogonal.* The system of equations (2.12) is then no longer coupled, and the flux through the edges can be determined by eliminating the edge values $\bar{u}_i$. This gives a two-point flux expression. If the grid is not $\boldsymbol{K}$-orthogonal, the edge values $\bar{u}_i$ may still be eliminated in each interaction volume. We then proceed in the following way.

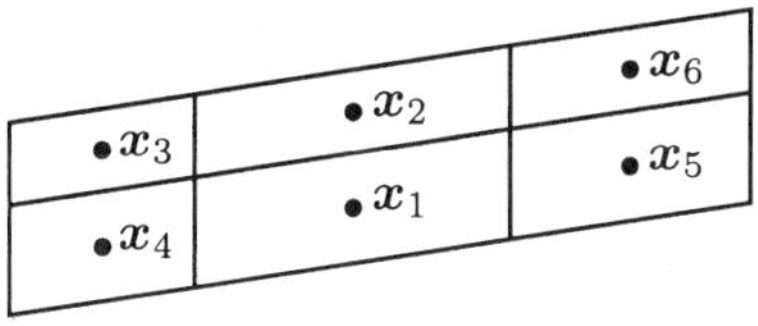

FIG. 5. *Flux stencil.*

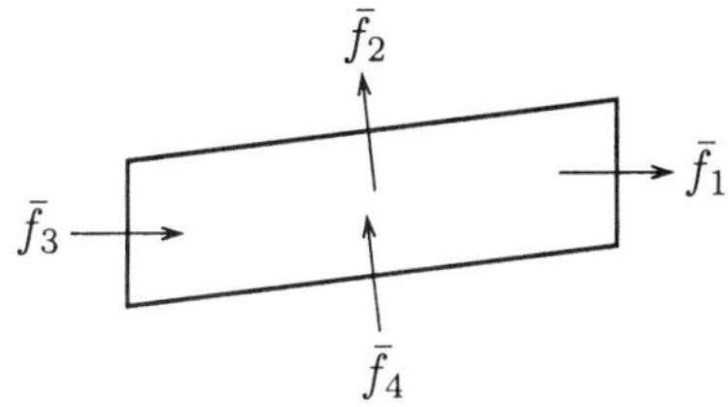

FIG. 6. *Flux through the cell edge of a cell.*

The fluxes of the system of equations (2.12) can be collected in the vector $\boldsymbol{f}$ defined by $\boldsymbol{f} = [f_1,\ f_2,\ f_3,\ f_4]^{\mathrm{T}}$. The system of equations further contains the potential values of the cell centers $\boldsymbol{u} = [u_1,\ u_2,\ u_3,\ u_4]^{\mathrm{T}}$ and the potential values at the midpoints of the cell edges $\boldsymbol{v} = [\bar{u}_1,\ \bar{u}_2,\ \bar{u}_3,\ \bar{u}_4]^{\mathrm{T}}$. The expressions on each side of the left equality sign of (2.12) can therefore be written on the form

$$\boldsymbol{f} = \boldsymbol{C}\boldsymbol{v} + \boldsymbol{F}\boldsymbol{u}. \tag{2.13}$$

The expressions on each side of the right equality sign in the system of equations (2.12) may after a reorganization be written in the form

$$\boldsymbol{A}\boldsymbol{v} = \boldsymbol{B}\boldsymbol{u}. \tag{2.14}$$

Hence, $\boldsymbol{v}$ may be eliminated by solving Eq. (2.14) with respect to $\boldsymbol{v}$ and putting $\boldsymbol{v} = \boldsymbol{A}^{-1}\boldsymbol{B}\boldsymbol{u}$ into (2.13). This gives the flux expression

$$\boldsymbol{f} = \boldsymbol{T}\boldsymbol{u}, \tag{2.15}$$

where

$$\boldsymbol{T} = \boldsymbol{C}\boldsymbol{A}^{-1}\boldsymbol{B} + \boldsymbol{F}. \tag{2.16}$$

The entries of the matrix $\boldsymbol{T}$ are called *transmissibility coefficients.* Equation (2.15) gives the flux through the half edges expressed by the potential values at the cell centers of an interaction volume.

Having determined the flux expression for all half edges, the two flux expressions of the two half edges which constitute an edge, can be added. This is shown in Fig. 5, where the cells 1, 2, 3, and 4 constitute one interaction volume, and the cells 1, 2, 5, and 6 constitute another. The flux stencil of the edge between cell 1 and 2 will therefore consist of the six cells of the figure. When the flux expressions have been found, these may be used in a discrete variant of Eq. (1.1). For the cell shown in Fig. 6 this yields the equation

$$\bar{f}_1 + \bar{f}_2 - \bar{f}_3 - \bar{f}_4 = VQ, \tag{2.17}$$

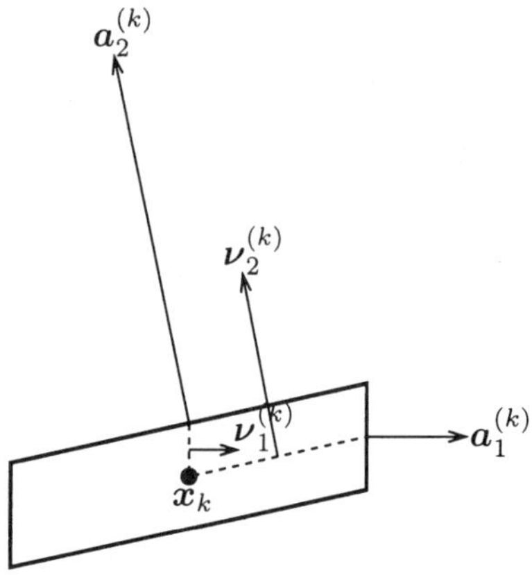

FIG. 7. *Normal vectors in parallelogram cells.*

where $\bar{f}_i$ is the flux through the entire edge i, V is the volume of the cell, and the source term Q has been approximated by a constant in the cell. This is a difference equation with u at the cell centers as the unknowns.

If two neighboring cells have vanishing conductivity, the corresponding row in the matrix $\boldsymbol{A}$ vanishes, and hence, the matrix $\boldsymbol{A}$ is singular. Because there is no need to determine the flux across the interfaces of cells with vanishing conductivity, the system may be reduced, and this will remove the singularity. However, it is more favorable to retain the system of unknowns and redefine the matrix $\boldsymbol{A}$ such that it becomes nonsingular. This is easiliy done by setting the diagonal elements of the vanishing rows in the matrix $\boldsymbol{A}$ equal to 1. The new system of equations has for the interfaces between cells with nonvanishing conductivity the same transmissibility coefficients as the reduced system.

For homogeneous media, test runs indicate that the matrix $\boldsymbol{A}$ is well conditioned, also for geometrically distorted cells. On the tested rough grids, the condition number satisfied $\text{cond}_2\, \boldsymbol{A} < 50$.

If cell k in Fig. 3 is a parallelogram, the expression for the matrix $\boldsymbol{G}_k$, Eq. (2.7), is simplified. For a parallelogram-shaped cell with index k, we denote the normal vectors of the edges with $\boldsymbol{a}_i^{(k)}$, $i = 1, 2$. These have length equal to the length of the edges. The normal vectors are shown in Fig. 7. Obviuosly, $\Gamma_i \boldsymbol{n}_i = \boldsymbol{a}_i^{(k)}/2$ and $\boldsymbol{\nu}_i^{(k)} = \boldsymbol{a}_i^{(k)}/2$. Further, $F_k = V_k/8$, where V_k is the area of cell k. It follows that for a parallelogram-shaped cell,

$$\boldsymbol{G}_k = \frac{1}{V_k} \left[\boldsymbol{a}_1^{(k)} \quad \boldsymbol{a}_2^{(k)}\right]^{\mathrm{T}} \boldsymbol{K}_k \left[\boldsymbol{a}_1^{(k)} \quad \boldsymbol{a}_2^{(k)}\right]. \tag{2.18}$$

Letting $\boldsymbol{J}_k = \left[\boldsymbol{a}_1^{(k)}, \, \boldsymbol{a}_2^{(k)}\right]$, one gets $V_k = |\det \boldsymbol{J}_k|$, and Eq. (2.18) becomes

$$\boldsymbol{G}_k = \frac{1}{|\det \boldsymbol{J}_k|} \boldsymbol{J}_k^{\mathrm{T}} \boldsymbol{K}_k \boldsymbol{J}_k. \tag{2.19}$$

Hence, for a parallelogram cell the tensor $\boldsymbol{G}_k$ is symmetric. Equation (2.19) is a congruence transformation. Thus, the tensor $\boldsymbol{G}_k$, as given by (2.18), is symmetric and positive definite if and only if $\boldsymbol{K}_k$ has these properties. If the tensor $\boldsymbol{G}_k$ is diagonal for all cell indices k, i.e., if

$$\left(\boldsymbol{a}_i^{(k)}\right)^{\mathrm{T}} \boldsymbol{K}_k \boldsymbol{a}_j^{(k)} = 0, \qquad i \neq j, \tag{2.20}$$

then the grid is $\boldsymbol{K}$-orthogonal.

In the matrix $\boldsymbol{G}_k$ it is sometimes useful to perform a splitting, such that anisotropy and grid skewness appears in one matrix and the mesh distances in another. If $\Delta\eta_k$ is the length of $\boldsymbol{a}_1^{(k)}$ and $\Delta\xi_k$ is the length of $\boldsymbol{a}_2^{(k)}$, then for a parallelogram grid,

$$\boldsymbol{G}_k = \frac{1}{\Delta\xi_k \Delta\eta_k} \boldsymbol{D}_k \boldsymbol{H}_k \boldsymbol{D}_k, \tag{2.21}$$

where

$$\begin{aligned} \boldsymbol{H}_k &= \frac{1}{\det[\boldsymbol{n}_1,\ \boldsymbol{n}_2]} \begin{bmatrix} \boldsymbol{n}_1 & \boldsymbol{n}_2 \end{bmatrix}^{\mathrm{T}} \boldsymbol{K}_k \begin{bmatrix} \boldsymbol{n}_1 & \boldsymbol{n}_2 \end{bmatrix} \\ &= \frac{1}{\det[\boldsymbol{n}_1,\ \boldsymbol{n}_2]} \begin{bmatrix} \boldsymbol{n}_1^{\mathrm{T}} \boldsymbol{K}_k \boldsymbol{n}_1 & \boldsymbol{n}_1^{\mathrm{T}} \boldsymbol{K}_k \boldsymbol{n}_2 \\ \boldsymbol{n}_2^{\mathrm{T}} \boldsymbol{K}_k \boldsymbol{n}_1 & \boldsymbol{n}_2^{\mathrm{T}} \boldsymbol{K}_k \boldsymbol{n}_2 \end{bmatrix}, \end{aligned} \tag{2.22}$$

and

$$\boldsymbol{D}_k = \operatorname{diag}(\Delta\eta_k,\ \Delta\xi_k). \tag{2.23}$$

Here, $\boldsymbol{n}_i$ is the unit normal vector which is parallel with $\boldsymbol{a}_i^{(k)}$, see Fig. 7. If $\boldsymbol{H}_k$ is diagonal, the grid is $\boldsymbol{K}$-orthogonal.

2.1. Extension to three dimensions. The principles of the MPFA O-method carry over to three dimensions. In three dimensions, an interaction volume contains 8 subcells and 12 interfaces, see Fig. 8. The linear functions in the eight cells are described by 32 coefficients. Eight of these are determined by the potential values at the cell centers. The rest of them are determined by the two continuity conditions at each of the 12 interfaces: the flux is required to be continuous at the interfaces, and the potential is required to be continuous at the interface midpoints.

The generalization of the equations of section 2 to three dimensions is straight forward. However, a three-dimensional cell described by its eight corners, generally does not have planar surfaces. The unit normal vector $\boldsymbol{n}_i$ of an interface is therefore not a constant across the interface. This can be accounted for by integrating the normal vector over the interface of the subcell in question. If a cell interface has corners $\boldsymbol{x}_k$, $k = 1, \ldots, 4$, see Fig. 9, the integrated normal vector over the interface of the subcell at vertex $\boldsymbol{x}_1$ is

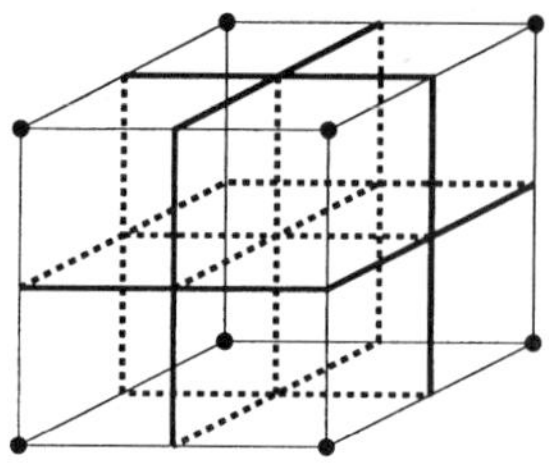

FIG. 8. *Tree-dimensional interaction volume (thin lines) with 8 subcells and 12 interfaces (thick lines).*

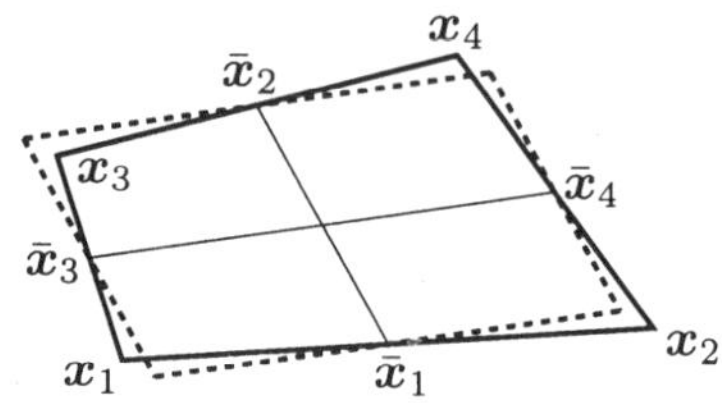

FIG. 9. *Replacing a quadrilateral (solid) by its associated parallelogram (dashed).*

$$\hat{\boldsymbol{n}} = \frac{1}{64}\Big[9(\boldsymbol{x}_2 - \boldsymbol{x}_1) \times (\boldsymbol{x}_3 - \boldsymbol{x}_1) + 3(\boldsymbol{x}_2 - \boldsymbol{x}_1) \times (\boldsymbol{x}_4 - \boldsymbol{x}_2) + 3(\boldsymbol{x}_4 - \boldsymbol{x}_3) \times (\boldsymbol{x}_3 - \boldsymbol{x}_1) + (\boldsymbol{x}_4 - \boldsymbol{x}_3) \times (\boldsymbol{x}_4 - \boldsymbol{x}_2)\Big]. \tag{2.24}$$

The vector $\hat{\boldsymbol{n}}$ has length equal to the area of the subcell interface, see [1] for details.

2.2. Symmetrization. The method described above yields a system of equations

$$\boldsymbol{M}\boldsymbol{u} = \boldsymbol{b}. \tag{2.25}$$

This is the discrete approximation of Eq. (1.1). Since the differential operator of Eq. (1.1) is self adjoint, one would like the matrix $\boldsymbol{M}$ of Eq. (2.25) to be symmetric. Further, the matrix $\boldsymbol{M}$ should be positive definite, to ensure that (2.25) approximates an elliptic equation.

Unfortunately, on a general quadrilateral grid the matrix $\boldsymbol{M}$ is not symmetric. However, if the matrices $\boldsymbol{G}_k$, given in Eq. (2.7), are symmetric, one may show that the matrix of coefficients $\boldsymbol{M}$ is symmetric [4]. Therefore, if all the cells are parallelograms (parallelepipeds in 3D), then the matrix of coefficients $\boldsymbol{M}$ is symmetric. For general quadrilaterals, this can be accomplished by replacing each cell with its associated parallelogram cell. This is shown for two dimensions in Fig. 9. The associated parallelogram is constructed as follows. Let $\bar{\boldsymbol{x}}_k$, $k = 1, \ldots, 4$, be the four midpoints of the edges of the quadrilateral, see Fig. 9. Draw the lines $\bar{\boldsymbol{x}}_1\bar{\boldsymbol{x}}_2$ and $\bar{\boldsymbol{x}}_3\bar{\boldsymbol{x}}_4$. Through each of the midpoints $\bar{\boldsymbol{x}}_1$ and $\bar{\boldsymbol{x}}_2$, lines parallel to $\bar{\boldsymbol{x}}_3\bar{\boldsymbol{x}}_4$ are drawn. Through each of the midpoints $\bar{\boldsymbol{x}}_3$ and $\bar{\boldsymbol{x}}_4$, lines parallel to $\bar{\boldsymbol{x}}_1\bar{\boldsymbol{x}}_2$ are drawn. The resulting quadrilateral (shown with dashed lines in Fig. 9) is the associated parallelogram.

Replacing each quadrilateral with its associated parallelogram yields a symmetric MPFA method. However, the order of convergence is generally lower, as shown in subsection 3.1. The described symmetric MPFA method is equivalent to the method which appears when each quadrilateral is transformed to a reference space with a bilinear mapping, and the flux is calculated in the reference space, using the Jacobian matrix evaluated at the cell center [1]. For a parallelogram, the matrix $\boldsymbol{J}^{-\mathrm{T}}$ of Eq. (2.19) equals the Jacobian matrix $\boldsymbol{dx}/\boldsymbol{d\xi}$ of the bilinear mapping.

3. Convergence. In this section we test the convergence properties of the MPFA O-method on quadrilateral grids by numerical experiments. In the derivation of the method, we made use of the cell center, without defining the location of this center. We will first test which location is the best in terms of convergence. Further, we investigate different grids for the same reference solution (on homogeneous media). We also compare the solutions obtained by discretizing on the physical quadrilaterals and discretizing on the associated parallelograms. Finally, we discuss the convergence rates on physical quadrilaterals for solutions with different smoothness. Most of the test runs are in 2D, but at the end we supplement with 3D test runs.

In this section, the potential u is termed the pressure as in reservoir simulation. Except for the test runs of subsection 3.2, the convergence rates are measured by the following discrete L^2 norms for both the pressures and the edge normal velocities [9],

$$\|u_h - u\|_{L^2} = \left(\sum_i V_i(u_{h,i} - u_i)^2\right)^{1/2}, \tag{3.1}$$

$$\|q_h - q\|_{L^2} = \left(\sum_j \tfrac{1}{4}(V_{j+} + V_{j-})(q_{h,j} - q_j)^2\right)^{1/2}. \tag{3.2}$$

Here, $q = \boldsymbol{q}\cdot\boldsymbol{n}$ is the edge normal velocity. Subscript h refers to the discrete solution. Further, V_i is the volume (area) of cell i, and $V_{j\pm}$ are the volumes of the two cells separated by edge j. The total volume of the simulated domains is for all test cases equal to unity.

3.1. 2D results in L^2 norm. Figure 10 shows some of the grids used in the test runs. One grid is constructed such that the entire grid has to account for an internal 120° grid line (Fig. 10.a). Another grid is a uniform parallelogram grid with internal acute angles of 45° (not shown in the figure). A third grid is a zig-zag parallelogram grid (Fig. 10.c).

A randomization may be performed for the grids [10, 11, 15]. By displacing the corners of the grid in Fig. 10.a by a random h^β pertubation, a grid with an arbitrary roughness appears. Such a rough grid is shown in Fig. 10.b. Finally, the grid shown in Fig. 10.d will be applied for a test case found in [8].

The first test cases are performed with the solution

$$u(x, y) = \cosh(\pi x)\cos(\pi y) \tag{3.3}$$

of the problem (1.1) on an isotropic, homogeneous medium. The boundary conditions are given by Dirichlet conditions, and are implemeted by interpolation [8, 9].

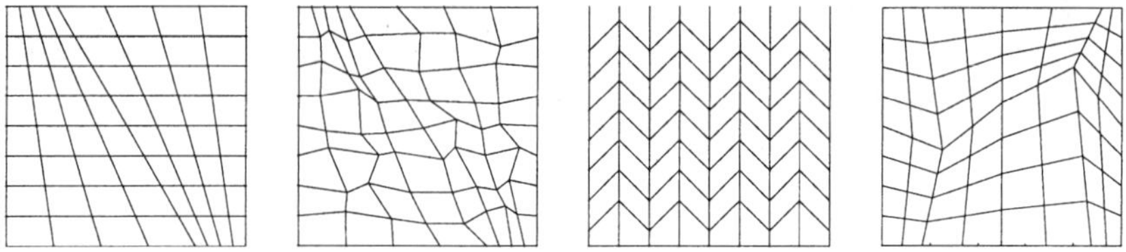

FIG. 10. *Grids used for simulations. From left ro right: (a): Smooth grid. (b): Random h^1 perturbution of the smooth grid. (c): Zig-zag parallelogram grid. (d): Grid used for simulation of* (3.7).

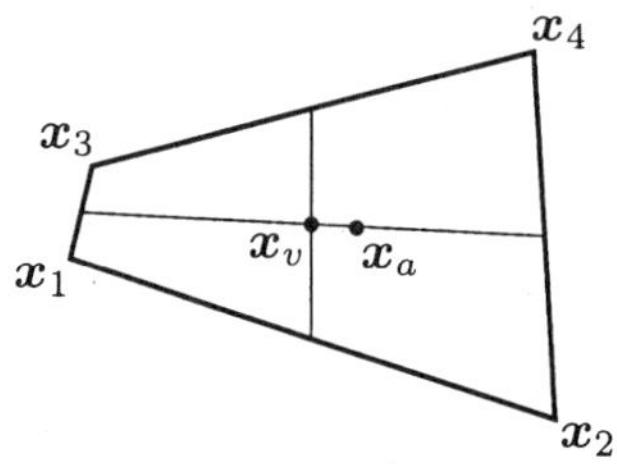

FIG. 11. *Vertex center $\boldsymbol{x}_v$ and area center $\boldsymbol{x}_a$ of a quadrilateral.*

We begin by testing different cell center locations. For quadrilaterals, there are two "natural" cell centers. The first is the vertex center

$$\boldsymbol{x}_v = \tfrac{1}{4}(\boldsymbol{x}_1 + \boldsymbol{x}_2 + \boldsymbol{x}_3 + \boldsymbol{x}_4), \tag{3.4}$$

where $\boldsymbol{x}_i$, $i = 1, \ldots, 4$, are the vertices of the quadrilateral. The second is the area center

$$\boldsymbol{x}_a = \frac{\int_V \boldsymbol{x}\, d\tau}{\int_V d\tau}, \tag{3.5}$$

where V is the quadrilateral. These centers are shown in Fig. 11. The area center is the barycenter of the area, whereas the vertex center is the barycenter of the vertices.

The use of these two cell centers is tested on the grid shown in Fig. 10.b for the solution (3.3) in a homogeneous medium. Since the grid is rough, the two different cell centers may deviate significantly.

As seen in Fig. 12, the convergence order is the same for both cases, but the normal-velocity error is smaller when using the vertex center compared to the use of the area center. In the following we therefore use the vertex center in all test runs.

Test runs on the different grids of Fig. 10 with the solution (3.3) are now performed. The convergence for the discretization in physical space is considered for these cases. As seen from Fig. 13.a, the convergence is second order for the pressure for the skew grid in Fig. 10.a, the uniform

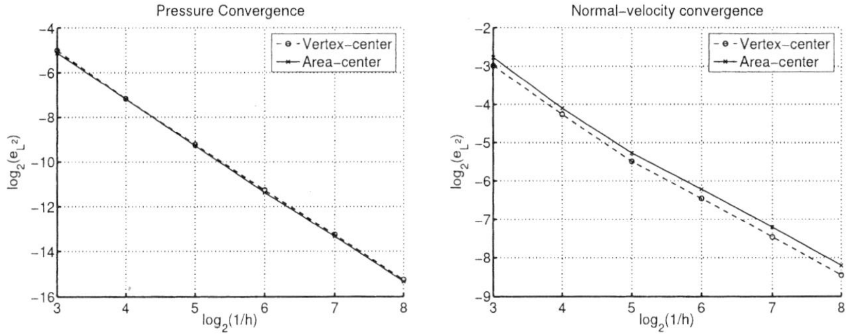

FIG. 12. *Test on cell center location with solution* (3.3) *on the grid in Fig. 10.b. Left (a): Pressure. Right (b): Edge normal velocities.*

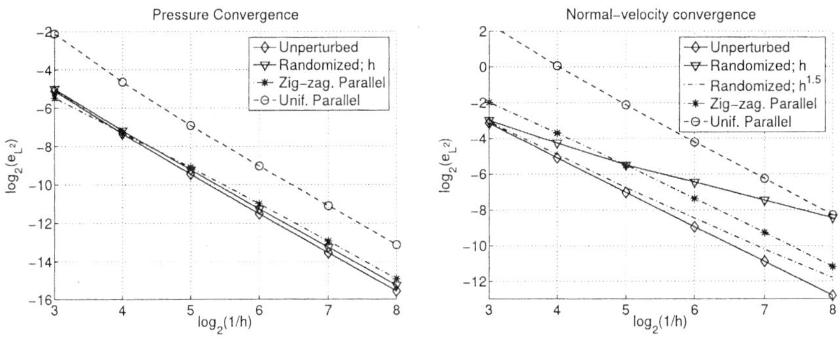

FIG. 13. *Convergence behavior for the solution* (3.3). *Left (a): Pressure. Right (b): Edge normal velocities.*

parallelogram grid, and the zig-zag parallelogram grid in Fig. 10.c. The velocity convergence is second order for both the parallelogram grids and the skew grid. Note that the domain of the uniform parallelogram grid is different from the domain in the other test cases. Therefore, only the order, and not the magnitude of the error, may be compared.

Figure 13 also shows the solutions on the rough grids shown in Fig. 10.b. For a random h^1 perturbation, the pressure is still seen to converge as h^2, whereas the convergence rate for the velocities gradually decreases to h^1, although almost $h^{1.5}$ is observed in the first refinement levels. If the perturbation is of order h^2, the velocity convergence is again of order h^2. Various h^β perturbations have been tested for $1 \leq \beta \leq 2$, and the specific case $\beta = 1.5$ is plotted in Fig. 13.b. Convergence of order $h^{1.75}$ is observed in the latest refinement step here.

Next, we compare the symmetrized version with the unsymmetric version, i.e., we compare the use of associated parallelograms with the use of

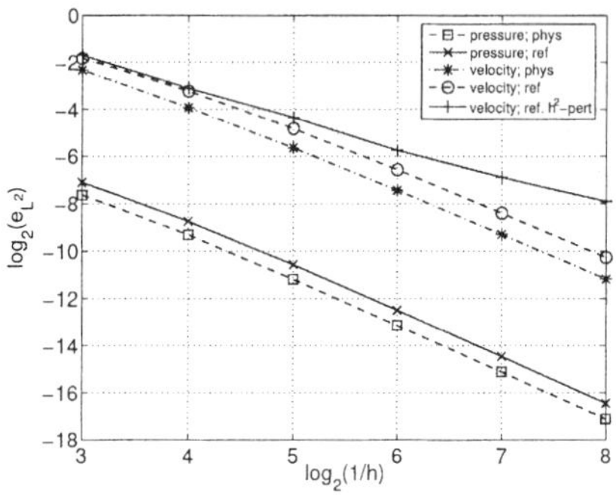

FIG. 14. *Convergence behavior of pressure and edge normal velocities for the piecewise quadratic solution* (3.7) *on the grid in Fig. 10.d.*

the physical quadrilaterals. Use of associated parallelograms may also be referred to as calculation in a reference space, since this method is identical to the method achieved by using the bilinear mapping of Sec. 2.2.

The example uses a reference solution which is a piecewise quadratic pressure, taken from [8] for a case where the medium is layered. The domain is $[0,1] \times [0,1]$, and the discontinuity line follows $x = 1/2$. Conductivities of the medium are specified by

$$\boldsymbol{K}_l = \begin{bmatrix} 50 & 0 \\ 0 & 1 \end{bmatrix}, \qquad \boldsymbol{K}_r = \begin{bmatrix} 1 & 0 \\ 0 & 10 \end{bmatrix}, \tag{3.6}$$

for which the following analytical solution holds

$$u(x,y) = \begin{cases} c_l x^2 + d_l y^2, & x < 1/2, \\ a_r + b_r x + c_r x^2 + d_r y^2, & x > 1/2. \end{cases} \tag{3.7}$$

The coefficients of (3.7) are comprised of the defined conductivities [8].

This case is simulated on the grid in Fig. 10.d, and the results are shown in Fig. 14. The asymptotic order of convergence again seems to be h^2 for both the pressure and normal velocities in physical space (the normal velocities converge as $h^{1.9}$ in the last refinement level). The order of convergence seems to be h^2 in the limit for both the pressure and normal velocities in computational space for an unperturbed grid, but initial errors are larger for the computational space discretization. When h^2-order perturbations are introduced for the corners of the grid, we see that the convergence of the normal velocities decreases to h^1, whereas h^2-order convergence is retained for the pressure. The velocity convergence in physical space is still close to h^2, and the curve will here almost coincide with the curve for the unperturbed grid. Increasing the perturbations to order h^1, our tests show that the pressure may converge slower than h^1, whereas the velocities may not even converge. For discretization in physical space, the

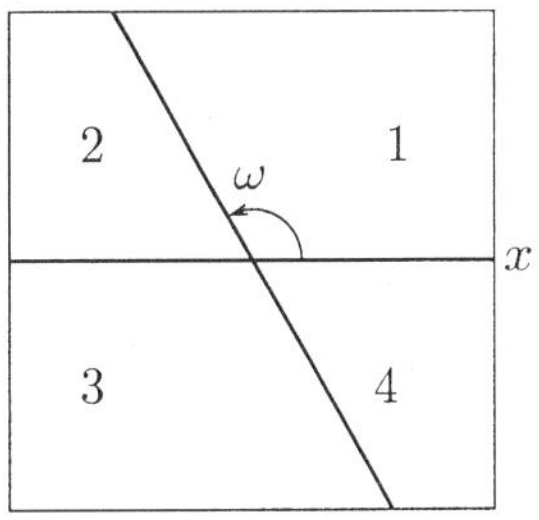

FIG. 15. *Corner with regions 1, 2, 3, and 4.*

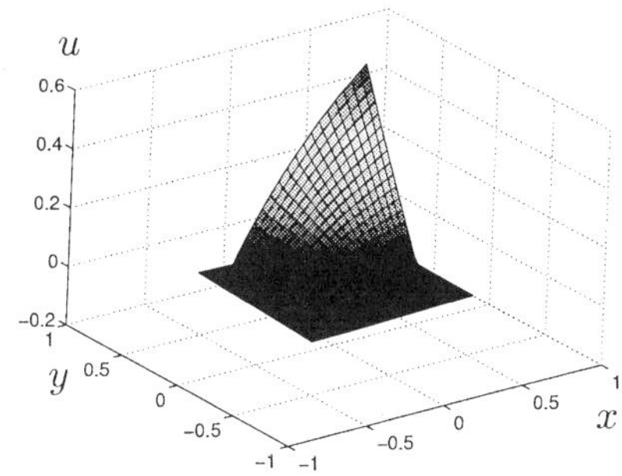

FIG. 16. *Solution* (3.8) *with* $\alpha \approx 1.4787$.

pressure converges as order h^2, whereas velocities converge as order h^1. Convergence on perturbed grids for the reference space discretization will be an issue for further research.

We conclude that although symmetry is achieved by this discretization, the price to pay is in general less accuracy, and even loss of convergence for the velocities.

The third test series consists of cases where internal corners of the medium exist and impose singularities on the velocity field [14, 19]. Such cases have been extensively tested for MPFA in [9]. Areas where grid cells with different conductivities meet, are encountered extensively for grids used in reservoir simulation, and may give rise to singular solutions. Nonorthogonal grid cells must also be used to handle the geo-description.

In Fig. 15, each of the regions labeled 1 to 4 may have different conductivities, and this may yield a singularity at the corner where the regions meet. We assume that the medium is isotropic, and use $\omega = 2\pi/3 = 120°$. Let the distance from the corner be r and the angle from the x-axis be θ. In the case where the conductivities in the regions 2, 3, and 4 are equal, there exists a solution of (1.1) of the form

$$u(r,\theta) = r^\alpha \begin{cases} \cos\alpha(\theta - \pi/3) & \text{for } \theta \in [0, 2\pi/3], \\ d\cos\alpha(4\pi/3 - \theta) & \text{for } \theta \in [2\pi/3, 2\pi], \end{cases} \tag{3.8}$$

where $\alpha = (3/\pi)\arctan\sqrt{1 + 2/\kappa}$ and $d = \cos(\alpha\pi/3)/\cos(2\alpha\pi/3)$. Also, $\kappa = k_1/k_2$ is the conductivity ratio. For $\kappa \geq 0$ one gets exponents $\alpha \in [0.75, 1.5]$. The solution (3.8) belongs to the space $H^{1+\alpha-\epsilon}$ for any $\epsilon > 0$.

The case $\kappa = 10^{-3}$ yields $\alpha \approx 1.4787$. This solution is shown in Fig. 16, and is simulated on the grids shown in Figs. 10.a and 10.b. As seen from Fig. 17, the convergence rate in L^2 norm is h^2 for the pressure on both grids. The convergence rate for the edge normal velocities is $h^{1.47}$ for the grid in Fig. 10.a and h^1 for the grid in Fig. 10.b. The convergence rates in maximum norm shown in the figures are discussed in subsection 3.2.

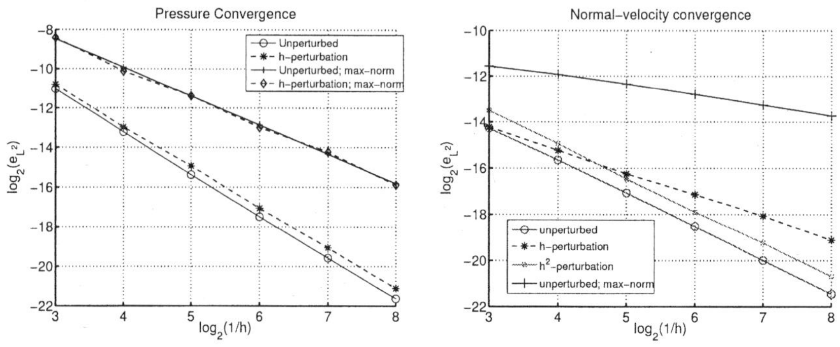

FIG. 17. *Convergence behavior for the $H^{2.47}$ solution* (3.8). *Left (a): Pressure. Right (b): Edge normal velocities.*

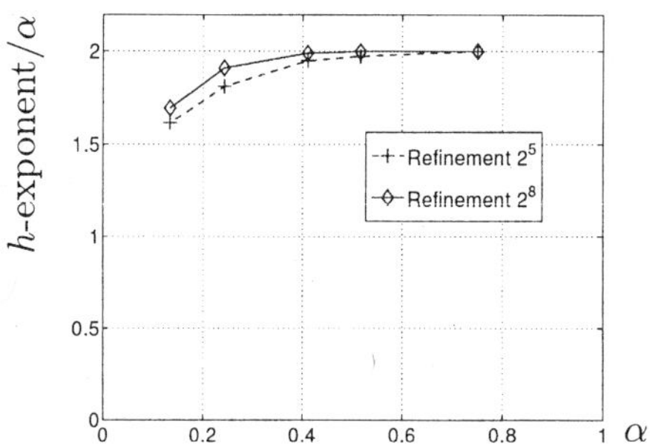

FIG. 18. *Convergence behavior of pressure for varying α in the refinement steps* $2^4 \to 2^5$ *(dashed) and* $2^7 \to 2^8$ *(solid). The ordinate is the h exponent divided by α. The asymptotic region is not reached for $\alpha < 0.4$ for the solid curve.*

The case $\kappa = 10^2$ yields the solution $\alpha \approx 0.7547$. A similar test as the above gives the convergence rates $h^{1.51}$ for the pressure and $h^{0.75}$ for the edge normal velocities, and is discussed in [9]. These rates hold on both grids (Figs. 10.a and 10.b).

If we let the conductivities in region 1 and 3 be equal, and likewise the conductivities in region 2 and 4 equal, solutions with $\alpha \in [0, 1.5]$ exist. The solution satisfies $u(r,\theta) = -u(r,\theta - \pi)$ with

$$u(r,\theta) = r^{\alpha} \begin{cases} \cos\alpha(\theta - \pi/3) & \text{for } \theta \in [0, 2\pi/3], \\ d\sin\alpha(5\pi/6 - \theta) & \text{for } \theta \in [2\pi/3, \pi]. \end{cases} \tag{3.9}$$

Here, $\alpha = (6/\pi)\arctan\bigl(1/\sqrt{1+2\kappa}\bigr)$ and $d = \cos(\alpha\pi/3)/\sin(\alpha\pi/6)$. As above, $\kappa = k_1/k_2$ is the conductivity ratio.

Several cases of the form (3.9) with varying regularity (described by α) are tested in [9] for the O-method in physical space for various nonorthogonal grids. Finite-element theory [16] yields pressure convergence of order

$h^{2\alpha}$. The convergence behavior of the pressure in two grid refinement steps for different values of α is plotted in Fig. 18. The refinement steps are given as the number of nodes in each direction, and are $2^4 \rightarrow 2^5$ for the dashed curve and $2^7 \rightarrow 2^8$ for the solid curve. As seen from the diagram, the convergence is not fully $h^{2\alpha}$ yet for the pressure when α decreases. However, as the curves show, the smaller α is, the later the asymptotic region of convergence is entered. The same trend is seen for mixed finite element formulations, see [9] for a specific implementation.

Normal velocities are seen to converge with order h^α for the same refinement levels for examples with $\alpha > 0.3$, and then experience a decrease for smaller α's which is similar to the decrease in pressure convergence.

In conclusion, the simulated runs indicate the following error bounds for discretization on arbitrary grids in physical space,

$$\|u_h - u\|_{L^2} \sim h^2, \tag{3.10}$$

$$\|q_h - q\|_{L^2} \sim h. \tag{3.11}$$

These error bounds require that $u \in H^2$. If $u \in H^3$ and the grid is a parallelogram grid or a smooth quadrilateral grid, the convergence order in (3.11) becomes h^2.

If $u \in H^{1+\alpha}$, $\alpha < 1$, the simulated runs indicate the following error bounds for discretization in physical space,

$$\|u_h - u\|_{L^2} \sim h^{2\alpha}, \tag{3.12}$$

$$\|q_h - q\|_{L^2} \sim h^{\alpha}. \tag{3.13}$$

Equation (3.13) still holds for $u \in H^{1+\alpha}$, $\alpha \in [1, 2]$, provided the grid is smooth.

3.2. 2D results in maximum norm. Figure 13 shows convergence results in L^2 norm for the solution (3.3). For these test examples we have also measured the error in maximum norm. These results are shown in Fig. 19. As can be seen in Fig. 19.a, the convergence rate for the pressure is h^2 for the grids shown in Figs. 10.a, 10.b, and 10.c, as well as for the uniform parallelogram grid. This is the same convergence rate as found for the error in L^2 norm.

Figure 19.b shows the error of the edge normal velocities in j-direction. Here, the convergence rate is h^1 for the grids shown in Figs. 10.a, 10.b, and 10.c. Only the uniform parallelogram grid gives second order convergence in the maximum norm. Thus, for the normal velocities, the L^2 error is second order for smooth grids, but the L^∞ error is second order only for uniform grids.

The error in maximum norm has also been measured for the $H^{2.47}$ solution shown in Fig. 17. As can be seen, the convergence rate for the pressure is roughly $h^{1.47}$ for the grids shown in Figs. 10.a and 10.b. The

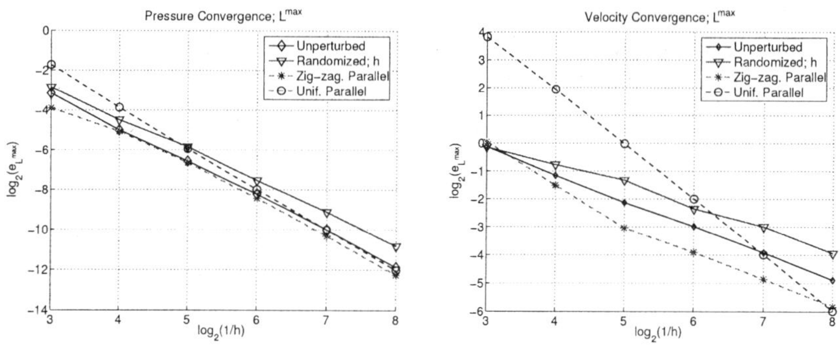

FIG. 19. *Convergence behavior in maximum norm for the solution* (3.3). *Left (a): Pressure. Right (b): Edge normal velocities.*

convergence rate for the edge normal velocities is roughly $h^{0.47}$ on the grid shown in Fig. 10.a. There was an oscillatory behavior for the edge normal velocities on the h-perturbed grid of Fig. 10.b.

The above results can be summarized as follows. For $u \in H^{1+\alpha}$, $\alpha > 0$, the error bound in the pressure seems to be

$$\|u_h - u\|_{L^\infty} \sim h^{\min\{2,\alpha\}}. \tag{3.14}$$

For smooth solutions, the found error bound for the edge normal velocities is h^1. The discussed test runs do not clarify the regularity required to achieve this convergence rate for rough grids. For smooth grids, however, the error bound in the edge normal velocities seems to be $h^{\min\{1,\alpha-1\}}$, provided $u \in H^{1+\alpha}$, $\alpha > 1$.

The L^∞ convergence rates which are indicated above, must be taken with precaution. For example, it is by no means clear that only the Sobolev-space regularity and the grid smoothness determine the L^∞ convergence rates. A property like the monotonicity of the method [13] might be important for these convergence rates.

3.3. 3D results on uniform flow. Numerical test runs are next performed on three-dimensional grids in physical space. When going from 2D to 3D, general positioning of corners of the control volumes implies that bilinear cell surfaces may arise. These surfaces may for some methods create additional difficulties for handling of fluxes across cell interfaces [12]. In particular, methods that rely on a transformation from the physical grid to an orthogonal reference grid, will not be able to reproduce uniform flow exactly.

This is not the case with the O-method discretization in physical space. As an example, a 3D grid created by random h perturbations of the corners in all directions of an initial orthogonal grid, is shown in Fig. 20. The numerical pressure is exact to working precision (10^{-16}) when uniform flow

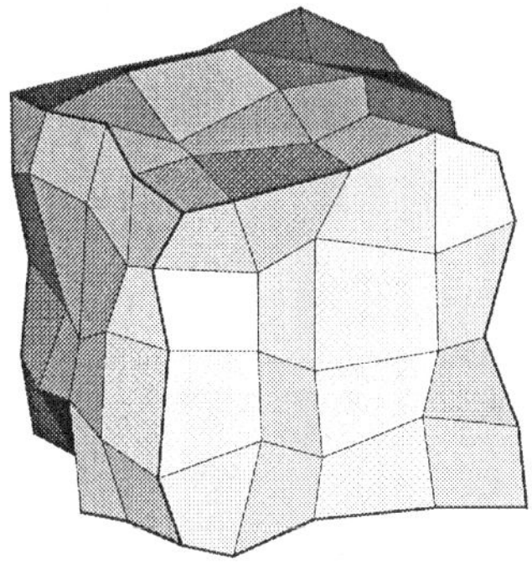

FIG. 20. *3D grid. All corners perturbed randomly in x, y, and z direction.*

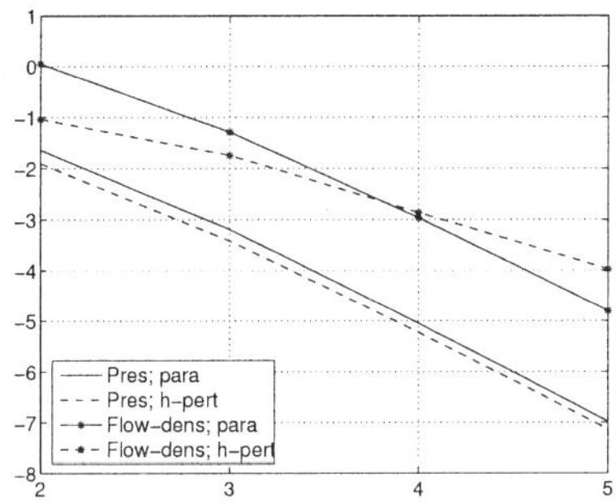

FIG. 21. *Convergence behavior for pressure and normal velocities of i-edges of 3D grids.*

is used as a reference case for all possible conforming grids that we have tested. This applies for both of the cell center definitions (3.4) and (3.5). The exact solution property is explained by the way the transmissibilities are derived in 3D in physical space. The term $-\int_S \boldsymbol{n}^{\mathrm{T}} \boldsymbol{K} \,\mathrm{grad}\, u \, d\sigma$ for each edge is discretized by the assumption of piecewise linear pressure as in 2D. The normal vector $\boldsymbol{n}$ is parametrized in (2.24) for general bilinear surfaces, and $-\int_S \boldsymbol{n} \, d\sigma$ is hence evaluated correctly. For linear pressure, $\mathrm{grad}\, u$ is constant, and the flux discretization that the transmissibility calculations apply, is therefore exact. Together with the uniform flow result for 2D cases in [9], this then implies that the pressure solution for the O-method in 3D is exact for uniform flow, and this is verified by our numerical results. This again shows the superiority of the discretization in physical space.

3.4. 3D results in L^2 norm. Next, a case of nonuniform flow is tested for our 3D implementation of the O-method. It is trivial to verify that the function

$$u(x, y, z) = \sin(\sqrt{2}\pi x) \sinh(\pi y) \sinh(\pi z) \tag{3.15}$$

is a solution to the problem (1.1) when the medium is isotropic and homogeneous. The convergence is examined for both the pressure and normal velocities for the set of i-edges in a parallelepiped grid and the grid in Fig. 20, and is depicted in Fig. 21. Dirichlet boundary conditions which correspond to the reference solution, are implemented by artificial layers of grid cells. The first test shows the numerical results for a uniform parallelepiped grid on a given domain. As is expected, both the pressure and normal velocities converge as h^2 when the grid is refined uniformly.

The second test shows the convergence behavior for an orthogonal initial grid for which all corners are arbitrarily perturbed by terms of order h in all directions, but for which nonoverlapping grid cells do not occur.

The pressure still converges as h^2, but the normal velocity convergence decreases to h as is expected from the results in 2D.

The 3D test runs agree with the estimated convergence rates in 2D.

REFERENCES

[1] I. AAVATSMARK, *Introduction to multipoint flux approximations for quadrilateral grids*, Comput. Geosci. **6** (2002), 405–432.

[2] I. AAVATSMARK, T. BARKVE, Ø. BØE, AND T. MANNSETH, *Discretization on non-orthogonal, quadrilateral grids for inhomogeneous, anisotropic media*, J. Comput. Phys. **127** (1996), 2–14.

[3] I. AAVATSMARK, T. BARKVE, Ø. BØE, AND T. MANNSETH, *A class of discretization methods for structured and unstructured grids in anisotropic, inhomogeneous media*, Proc. of the 5th European Conf. on the Mathematics of Oil Recovery, eds. Z.E. Heinemann and M. Kriebernegg, Leoben, Austria, 1996, pp. 157–166.

[4] I. AAVATSMARK, T. BARKVE, AND T. MANNSETH, *Control-volume discretization methods for 3D quadrilateral grids in inhomogeneous, anisotropic reservoirs*, SPE J. **3** (1998), 146–154.

[5] I. AAVATSMARK, T. BARKVE, Ø. BØE, AND T. MANNSETH, *Discretization on unstructured grids for inhomogeneous, anisotropic media. Part I: Derivation of the methods*, SIAM J. Sci. Comput. **19** (1998), 1700–1716.

[6] I. AAVATSMARK, T. BARKVE, Ø. BØE, AND T. MANNSETH, *Discretization on unstructured grids for inhomogeneous, anisotropic media. Part II: Discussion and numerical results*, SIAM J. Sci. Comput. **19** (1998), 1717–1736.

[7] M.G. EDWARDS, *Unstructured, control-volume distributed, full-tensor finite-volume schemes with flow based grids*, Comput. Geosci. **6** (2002), 433–452.

[8] M.G. EDWARDS AND C.F. ROGERS, *Finite volume discretization with imposed flux continuity for the general tensor pressure equation*, Comput. Geosci. **2** (1998), 259–290.

[9] G.T. EIGESTAD AND R.A. KLAUSEN, *On the convergence of the multi-point flux approximation O-method; Numerical experiments for discontinuous permeability*, To appear in Numer. Methods Partial Diff. Eqns. **21** (2005).

[10] J. HYMAN, M. SHASHKOV, AND S. STEINBERG, *The numerical solution of diffusion problems in strongly heterogeneous non-isotropic materials*, J. Comput. Phys. **132** (1997), 130–148.

[11] I.D. MISHEV, *Nonconforming finite volume methods*, Comput. Geosci. **6** (2002), 253–268.

[12] R.L. NAFF, T.F. RUSSEL, AND J.D. WILSON, *Shape functions for velocity interpolation in general hexahedral cells*, Comput. Geosci. **6** (2002), 285–314.

[13] J.M. NORDBOTTEN AND I. AAVATSMARK, *Monotonicity conditions for control volume methods on uniform parallelogram grids in homogeneous media*, Comput. Geosci. **9** (2005), 61–72.

[14] B. RIVIÈRE, M.F. WHEELER, AND K. BANAŚ, *Discontinuous Galerkin method applied to a single phase flow in porous media*, Comput. Geosci. **4** (2000) 337–349.

[15] M. SHASHKOV AND S. STEINBERG, *Solving diffusion equations with rough coefficients in rough grids*, J. Comput. Phys. **129** (1996), 383–405.

[16] G. STRANG AND G.J. FIX, *An analysis of the finite element method*, Wiley, New York, 1973.

[17] S. VERMA AND K. AZIZ, *Two- and three-dimensional flexible grids for reservoir simulation*, Proc. of the 5th European Conf. on the Mathematics of Oil Recovery, eds. Z.E. Heinemann and M. Kriebernegg, Leoben, Austria, 1996, pp. 143–156.

[18] A.F. WARE, A.K. PARROTT, AND C. ROGERS, *A finite volume discretization for porous media flows governed by non-diagonal permeability tensors*, Proc. of

CFD95, eds. P.A. Thibault and D.M. Bergeron, Banff, Canada, 1995, pp. 357–364.

[19] J.A. Wheeler, M.F. Wheeler, and I. Yotov, *Enhanced velocity mixed finite element methods for flow in multiblock domains*, Comput. Geosci. **6** (2002), 315–332.

DIFFERENTIAL COMPLEXES AND STABILITY OF FINITE ELEMENT METHODS I. THE DE RHAM COMPLEX

DOUGLAS N. ARNOLD*, RICHARD S. FALK†, AND RAGNAR WINTHER‡

Abstract. In this paper we explain the relation between certain piecewise polynomial subcomplexes of the de Rham complex and the stability of mixed finite element methods for elliptic problems.

Key words. Mixed finite element method, de Rham complex, stability.

AMS(MOS) subject classifications. 65N12, 65N30.

1. Introduction. Many standard finite element methods are based on extremal variational formulations. Typically, the desired solution is characterized as the minimum of some functional over an appropriate trial space of functions, and the discrete solution is then taken to be the minimum of the same functional restricted to a finite dimensional subspace of the trial space consisting of piecewise polynomials with respect to a triangulation of the domain of interest. For such methods, stability is often a simple consideration. For mixed finite element methods, which are based on saddle-point variational principles, the situation is very different: stability is generally a subtle matter and the development of stable mixed finite element methods very challenging. In recent years, a new approach has added greatly to our understanding of stability of mixed methods and enabled the development of stable methods for a number of previously intractable problems. This approach is homological, involving differential complexes related to the problem to be solved, discretizations of these complexes obtained by restricting the differential operators to finite dimensional subspaces, and commutative diagrams relating the two. See, e.g., [1, 2, 14, 18]. In this paper we will survey these ideas. While the presentation aims to be relatively self-contained, it is directed primarily at readers familiar with the classical theory of mixed finite element methods as exposed in, for instance, [12].

We will concentrate first on the problem of steady state heat conduction. In this problem we seek a scalar temperature field u and a vector flux field σ defined on the domain of interest $\Omega \subset \mathbb{R}^n$. These satisfy Fourier's

*Institute for Mathematics and its Applications, University of Minnesota, Minneapolis, MN 55455. The work of the first author was supported in part by NSF grant DMS-0411388.

†Department of Mathematics, Rutgers University, Hill Center, 110 Frelinghuysen Rd, Rutgers University, Piscataway, NJ 08854-8019. The work of the second author was supported in part by NSF grant DMS03-08347.

‡Centre of Mathematics for Applications and Department of Informatics, University of Oslo, P.O. Box 1053, 0316 Oslo, NORWAY. The work of the third author was supported by the Norwegian Research Council.

law of heat conduction

$$A\sigma + \operatorname{grad} u = 0 \text{ on } \Omega \tag{1.1}$$

and the equation of thermal equilibrium

$$\operatorname{div} \sigma = f \text{ on } \Omega. \tag{1.2}$$

Here A is the thermal resistivity tensor (the inverse of the thermal conductivity tensor), an $n \times n$ symmetric positive-definite matrix field (scalar for an isotropic material), and f the given rate of heat generated per unit volume. To obtain a well-posed problem, these differential equations must be supplemented by suitable boundary conditions, for example, the Dirichlet condition $u = 0$ on $\partial\Omega$.

Multiplying the constitutive equation by a test field τ and integrating by parts over Ω (taking into account the homogeneous Dirichlet boundary condition), we obtain

$$\int_\Omega A\sigma \cdot \tau \, dx - \int_\Omega u \operatorname{div} \tau \, dx = 0 \quad \forall \tau \in H(\operatorname{div}, \Omega; \mathbb{R}^n), \tag{1.3}$$

while from the equilibrium equation we obtain

$$\int_\Omega \operatorname{div} \sigma v \, dx = \int_\Omega f v \, dx \quad \forall v \in L^2(\Omega). \tag{1.4}$$

The space $H(\operatorname{div}, \Omega; \mathbb{R}^n)$ consists of all vector fields $\tau : \Omega \to \mathbb{R}^n$ which are square integrable and for which the divergence $\operatorname{div} \tau$ is also square integrable. The pair of spaces $H(\operatorname{div}, \Omega; \mathbb{R}^n)$, $L^2(\Omega)$ are the natural ones for this problem. Indeed, it can be shown that for any $f \in L^2(\Omega)$, there is a unique pair $(\sigma, u) \in H(\operatorname{div}, \Omega; \mathbb{R}^n) \times L^2(\Omega)$ satisfying (1.3) and (1.4), and so providing a (weak) solution to (1.1) and (1.2) and the boundary conditions.

Equivalent to the weak formulation is a saddle-point variational formulation, namely

$$(\sigma, u) = \operatorname*{argcrit}_{(\tau,v)\in H(\operatorname{div})\times L^2} \left[\int (\frac{1}{2} A\tau \cdot \tau - v \operatorname{div} \tau) \, dx + \int f v \, dx \right]. \tag{1.5}$$

A more familiar variational characterization of the solution of the heat conduction problem is Dirichlet's principle, which involves the temperature field alone:

$$u = \operatorname*{argmin}_{v \in \mathring{H}^1(\Omega)} \left(\frac{1}{2} \int A^{-1} \operatorname{grad} v \cdot \operatorname{grad} v \, dx - \int f v \, dx \right).$$

This is connected to the second order scalar elliptic equation

$$-\operatorname{div} A^{-1} \operatorname{grad} u - f \text{ on } \Omega$$

and its natural weak formulation. A standard finite element methods uses a finite element subspace V_h of $\mathring{H}^1(\Omega)$ and defines the approximate solution

$$u_h = \operatorname*{argmin}_{v \in V_h} \left(\frac{1}{2} \int A^{-1} \operatorname{grad} v \cdot \operatorname{grad} v \, dx - \int f v \, dx \right).$$

Such a method is automatically stable with respect to the H^1 norm, and consequently the quasioptimal estimate

$$\|u - u_h\|_{H^1} \leq C \inf_{v \in V_h} \|u - v\|_{H^1}$$

holds (with C depending only on A and Ω).

Returning to the saddle-point formulation, a mixed finite element method defines an approximation solution σ_h, u_h belonging to finite element subspaces $\Sigma_h \subset H(\operatorname{div}, \Omega; \mathbb{R}^n)$, $V_h \subset L^2(\Omega)$, by

$$(\sigma_h, u_h) = \operatorname*{argcrit}_{(\tau, v) \in \Sigma_h \times V_h} \left[\int (\frac{1}{2} A\tau \cdot \tau - v \operatorname{div} \tau) \, dx + \int f v \, dx \right]. \qquad (1.6)$$

The corresponding quasioptimal estimate

$$\|\sigma - \sigma_h\|_{H(\operatorname{div})} + \|u - u_h\|_{L^2} \leq C(\inf_{\tau \in \Sigma_h} \|\sigma - \tau\|_{H(\operatorname{div})} + \inf_{v \in V_h} \|u - v\|_{L^2})$$

will, however, not hold in general. This requires *stability*, which holds only for very special choices of the finite element spaces. The method (1.6) falls into a well-studied class of saddle-point discretizations for which sufficient (and nearly necessary) conditions for stability can be given [8, 12]. Namely the discretization will be stable if there exist constants c_1 and c_2, independent of the discretization parameter h, such that

(A1) $\|\tau\|_{H(\operatorname{div})} \leq c_1 \|\tau\|_{L^2}$ whenever $\tau \in \Sigma_h$ satisfies $\int_\Omega v \operatorname{div} \tau \, dx = 0$ for all $v \in V_h$.

(A2) For all nonzero $v \in V_h$, there exists nonzero $\tau \in \Sigma_h$ with $\int_\Omega v \operatorname{div} \tau \, dx \geq c_2 \|\tau\|_{H(\operatorname{div})} \|v\|_{L^2}$.

The development of finite element methods satisfying these stability conditions is quite subtle. In the next section, we illustrate this in the simplest case of 1 dimension. In Section 3, we review the two main families of stable finite element spaces for this mixed problem in arbitrary dimensions. Section 4 is a concise review of the main relevant concepts of exterior algebra and exterior calculus, particularly the de Rham complex, the Hodge Laplacian, and the Koszul complex. With these preliminaries, we develop families of finite element discretizations of differential forms of all orders in all dimensions, and show how to combine them into piecewise polynomial subcomplexes of the de Rham complex, obtaining 2^{n-1} such subcomplexes in dimension n for each polynomial degree. The finite element spaces involved in these subcomplexes provide most of the stable finite elements that have been derived for mixed problems closely related to a Hodge Laplacian. In the final section, we show how to use these subcomplexes and the commutative diagrams relating them to the de Rham complex to obtain stability of mixed finite element methods. For reasons of space, many results are stated in this paper without proof. Proofs for most of the assertions can be found in the cited references, while for the material new here (the 2^{n-1} subcomplexes and the degrees of freedom in (5.1)), a more complete presentation will appear elsewhere.

2. Some 1-dimensional examples. The subtle nature of stability of finite elements for this problem arises already in the simplest case of one-dimension, with $A \equiv 1$. Thus we are approximating the problem

$$\sigma + u' = 0, \quad \sigma' = f \text{ on } (-1,1), \quad u(\pm 1) = 0.$$

We shall present some examples to illustrate both stable and unstable choices of elements for this problem. Although in this simple 1-dimensional context, these results can be fully analyzed theoretically, we will limit ourselves to displaying numerical results.

A stable choice of elements in this case consists of continuous piecewise linear functions for the flux and piecewise constant functions for the temperature, which we shall refer to as the $\mathcal{P}_1^{\text{cont}}$–$\mathcal{P}_0$ method. The exact and numerical solution are shown in Figure 1 for uniform meshes of 10, 20, and 40 subintervals, first in the case where $u(x) = 1 - |x|^{7/2}$, and then for the rougher solution $u(x) = 1 - |x|^{5/4}$. In the first case, $u \in H^3$ and $\sigma \in H^2$, but in the second case, $u \in H^s$ and $\sigma \in H^{s-1}$ for $s < 7/4$ but not larger, which limits the order of convergence of the best approximation by piecewise linears to σ. In the first part of Table 1, we see clearly that $\|\sigma - \sigma_h\|_{L^2} = O(h^2)$ and $\|u - u_h\|_{L^2} = O(h)$, both of which orders are optimal. In the second part of the table, the order of convergence is lowered due to the lowered smoothness of the solution, but the convergence order is as high as that of the best approximation, illustrating the stability of this method.

By contrast, if we use continuous piecewise linear elements for both σ and u (e.g., in the hope of improving the order of convergence to u), the method is not stable. For the smoother problem, $u(x) = 1 - |x|^{7/2}$, we again have second order convergence for σ and first order (not second order) convergence for u. But for $u(x) = 1 - |x|^{5/4}$, the convergence is clearly well-below that of the best approximation, a manifestation of instability, which is plainly visible in Figure 2 and Table 2. This example has been analyzed in detail by Babuška and Narasimhan [5].

In Figure 3, we show the result of using continuous piecewise quadratic elements for σ and piecewise constant elements for u (e.g., in hope of improving the order of convergence to σ). The method, which was analyzed in [9], is clearly unstable as well.

Although this $\mathcal{P}_2^{\text{cont}}$–$\mathcal{P}_0$ method is not stable, the $\mathcal{P}_2^{\text{cont}}$–$\mathcal{P}_1$ method is. In fact, in one dimension the $\mathcal{P}_r^{\text{cont}}$–$\mathcal{P}_{r-1}$ method (continuous piecewise

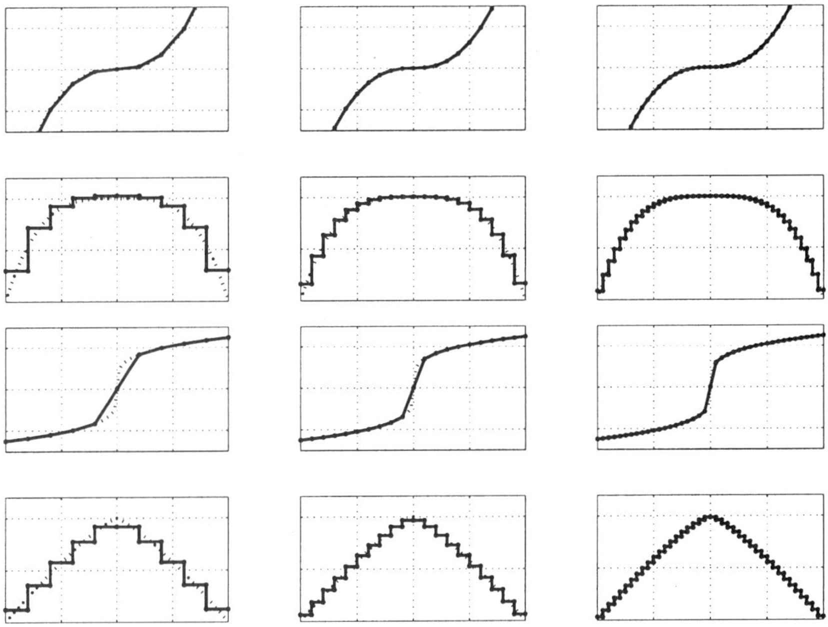

FIG. 1. *The $\mathcal{P}_1^{cont}$–$\mathcal{P}_0$ mixed method (which is stable) for meshes of 10, 20, and 40 elements. Lines 1 and 2: σ_h versus σ and u_h versus u for the $\mathcal{P}_1^{cont}$–$\mathcal{P}_0$ mixed method when $u(x) = 1 - |x|^{7/2}$. Lines 3 and 4: same with $u(x) = 1 - |x|^{5/4}$.*

polynomials of degree r for the flux and arbitrary piecewise polynomials of degree $r-1$ for the temperature) is stable for any $r \geq 1$.

3. Basic mixed finite elements in higher dimensions. Consider now the saddle point problem (1.5) in n dimensions and its discretization (1.6) using finite element spaces Σ_h and V_h consisting of piecewise polynomials with respect to a simplicial decomposition $\mathcal{T}_h$ of Ω. A simple choice of elements, which we saw to be stable in one dimension, is $\mathcal{P}_1^{cont}$–$\mathcal{P}_0$:

$$\begin{aligned} \Sigma_h &= \{\, \tau \in H^1(\Omega;\mathbb{R}^n) \mid \tau|_T \in \mathcal{P}_1(T) \quad \forall T \in \mathcal{T}_h \,\}, \\ V_h &= \{\, v \in L^2(\Omega) \mid v|_T \in \mathcal{P}_0(T) \quad \forall T \in \mathcal{T}_h \,\}. \end{aligned} \tag{3.1}$$

However, for $n > 1$, this choice is highly unstable. In fact, on generic triangular meshes the discrete problem is singular and u_h is undetermined. And even if σ_h could be determined, it would belong to the space of divergence-free continuous piecewise linear functions, which reduces to the space of global constants on many triangular meshes, so does not offer any approximation.

However, there are several stable choice of elements in higher dimensions that may be regarded as natural extensions of the simple $\mathcal{P}_1^{\mathrm{cont}}$–$\mathcal{P}_0$

TABLE 1
Errors and orders of convergence for the $\mathcal{P}_1^{cont}$–$\mathcal{P}_0$ mixed method.

$u = 1 - |x|^{7/2}$

n	$\|\sigma - \sigma_h\|_{L^2}$ err.	%	order	$\|u - u_h\|_{L^2}$ err.	%	order
10	4.78e−02	3.348		1.18e−01	10.141	
20	1.20e−02	0.838	2.00	5.87e−02	5.027	1.01
40	2.99e−03	0.210	2.00	2.93e−02	2.508	1.00
80	7.49e−04	0.052	2.00	1.46e−02	1.253	1.00
160	1.87e−04	0.013	2.00	7.31e−03	0.627	1.00

$u = 1 - |x|^{5/4}$

n	$\|\sigma - \sigma_h\|_{L^2}$ err.	%	order	$\|u - u_h\|_{L^2}$ err.	%	order
10	1.75e−01	17.102		8.47e−02	9.503	
20	1.04e−01	10.169	0.75	4.20e−02	4.712	1.01
40	6.17e−02	6.047	0.75	2.09e−02	2.349	1.00
80	3.67e−02	3.595	0.75	1.04e−02	1.173	1.00
160	2.18e−02	2.138	0.75	5.22e−03	0.586	1.00

element. First, we consider the first order Brezzi–Douglas–Marini elements developed in [11] in two dimensions and [20] and [10] in three dimensions:

$$\begin{aligned} \Sigma_h &= \{\, \tau \in H(\operatorname{div}, \Omega; \mathbb{R}^n) \mid \tau|_T \in \mathcal{P}_1(T; \mathbb{R}^n) \quad \forall T \in \mathcal{T}_h \,\}, \\ V_h &= \{\, v \in L^2(\Omega) \mid v|_T \in \mathcal{P}_0(T) \quad \forall T \in \mathcal{T}_h \,\}. \end{aligned} \tag{3.2}$$

The difference from the previous choice is that for (3.1), the trial functions for flux are restricted to $H^1(\Omega; \mathbb{R}^n)$, which means that full interelement continuity must be imposed (a piecewise polynomial belongs to H^1 if and only if it is continuous). But for the stable choice (3.2), the flux functions need only belong to $H(\operatorname{div})$, which requires only interelement continuity of the normal component (a piecewise polynomial vector field belongs to $H(\operatorname{div})$ if and only if its normal component is continuous across each $(n-1)$-dimensional face shared by two elements).

In order that the spaces given in (3.2) are implementable via the standard finite element assembly procedure—in fact, in order that they that are finite element spaces in the sense of [13]—we must be able to specify degrees of freedom for the local shape function spaces $\mathcal{P}_1(T; \mathbb{R}^n)$ and $\mathcal{P}_0(T)$ which enforce exactly the required interelement continuity. For the former, we choose the moments of degree at most 1 of the normal component of

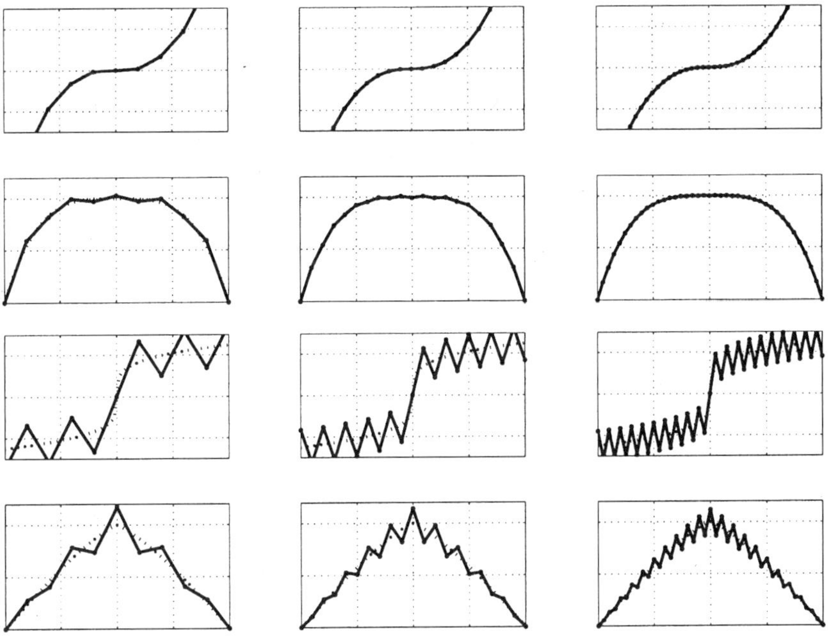

FIG. 2. $\mathcal{P}_1^{cont}$–$\mathcal{P}_1^{cont}$ *mixed method (unstable) for meshes of 10, 20, and 40 elements.*

the field on each face of the element:

$$\tau \mapsto \int_f (\tau \cdot \nu)p, \quad p \in \mathcal{P}_1(f), \quad f \in \Delta_{n-1}(T). \tag{3.3}$$

(We use the notation $\Delta_k(T)$ to denote the set of subsimplices of dimension k of the simplex T, i.e., the set of vertices for $k = 0$, edges for $k = 1$, etc.) Since the normal component of the field is itself linear, these functionals exactly impose the desired continuity of the normal component. Choosing a basis for each of the n-dimensional spaces $\mathcal{P}_1(f)$ for each of $n+1$ faces $f \in \Delta_{n-1}(T)$, we obtain $n(n+1) = \dim P_1(T; \mathbb{R}^n)$ degrees of freedom. These degrees of freedom are clearly unisolvent, since if they all vanish for some $\tau \in \mathcal{P}_1(T)$, then at each vertex $\tau \cdot n$ vanishes for the vector n normal to each face meeting at the vertex. These normal vectors span $\mathbb{R}^n$, so τ itself vanishes at each vertex, and therefore vanishes on all of T. Since the space V_h involves no interelement continuity, we make the obvious choice of degree of freedom for $\mathcal{P}_0(T)$:

$$v \mapsto \int_T v. \tag{3.4}$$

The moments (3.3) determine a projection operator $\Pi_{\Sigma_h} : H^1(\Omega; \mathbb{R}^n) \to \Sigma_h$

TABLE 2
Errors and orders of convergence for the $\mathcal{P}_1^{cont}$–$\mathcal{P}_1^{cont}$ mixed method.

$u = 1 - |x|^{7/2}$

n	$\|\sigma - \sigma_h\|_{L^2}$ err.	%	order	$\|u - u_h\|_{L^2}$ err.	%	order
10	2.09e−02	1.464		2.38e−01	20.429	
20	5.07e−03	0.355	2.04	1.17e−01	10.066	1.02
40	1.25e−03	0.088	2.02	5.85e−02	5.011	1.01
80	3.11e−04	0.022	2.01	2.92e−02	2.502	1.00
160	7.76e−05	0.005	2.00	1.46e−02	1.251	1.00

$u = 1 - |x|^{5/4}$

n	$\|\sigma - \sigma_h\|_{L^2}$ err.	%	order	$\|u - u_h\|_{L^2}$ err.	%	order
10	3.96e−01	38.769		2.24e−01	25.182	
20	3.36e−01	32.875	0.24	1.42e−01	15.974	0.66
40	2.83e−01	27.759	0.24	1.04e−01	11.663	0.45
80	2.39e−01	23.391	0.25	8.23e−02	9.243	0.34
160	2.01e−01	19.689	0.25	6.77e−02	7.601	0.28

onto the first-order Brezzi–Douglas–Marini space, given by

$$\int_f (\Pi_{\Sigma_h}\tau)\cdot\nu\, p = \int_f (\tau\cdot\nu)p, \quad p \in \mathcal{P}_1(f), \quad f \in \Delta_{n-1}(T),$$

while the projection operator $\Pi_{V_h} : L^2(\Omega) \to V_h$ determined by the degrees of freedom (3.4) is simply the usual L^2-projection. An important relation between these operators is expressed by the commutativity of the following diagram:

$$\begin{array}{ccc} H^1(\Omega;\mathbb{R}^n) & \xrightarrow{\text{div}} & L^2(\Omega) \\ \downarrow \Pi_{\Sigma_h} & & \downarrow \Pi_{V_h} \\ \Sigma_h & \xrightarrow{\text{div}} & V_h \end{array} \tag{3.5}$$

(This can be verified via integration by parts.) Note that we have taken $H^1(\Omega;\mathbb{R}^n)$, rather than $H(\text{div},\Omega;\mathbb{R}^n)$, as the domain of Π_{Σ_h}. This is because Π_{Σ_h} defines a bounded operator $H^1(\Omega;\mathbb{R}^n) \to L^2(\Omega;\mathbb{R}^n)$. In fact, it is bounded uniformly in the mesh size h if we restrict to a shape-regular family of triangulations. However, Π_{Σ_h} does not extend to a bounded operator on all of $H(\text{div},\Omega;\mathbb{R}^n)$, because of lack of sufficiently regular traces.

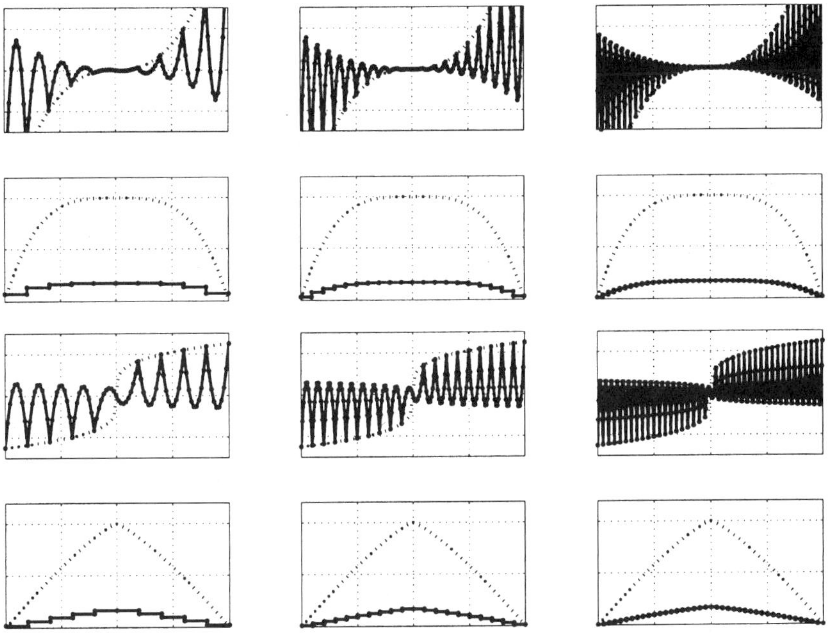

FIG. 3. $\mathcal{P}_2^{cont}$–$\mathcal{P}_0$ *mixed method (unstable).*

The commutative diagram (3.5) encapsulates the properties of the spaces Σ_h and V_h needed to verify the stability conditions (A1) and (A2). First, since $\operatorname{div}\Sigma_h \subset V_h$, any $\tau \in \Sigma_h$ satisfying $\int_\Omega v \operatorname{div}\tau\,dx = 0$ for all $v \in V_h$ is in fact divergence-free, and so (A1) holds. In order to verify (A2), let $v \in V_h$ be given. Since div maps $H^1(\Omega;\mathbb{R}^n)$ onto $L^2(\Omega)$ and admits a bounded right inverse, c.f. [16], we can find $\tilde\tau \in H^1(\Omega;\mathbb{R}^n)$ with $\operatorname{div}\tilde\tau = v$ and $\|\tilde\tau\|_{H^1} \le c\|v\|_{L^2}$. Now let $\tau = \Pi_{\Sigma_h}\tilde\tau$. From the commutative diagram (3.5) we see that

$$\operatorname{div}\tau = \operatorname{div}\Pi_{\Sigma_h}\tilde\tau = \Pi_{V_h}\operatorname{div}\tilde\tau = \Pi_{V_h}v = v.$$

Invoking also the $H^1(\Omega;\mathbb{R}^n) \to L^2(\Omega)$ boundedness of Π_{Σ_h}, we obtain

$$\int_\Omega v \operatorname{div}\tau\,dx = \|v\|_{L^2}^2, \quad \|\tau\|_{H(\operatorname{div})} \le c'\|v\|_{L^2},$$

from which (A2) follows. Thus the first-order Brezzi–Douglas–Marini elements (3.2) are stable in n dimensions.

Note that (3.2) coincides with (3.1) in the case $n = 1$, so these elements are indeed a generalization to higher dimensions of the simple $\mathcal{P}_1^{\text{cont}}$–$\mathcal{P}_0$ elements. Moreover, they can be viewed as the lowest order case of a family of stable elements of arbitrary order:

$$\begin{aligned}\Sigma_h &= \{\, \tau \in H(\operatorname{div}, \Omega; \mathbb{R}^n) \mid \tau|_T \in \mathcal{P}_r(T; \mathbb{R}^n) \quad \forall T \in \mathcal{T}_h \,\}, \\ V_h &= \{\, v \in L^2(\Omega) \mid v|_T \in \mathcal{P}_{r-1}(T) \quad \forall T \in \mathcal{T}_h \,\}.\end{aligned} \tag{3.6}$$

The interelement continuity for Σ_h can be specified by continuity of the moments

$$\tau \mapsto \int_f (\tau \cdot n) p, \quad p \in \mathcal{P}_r(f), \quad f \in \Delta_{n-1}(T),$$

and a set of degrees of freedom determined by these moments with p restricted to a basis in each $\mathcal{P}_r(f)$, together with and an additional $(r-1)\binom{n+r-1}{r}$ moments over T, about which more will be said below. In one dimension, this is just the $\mathcal{P}_r^{\text{cont}}$–$\mathcal{P}_{r-1}$ element discussed at the end of the last section. The first two elements in the Brezzi–Douglas–Marini family in two dimensions are pictured in Figure 4.

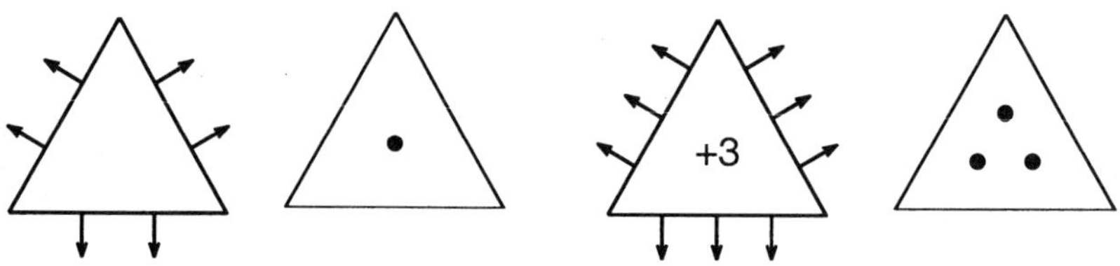

FIG. 4. *Brezzi–Douglas–Marini element pairs for* $r = 1$ *and* 2 *in two dimensions.*

Although the Brezzi–Douglas–Marini family of elements provide a natural analogue of the $\mathcal{P}_r^{\text{cont}}$–$\mathcal{P}_{r-1}$ family of elements to higher dimensions, it is not the only such analogue. Another is the Raviart–Thomas family introduced in [21] and improved and extended from two to three dimensions in [19]. To describe it, we define for $T \subset \mathbb{R}^n$ and integer $r \geq 0$,

$$\mathcal{RT}_r(T) = \{\tau : T \to \mathbb{R}^n \,|\, \tau(x) = \alpha(x) + x\beta(x),\ \alpha \in \mathcal{P}_r(T; \mathbb{R}^n), \beta \in \mathcal{P}_r(T)\}. \tag{3.7}$$

Then the Raviart–Thomas elements of index $r \geq 0$ are

$$\begin{aligned}\Sigma_h &= \{\, \tau \in H(\operatorname{div}, \Omega; \mathbb{R}^n) \mid \tau|_T \in \mathcal{RT}_r(T) \quad \forall T \in \mathcal{T}_h \,\}, \\ V_h &= \{\, v \in L^2(\Omega) \mid v|_T \in \mathcal{P}_r(T) \quad \forall T \in \mathcal{T}_h \,\},\end{aligned} \tag{3.8}$$

with some element diagrams shown in Figure 5. These elements are defined in all dimensions. In dimension one, the Raviart–Thomas elements (3.8) coincide with the Brezzi–Douglas–Marini elements (3.6) with r replaced by $r + 1$. But for $n \geq 2$, these families are distinct.

4. Exterior calculus. The finite elements described above, and others, can better be understood with the help of differential forms and exterior calculus. We begin by recalling the basic notions of exterior algebra. (For

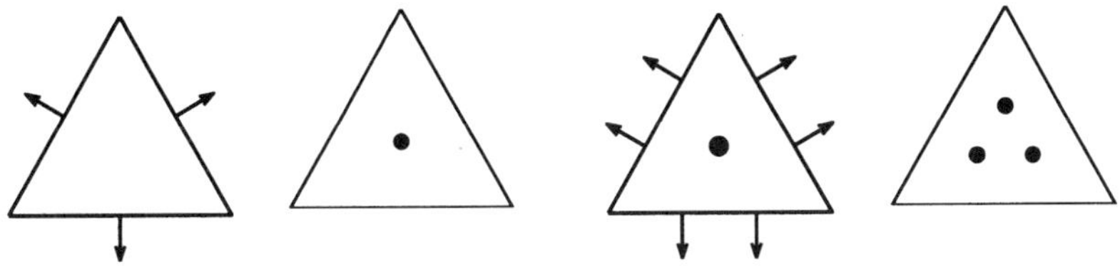

FIG. 5. *Raviart–Thomas element pairs for $r = 0$ and 1 in two dimensions.*

details see, e.g., [3], Ch. 7. Let V be a vector space of dimension n. We denote by $\mathrm{Alt}^k V$ the space of exterior k-forms on V, i.e., of alternating k-linear maps $V \times \cdots \times V \to \mathbb{R}$. That is, an element of $\mathrm{Alt}^k V$ assigns a real number to k elements of V, is linear in each argument, and reverses sign when two arguments are interchanged. In particular, $\mathrm{Alt}^1 V$ is simply the dual space V^* and $\mathrm{Alt}^0 V$ may be identified with $\mathbb{R}$. For $k > n$, $\mathrm{Alt}^k V = 0$, while for all k we have

$$\dim \mathrm{Alt}^k V = \binom{n}{k}.$$

A form in the one-dimensional space $\mathrm{Alt}^n V$ is uniquely determined by its value on any one coordinate frame (i.e., ordered basis). The value of the form on any other ordered n-tuple of vectors can be obtained by expanding the vectors in the coordinate frame to obtain a matrix, and multiplying the value of the form on the coordinate frame by the determinant of the matrix.

An inner product on V determines an inner product on each $\mathrm{Alt}^k V$ by the formula

$$\langle \omega, \eta \rangle = \sum_{1 \le \sigma_1 < \cdots < \sigma_k \le n} \omega(v_{\sigma_1}, \ldots, v_{\sigma_k}) \eta(v_{\sigma_1}, \ldots, v_{\sigma_k}), \quad \omega, \eta \in \mathrm{Alt}^k V \quad (4.1)$$

for any orthonormal basis $v_1, \ldots, v_n$ (the right hand side is independent of the choice of orthonormal basis).

We recall also the *exterior product* $\wedge : \mathrm{Alt}^j V \times \mathrm{Alt}^k V \to \mathrm{Alt}^{j+k} V$ defined by

$$(\omega \wedge \eta)(v_1, \ldots, v_{j+k}) = \sum_{\sigma \in \Sigma(j, j+k)} (\operatorname{sign} \sigma) \omega(v_{\sigma_1}, \ldots, v_{\sigma_j}) \eta(v_{\sigma_{j+1}}, \ldots, v_{\sigma_{j+k}}),$$
$$\omega \in \mathrm{Alt}^j V, v_i \in V,$$

where $\Sigma(j, j+k)$ is the set of all permutations of $\{1, \ldots, j+k\}$, for which $\sigma_1 < \sigma_2 < \cdots \sigma_j$ and $\sigma_{j+1} < \sigma_{j+2} < \cdots \sigma_{j+k}$.

In the case $V = \mathbb{R}^n$, there is a canonical basis, and we denote by $dx^1, \ldots, dx^n$ the elements of the dual basis, which form a canonical basis for $\mathrm{Alt}^1 \mathbb{R}^n$. Then a canonical basis for $\mathrm{Alt}^k \mathbb{R}^n$ consists of the forms $dx^{\sigma_1} \wedge \cdots \wedge dx^{\sigma_k}$, where $1 \le \sigma_1 < \cdots < \sigma_k \le n$.

For readers less familiar with exterior algebra, it is worthwhile to examine in detail the example $V = \mathbb{R}^3$, endowed with the usual inner product and orientation. In this case

- The general element of $\mathrm{Alt}^0\mathbb{R}^3$ is c, $c \in \mathbb{R}$.
- The general element of $\mathrm{Alt}^1\mathbb{R}^3$ is $\langle u, \cdot \rangle$, or, equivalently, $u_1 dx^1 + u_2 dx^2 + u_3 dx^3$, $u \in \mathbb{R}^3$.
- The general element of $\mathrm{Alt}^2\mathbb{R}^3$ is $\langle w, \cdot \times \cdot \rangle$, or, equivalently, $w_1 dx^2 \wedge dx^3 - w_2 dx^1 \wedge dx^3 + w_3 dx^1 \wedge dx^2$, $w \in \mathbb{R}^3$.
- The general element of $\mathrm{Alt}^3\mathbb{R}^3$ is $g\langle \cdot, \cdot \times \cdot \rangle$, or, equivalently, $g dx^1 \wedge dx^2 \wedge dx^3$, $g \in \mathbb{R}$.

Thus we may identify $\mathrm{Alt}^0\mathbb{R}^3$ and $\mathrm{Alt}^3\mathbb{R}^3$ with $\mathbb{R}$ and $\mathrm{Alt}^1\mathbb{R}^3$ and $\mathrm{Alt}^2\mathbb{R}^3$ with $\mathbb{R}^3$.

Next we identify the exterior product $\mathrm{Alt}^j\mathbb{R}^3 \times \mathrm{Alt}^k\mathbb{R}^3 \to \mathrm{Alt}^{j+k}\mathbb{R}^3$ for $0 \le j \le k$, $j + k \le 3$. (The exterior product for other values of j, k either follows from these or is identically zero.) If $j = 0$, the exterior product is the ordinary scalar multiplication. The exterior product $\mathrm{Alt}^1\mathbb{R}^3 \times \mathrm{Alt}^1\mathbb{R}^3 \to \mathrm{Alt}^2\mathbb{R}^3$ corresponds under our identifications to the usual cross product $\mathbb{R}^3 \times \mathbb{R}^3 \to \mathbb{R}^3$. Finally, the exterior product $\mathrm{Alt}^1\mathbb{R}^3 \times \mathrm{Alt}^2\mathbb{R}^3 \to \mathrm{Alt}^3\mathbb{R}^3$ corresponds to the usual inner product $\mathbb{R}^3 \times \mathbb{R}^3 \to \mathbb{R}$. It is straightforward to check that given the identifications mentioned, the inner product defined above on $\mathrm{Alt}^k\mathbb{R}^3$ is the usual product in $\mathbb{R}$ for $k = 0$ or 3, and the Euclidean product in $\mathbb{R}^3$ for $k = 1$ or 2.

Having reviewed the basic definitions of exterior algebra, we now turn to exterior calculus. If Ω is any smooth manifold, we define a smooth differential k-form on Ω as a mapping ω which assigns to each $x \in \Omega$ an alternating linear form $\omega_x \in \mathrm{Alt}^k(T_x\Omega)$ on the tangent space $T_x\Omega$ to Ω at x. We denote the space of all smooth differential k-forms on Ω by $\Lambda^k(\Omega)$. We write $C^0\Lambda^k(\Omega)$ to denote the larger space of all continuous differential forms, and use a similar notation for other functional spaces. For example, if Ω is a Riemannian manifold, we can talk about $L^2\Lambda^k(\Omega)$, etc.

Differential forms can be integrated and differentiated without the need for any additional structure, such as a measure or metric, on the manifold Ω. If $0 \le k \le n$ is an integer, f is an oriented piecewise smooth k-dimensional submanifold of Ω, and ω is a k-form, then the *integral* $\int_f \omega \in \mathbb{R}$ is well-defined. Thus, for example, 0-forms can be evaluated at points, 1-forms can be integrated on curves, and 2-forms can be integrated over surfaces. Also, for each such k, the *exterior derivative* d_k is a linear operator mapping $\Lambda^k(\Omega)$ into $\Lambda^{k+1}(\Omega)$. The de Rham complex of Ω is the sequence of maps

$$\mathbb{R} \hookrightarrow \Lambda^0(\Omega) \xrightarrow{d} \Lambda^1(\Omega) \xrightarrow{d} \cdots \xrightarrow{d} \Lambda^n(\Omega) \to 0 \tag{4.2}$$

where we have followed the usual convention of suppressing the subscript on the d_k. This is a complex in the sense that the composition of two consecutive maps is zero ($d_k d_{k-1} = 0$), and we can consider the kth de Rham cohomology space, defined to be the quotient of the null space of d_k modulo the range of d_{k-1}. If the manifold is contractible, this complex is exact in the sense that the cohomology spaces all vanish, or, equivalently, the range of each map is precisely equal to (and not just contained in) the null space of the succeeding map.

Assuming that Ω is a Riemannian manifold, so each tangent space $T_x\Omega$ is endowed with an inner product, we have an inner product on each $\Lambda^k(\Omega)$ which can be completed to a Hilbert space $L^2\Lambda^k(\Omega)$. We can define the Sobolev space of differential k-forms:

$$H\Lambda^k(\Omega) = \{\, \omega \in L^2\Lambda^k(\Omega) \,|\, d\omega \in L^2\Lambda^{k+1}(\Omega) \,\}.$$

The L^2 *de Rham complex*

$$\mathbb{R} \hookrightarrow H\Lambda^0(\Omega) \xrightarrow{d} H\Lambda^1(\Omega) \xrightarrow{d} \cdots \xrightarrow{d} H\Lambda^n(\Omega) \to 0$$

has the same cohomology as the smooth de Rham complex.

Viewing d_k as a (closed, densely-defined) unbounded linear operator mapping $L^2\Lambda^k(\Omega)$ to $L^2\Lambda^{k+1}(\Omega)$ with domain $H\Lambda^k(\Omega)$, we may use the inner product of differential forms to define the adjoint d_k^* which maps a dense subspace of $L^2\Lambda^{k+1}(\Omega)$ to $L^2\Lambda^k(\Omega)$. Namely, $\omega \in L^2\Lambda^{k+1}(\Omega)$ belongs to the domain of d_k^* if there exists $d_k^*\omega \in L^2\Lambda^k(\Omega)$ such that

$$\langle d^*\omega, \eta\rangle_{L^2\Lambda^k} = \langle \omega, d\eta\rangle_{L^2\Lambda^{k+1}}, \qquad \eta \in H\Lambda^k(\Omega).$$

The *Hodge Laplacian* is then the map $d^*d + dd^*$ (or, more precisely, $d_k^*d_k + d_{k-1}d_{k-1}^*$) which maps a part of $L^2\Lambda^k(\Omega)$ into $L^2\Lambda^k(\Omega)$.

In case Ω is an open subset of $\mathbb{R}^n$, every differential k-form may be written uniquely in the form

$$\omega_x = \sum_{i_1<\cdots<i_k} a_{i_1\ldots i_k}(x)dx^{i_1} \wedge \cdots \wedge dx^{i_k}, \tag{4.3}$$

for some smooth functions $a_{i_1\ldots i_k} : \Omega \to \mathbb{R}$. This is useful for computing the exterior derivative since:

$$d(a\, dx^{i_1} \wedge \cdots \wedge dx^{i_k}) = \sum_{j=1}^{n} \frac{\partial a}{\partial x^j}\, dx^j \wedge dx^{i_1} \wedge \cdots \wedge dx^{i_k}.$$

For use later in the paper, we introduce the Sobolev space $H^s\Lambda^k(\Omega)$ consisting of differential forms of the form (4.3) for which the coefficients $a_{i_1\ldots i_k} \in H^s(\Omega)$. The corresponding norm is given by $\|\omega\|_s = (\sum \|a_{i_1\ldots i_k}\|_s^2)^{1/2}$, which we write simply as $\|\omega\|$ if $s = 0$.

For $\Omega \subset \mathbb{R}^n$, we can also define the notion of polynomial differential forms. Namely, we say that $\omega \in \Lambda^k(\Omega)$ is a (homogeneous) polynomial k-form of degree r if for any choice $v^1, \ldots, v^k \in \mathbb{R}^n$, the map

$$x \mapsto \omega_x(v^1, \ldots, v^k)$$

is the restriction to Ω of a (homogeneous) polynomial of degree r. For ω given by (4.3), this is equivalent to saying that each of the coefficients $a_{i_1\dots i_k}$ is a (homogeneous) polynomial of degree r. We denote the spaces of polynomial k-forms of degree r and homogeneous k-forms of degree r by $\mathcal{P}_r\Lambda^k(\Omega)$ and $\mathcal{H}_r\Lambda^k(\Omega)$, respectively. We shall verify below that the *polynomial de Rham complex*

$$\mathbb{R} \hookrightarrow \mathcal{P}_r\Lambda^0(\Omega) \xrightarrow{d} \mathcal{P}_{r-1}\Lambda^1(\Omega) \xrightarrow{d} \cdots \xrightarrow{d} \mathcal{P}_{r-n}\Lambda^n(\Omega) \to 0 \qquad (4.4)$$

is exact for every $r \geq 0$ (with the understanding that $\mathcal{H}_m = \mathcal{P}_m = 0$ for $m < 0$). The same is true for the homogeneous polynomial de Rham sequence

$$R \hookrightarrow \mathcal{H}_r\Lambda^0(\Omega) \xrightarrow{d} \mathcal{H}_{r-1}\Lambda^1(\Omega) \xrightarrow{d} \cdots \xrightarrow{d} \mathcal{H}_{r-n}\Lambda^n(\Omega) \to 0 \qquad (4.5)$$

where $R = \mathbb{R}$ if $r = 0$ and $R = 0$ otherwise.

Finally, still in the case $\Omega \subset \mathbb{R}^n$, we introduce the *Koszul differential* $\kappa = \kappa_k : \Lambda^k \to \Lambda^{k-1}$, defined by

$$(\kappa\omega)_x(v^1, \dots, v^{k-1}) = \omega_x(x, v^1, \dots, v^{k-1}). \qquad (4.6)$$

Note that $\kappa_{k-1}\kappa_k = 0$. Also, κ maps $\mathcal{H}_r\Lambda^k(\Omega)$ into $\mathcal{H}_{r+1}\Lambda^{k-1}(\Omega)$, i.e., the Koszul differential increases polynomial degree and decreases the order of the differential form, exactly the opposite of exterior differentiation, which maps $\mathcal{H}_{r+1}\Lambda^{k-1}(\Omega)$ into $\mathcal{H}_r\Lambda^k(\Omega)$. The two operations are connected by the formula

$$(d\kappa + \kappa d)\omega = (r + k)\omega, \quad \omega \in \mathcal{H}_r\Lambda^k(\Omega). \qquad (4.7)$$

This can be used to establish exactness of the homogeneous polynomial de Rham sequence (4.5), and also of the homogeneous *Koszul complex*

$$0 \to \mathcal{H}_{r-n}\Lambda^n(\Omega) \xrightarrow{\kappa} \mathcal{H}_{r-n+1}\Lambda^{n-1}(\Omega) \xrightarrow{\kappa} \cdots \xrightarrow{\kappa} \mathcal{H}_r\Lambda^0(\Omega) \to R \to 0$$

where again $R = \mathbb{R}$ if $r = 0$ and $R = 0$ otherwise. Adding over polynomial degrees we get the exactness of (4.4) and of the Koszul complex

$$0 \to \mathcal{P}_{r-n}\Lambda^n(\Omega) \xrightarrow{\kappa} \mathcal{P}_{r-n+1}\Lambda^{n-1}(\Omega) \xrightarrow{\kappa} \cdots \xrightarrow{\kappa} \mathcal{P}_r\Lambda^0(\Omega) \to \mathbb{R} \to 0$$

We use the Koszul differential to define an important space of polynomial forms on a domain $T \subset \mathbb{R}^n$:

$$\mathcal{P}_r^+\Lambda^k(T) = \mathcal{P}_r\Lambda^k(T) + \kappa\mathcal{P}_r\Lambda^{k+1}(T),$$

where κ is the Koszul differential defined in (4.6). Clearly

$$\mathcal{P}_r^+\Lambda^k(T) - \mathcal{P}_r\Lambda^k(T) + \kappa\mathcal{H}_r\Lambda^{k+1}(T)$$

and, in view of (4.7),

$$\mathcal{P}_r^+\Lambda^k(T) + d\mathcal{H}_{r+2}\Lambda^{k-1} = \mathcal{P}_{r+1}\Lambda^k(T).$$

For 0-forms and n-forms, the $\mathcal{P}^+$ spaces are nothing new:

$$\mathcal{P}_r^+\Lambda^0(T) = \mathcal{P}_{r+1}\Lambda^0(T), \quad \mathcal{P}_r^+\Lambda^n(T) = \mathcal{P}_r\Lambda^n(T).$$

However, for $0 < k < n$

$$\mathcal{P}_r\Lambda^k(T) \subsetneq \mathcal{P}_r^+\Lambda^k(T) \subsetneq \mathcal{P}_{r+1}\Lambda^k(T).$$

If we identify $\Lambda^{n-1}(T)$ with $C^\infty(T;\mathbb{R}^n)$, then $\mathcal{P}_r^+\Lambda^{n-1}(T)$ corresponds exactly to the space of Raviart–Thomas polynomial fields defined in (3.7). In the general case, we may compute their dimensions:

$$\dim \mathcal{P}_r^+\Lambda^k(T) = \binom{n+r}{n}\binom{n}{k} + \binom{n+r}{n-k-1}\binom{r+k}{k},$$

while

$$\dim \mathcal{P}_r\Lambda^k(T) = \binom{n+r}{n}\binom{n}{k}.$$

Finally, we specialize to the case $\Omega \subset \mathbb{R}^3$. Then $T_x\mathbb{R}^3 \cong \mathbb{R}^3$ and $\mathrm{Alt}^k T_x\mathbb{R}^3$ may be identified with $\mathbb{R}$ for $k = 0, 3$ and with $\mathbb{R}^3$ for $k = 1, 2$. We can then interpret the integral in the sense of differential forms as follows. If ω is a 0-form, and v a point in Ω, then $\int_v \omega = \omega(v)$. If ω is a function on Ω which we identify with a 1-form and e is an oriented curve in Ω, then the differential form integral $\int_e \omega = \int_e \omega \cdot t\, d\mathfrak{H}_1$ where t is the unit tangent to e (determined uniquely by the orientation) and $\mathfrak{H}_1$ is 1-dimensional Hausdorff measure. If ω is a function on Ω which we identify with a 2-form and f is an oriented surface in Ω, then the differential form integral $\int_f \omega = \int_f \omega \cdot \nu\, d\mathfrak{H}_2$ where ν is the unit normal to f and $\mathfrak{H}_2$ is 2-dimensional Hausdorff measure. Finally, if T is an open subset of Ω and ω a 3-form, then $\int_T \omega$ is equal to the usual integral of the corresponding function with respect to Lebesgue measure.

Continuing with the identification of forms on $\mathbb{R}^3$ with functions and vector fields, we find that $d_0 = \mathrm{grad}$, $d_1 = \mathrm{curl}$, $d_2 = \mathrm{div}$, κ_3 is multiplication of a scalar field by x to get a vector field, κ_2 takes the cross product of a vector field with x to produce another vector field, and κ_1 takes the dot product of a vector field with x. Thus the differential complexes discussed above can be written as follows.

The smooth de Rham complex:

$$\mathbb{R} \hookrightarrow C^\infty(\Omega) \xrightarrow{\mathrm{grad}} C^\infty(\Omega;\mathbb{R}^3) \xrightarrow{\mathrm{curl}} C^\infty(\Omega;\mathbb{R}^3) \xrightarrow{\mathrm{div}} C^\infty(\Omega) \to 0.$$

The L^2 de Rham complex:

$$\mathbb{R} \hookrightarrow H^1(\Omega) \xrightarrow{\mathrm{grad}} H(\mathrm{curl},\Omega;\mathbb{R}^3) \xrightarrow{\mathrm{curl}} H(\mathrm{div},\Omega;\mathbb{R}^3) \xrightarrow{\mathrm{div}} L^2(\Omega) \to 0.$$

The polynomial de Rham complex:

$$\mathbb{R} \hookrightarrow \mathcal{P}_r(\Omega) \xrightarrow{\text{grad}} \mathcal{P}_{r-1}(\Omega;\mathbb{R}^3) \xrightarrow{\text{curl}} \mathcal{P}_{r-2}(\Omega;\mathbb{R}^3) \xrightarrow{\text{div}} \mathcal{P}_{r-3}(\Omega) \to 0.$$

The Koszul complex:

$$0 \to \mathcal{P}_{r-3}(\Omega) \xrightarrow{x} \mathcal{P}_{r-2}(\Omega;\mathbb{R}^3) \xrightarrow{\times x} \mathcal{P}_{r-1}(\Omega;\mathbb{R}^3) \xrightarrow{\cdot x} \mathcal{P}_r(\Omega) \to \mathbb{R} \to 0.$$

The Hodge Laplacian on 0-forms and 3-forms is the ordinary Laplacian $\Delta = \operatorname{div}\operatorname{grad}$ viewed as an unbounded operator on $L^2(\Omega)$ with certain boundary conditions imposed in its domain (basically, Neumann conditions in the case of 0-forms and Dirichlet conditions in the case of 3-forms). Similarly, the Hodge Laplacian on 1-forms and 2-forms gives the vector Laplacian $\operatorname{curl}\operatorname{curl} - \operatorname{grad}\operatorname{div}$ with two different sets of boundary conditions. We will say more on this in Section 6.

5. Piecewise polynomial differential forms. Let $\mathcal{T}$ be a triangulation by simplices of a domain $\Omega \subset \mathbb{R}^n$. In this section we define, in a unified fashion, a variety of finite-dimensional spaces of differential forms on Ω which are piecewise polynomials with respect to the triangulation $\mathcal{T}$. In the cases where we can identify differential forms with functions and vector fields on Ω, these spaces correspond to well-known finite element spaces, such as the Lagrange space, the Brezzi–Douglas–Marini spaces, the Raviart–Thomas spaces, and the Nedelec spaces of [19, 20].

We begin by describing a set of degrees of freedom for the polynomial spaces $\mathcal{P}_r\Lambda^k(T)$ and $\mathcal{P}_r^+\Lambda^k(T)$, which will reveal a strong connection between the $\mathcal{P}_r$ and $\mathcal{P}_r^+$ spaces. For T a simplex in $\mathbb{R}^n$, $0 \le k \le n$, and $r > 0$, an element $\omega \in \mathcal{P}_r\Lambda^k(T)$ is uniquely determined by the following quantities (see [20] for the case $n = 3$):

$$\int_f \omega \wedge \zeta, \quad \zeta \in \mathcal{P}^+_{r-d-1+k}\Lambda^{d-k}(f), \quad f \in \Delta_d(T), \quad k \le d \le n. \tag{5.1}$$

(For $r < 0$, we interpret $\mathcal{P}_r^+\Lambda^k(T) = \mathcal{P}_r\Lambda^k(T) = 0$.) Note that for $\omega \in \Lambda^k(T)$, ω naturally restricts on the face f to an element of $\Lambda^k(f)$. Therefore, for $\zeta \in \Lambda^{d-k}(f)$, the wedge product $\omega \wedge \zeta$ belongs to $\Lambda^d(f)$ and hence the integral of $\omega \wedge \zeta$ on the d-dimensional face f of T is a well-defined and natural quantity. A set of degrees of freedom for $\mathcal{P}_r\Lambda^k(T)$ is obtained from the quantities in (5.1) by restricting the weighting forms ζ to bases of the spaces $\mathcal{P}^+_{r-d-1+k}\Lambda^{d-k}(f)$. Notice that the degrees of freedom for a $\mathcal{P}\Lambda$ type space involve the moments on faces weighted by elements of $\mathcal{P}^+\Lambda$ type spaces. The reverse is true as well. The degrees of freedom for $\omega \in \mathcal{P}_r^+\Lambda^k(T)$ are obtained in a similar way (by selecting bases) from the moments

$$\int_f \omega \wedge \zeta, \quad \zeta \in \mathcal{P}_{r-d+k}\Lambda^{d-k}(f), \quad f \in \Delta_d(T), \quad k \le d \le n. \tag{5.2}$$

This set of degrees of freedom was presented in [17]. See also [19].

In this way, we obtain two families of piecewise polynomial k-forms, each indexed by polynomial degree r:

$$\mathcal{P}_r\Lambda^k(\mathcal{T}) = \{\,\omega \in H\Lambda^k(\Omega) \mid \omega_{|T} \in \mathcal{P}_r\Lambda^k(T) \quad \forall T \in \mathcal{T}\,\},$$
$$\mathcal{P}_r^+\Lambda^k(\mathcal{T}) = \{\,\omega \in H\Lambda^k(\Omega) \mid \omega_{|T} \in \mathcal{P}_r^+\Lambda^k(T) \quad \forall T \in \mathcal{T}\,\}.$$

We believe these should be regarded as the most natural finite element approximations of the Sobolev differential form spaces $H\Lambda^k(\Omega)$. This is certainly true in $n = 1$ dimension, where for each $r \ge 0$ and partition of the domain, we obtain a unique finite element discretization of $H^1(\Omega)$ and of $L^2(\Omega)$: the $\mathcal{P}_{r+1}^{\text{cont}}$ space of continuous piecewise polynomials of degree $r+1$ and the $\mathcal{P}_r$ space of all piecewise polynomials of degree r, respectively. For $k = 0$ in any number of dimensions, then

$$\mathcal{P}_r^+\Lambda^0(\mathcal{T}) = \mathcal{P}_{r+1}\Lambda^0(\mathcal{T})$$

is the usual Lagrange space of all continuous piecewise polynomials of degree $r+1$, the most natural discretization of $H^1(\Omega) \cong H\Lambda^0(\Omega)$. For $k = n$, we get

$$\mathcal{P}_r^+\Lambda^n(\mathcal{T}) = \mathcal{P}_r\Lambda^n(\mathcal{T})$$

is the space of all piecewise polynomials of degree r, the most natural discretization of $L^2(\Omega) \cong H\Lambda^n(\Omega)$. For $k = n-1$, we may identify $H\Lambda^{n-1}(\Omega)$ with $H(\text{div}, \Omega; \mathbb{R}^n)$, and $\mathcal{P}_r^+\Lambda^{n-1}(\mathcal{T})$ is the Raviart–Thomas space of index r and $\mathcal{P}_r\Lambda^{n-1}(\mathcal{T})$ the Brezzi–Douglas–Marini space of index r, the best known discretizations of $H(\text{div})$. Finally, for $k = 1$, $n = 3$, $H\Lambda^1(\Omega)$ can be identified with $H(\text{curl}, \Omega; \mathbb{R}^3)$ and $\mathcal{P}_r^+\Lambda^1(\mathcal{T})$ and $\mathcal{P}_r\Lambda^1(\mathcal{T})$ are the Nedelec finite element spaces of the first and second kind, respectively, the best known spaces of $H(\text{curl})$ elements, illustrated in Figure 6.

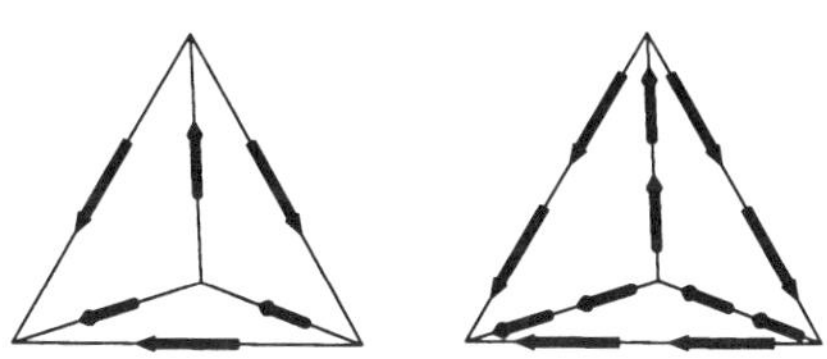

FIG. 6. *Lowest order Nedelec H(curl) elements of the first kind and the second kind.*

These spaces fit together to provide a number of piecewise polynomial analogues of the de Rham complex. For any $r \ge 0$, we have the complex

$$\mathbb{R} \hookrightarrow \mathcal{P}_r^+\Lambda^0(\mathcal{T}) \xrightarrow{d} \mathcal{P}_r^+\Lambda^1(\mathcal{T}) \xrightarrow{d} \cdots \xrightarrow{d} \mathcal{P}_r^+\Lambda^n(\mathcal{T}) \to 0. \tag{5.3}$$

In case $r = 0$, this is the complex of piecewise polynomial forms introduced by Whitney to calculate de Rham cohomology [22]. It has the same cohomology spaces as the smooth de Rham complex, so, in particular, is exact if Ω is contractible. The connection betweens Whitney's forms and mixed finite elements was recognized by Bossavit [7]. Using an element diagram to stand in for the corresponding finite element space, in $n = 3$ dimensions the complex of Whitney forms may be represented

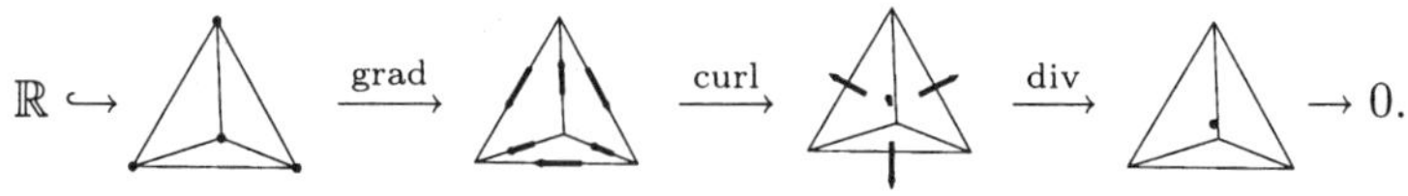

The degrees of freedom in (5.1) and (5.2) determine projection operators $\Pi_r^k : \Lambda^k(\Omega) \to \mathcal{P}_r\Lambda^k(\mathcal{T})$ and $\Pi_{r+}^k : \Lambda^k(\Omega) \to \mathcal{P}_r^+\Lambda^k(\mathcal{T})$ respectively. These may be used to relate the smooth de Rham complex (4.2) to the piecewise polynomial de Rham complex (5.3). Namely, the following diagram commutes:

$$\begin{array}{ccccccccc}
\mathbb{R} \hookrightarrow & \Lambda^0(\Omega) & \xrightarrow{d} & \Lambda^1(\Omega) & \xrightarrow{d} \cdots \xrightarrow{d} & \Lambda^n(\Omega) & \longrightarrow 0 \\
& \Big\downarrow \Pi_{r+}^0 & & \Big\downarrow \Pi_{r+}^1 & & \Big\downarrow \Pi_{r+}^n & \\
\mathbb{R} \hookrightarrow & \mathcal{P}_r^+\Lambda^0(\mathcal{T}) & \xrightarrow{d} & \mathcal{P}_r^+\Lambda^1(\mathcal{T}) & \xrightarrow{d} \cdots \xrightarrow{d} & \mathcal{P}_r^+\Lambda^n(\mathcal{T}) & \longrightarrow 0.
\end{array}$$

Another piecewise polynomial differential complex with the same cohomology uses the $\mathcal{P}_r\Lambda^k$ spaces:

$$\mathbb{R} \hookrightarrow \mathcal{P}_{r+n}\Lambda^0(\mathcal{T}) \xrightarrow{d} \mathcal{P}_{r+n-1}\Lambda^1(\mathcal{T}) \xrightarrow{d} \cdots \xrightarrow{d} \mathcal{P}_r\Lambda^n(\mathcal{T}) \to 0. \quad (5.4)$$

The complex (5.3) is a subcomplex of (5.4), in the sense that each space occurring in the former complex is a subspace of the corresponding space in the latter complex. The complex (5.4) appears, generalized to the case of degree varying by the element, in [14]. Note that this complex ends with the same space $\mathcal{P}_r\Lambda^n(\Omega) = \mathcal{P}_r^+\Lambda^n(\Omega)$ as (5.4), but in contrast with (5.3) the degree index r decreases with increasing differential form order k.

In one dimension the two complexes (5.3), (5.4) coincide, but in two dimensions they are distinct. In $n > 2$ dimensions there are additional piecewise polynomial complexes which can be built from the same $\mathcal{P}_r\Lambda^k$ and $\mathcal{P}_r\Lambda^{k+1}$ spaces, have the same cohomology, and end in the same space $\mathcal{P}_r\Lambda^0(\mathcal{T})$. These are intermediate between (5.3) and (5.4), and strictly ordered by the subcomplex relationship. Specifically, there are 2^{n-1} such piecewise polynomials complexes in n dimensions. In three dimensions the other two are

$$\mathbb{R} \hookrightarrow \mathcal{P}_{r+2}\Lambda^0(\mathcal{T}) \xrightarrow{d} \mathcal{P}_{r+1}\Lambda^1(\mathcal{T}) \xrightarrow{d} \mathcal{P}_r^+\Lambda^2(\mathcal{T}) \xrightarrow{d} \mathcal{P}_r\Lambda^3(\mathcal{T}) \to 0$$

and

$$\mathbb{R} \hookrightarrow \mathcal{P}_{r+2}\Lambda^0(\mathcal{T}) \xrightarrow{d} \mathcal{P}_{r+1}^+\Lambda^1(\mathcal{T}) \xrightarrow{d} \mathcal{P}_{r+1}\Lambda^2(\mathcal{T}) \xrightarrow{d} \mathcal{P}_r\Lambda^3(\mathcal{T}) \to 0.$$

6. Differential complexes and stability. Let Ω be a contractible subdomain of $\mathbb{R}^n$ and $0 \le k \le n$ an integer. Given $f \in L^2\Lambda^k(\Omega)$, define $\mathcal{L} : H\Lambda^{k-1}(\Omega) \times H\Lambda^k(\Omega) \to \mathbb{R}$ by

$$\mathcal{L}(\tau, v) = \int_\Omega \left(\frac{1}{2}\langle \tau, \tau \rangle - \langle d\tau, v \rangle - \frac{1}{2}\langle dv, dv \rangle + \langle f, v \rangle \right) dx,$$

where the angular brackets indicate the inner product of forms as defined in (4.1). Then $\mathcal{L}$ admits a unique critical point, $(\sigma, u) \in H\Lambda^{k-1}(\Omega) \times H\Lambda^k(\Omega)$ determined by the equations

$$\int_\Omega \langle \sigma, \tau \rangle \, dx = \int_\Omega \langle d\tau, u \rangle \, dx \quad \forall \tau \in H\Lambda^{k-1}(\Omega), \tag{6.1}$$

$$\int_\Omega \langle d\sigma, v \rangle \, dx + \int_\Omega \langle du, dv \rangle \, dx = \int_\Omega \langle f, v \rangle \, dx \quad \forall v \in H\Lambda^k(\Omega). \tag{6.2}$$

Note that this critical point is a saddle point—a minimizer with respect to σ and a maximizer with respect to u—but is not generally obtained from a constrained minimization problem for σ via introduction of a Lagrange multiplier. Equations (6.1) and (6.2) are weak formulations of the equations

$$\sigma = d^* u, \quad d\sigma + d^* du = f,$$

respectively, and hence together, give the Hodge Laplacian problem $(dd^* + d^*d)u = f$. Implied as well are the natural boundary conditions that the trace of u as a k-form on $\partial\Omega$ and the trace of du as a $(k+1)$-form on $\partial\Omega$ both must vanish.

Let us consider more concretely the situation in $n = 3$ dimensions, identifying the spaces $L^2\Lambda^k(\Omega)$ with function spaces as described at the end of Section 4. For $k = 3$, (6.1), (6.2) become: find $\sigma \in H(\operatorname{div}, \Omega; \mathbb{R}^3)$, $u \in L^2(\Omega)$ such that

$$\int_\Omega \sigma \cdot \tau \, dx = \int_\Omega \operatorname{div} \tau u \, dx \quad \forall \tau \in H(\operatorname{div}, \Omega; \mathbb{R}^3), \tag{6.3}$$

$$\int_\Omega \operatorname{div} \sigma v \, dx = \int_\Omega f v \, dx \quad \forall v \in L^2(\Omega), \tag{6.4}$$

i.e., the weak formulation of the steady state heat conduction problem (with unit resistivity) discussed in Section 1. This is the standard mixed formulation for the Dirichlet problem for the Poisson equation: (6.3) is equivalent to the differential equation $\sigma = -\operatorname{grad} u$ and the boundary condition $u = 0$, while (6.4) is equivalent to $\operatorname{div} \sigma = f$.

For $k = 2$, the unknowns $\sigma \in H(\operatorname{curl}, \Omega; \mathbb{R}^3)$ and $u \in H(\operatorname{div}, \Omega; \mathbb{R}^3)$ satisfy the differential equations

$$\sigma = \operatorname{curl} u, \quad \operatorname{curl} \sigma - \operatorname{grad} \operatorname{div} u = f,$$

and the boundary conditions $u \times \nu = 0$, $\operatorname{div} u = 0$ on $\partial\Omega$, so this is a mixed formulation for the vectorial Poisson equation

$$(\operatorname{curl}\operatorname{curl} - \operatorname{grad}\operatorname{div})u = f \tag{6.5}$$

with the auxiliary variable $\sigma = \operatorname{curl} u$. For $k = 1$, (6.1), (6.2) is a different mixed formulation of the vectorial Poisson equation (6.5). Now $\sigma \in H^1(\Omega)$ and $u \in H(\operatorname{curl}, \Omega; \mathbb{R}^3)$ satisfy

$$\sigma = -\operatorname{div} u, \quad (\operatorname{grad}\sigma + \operatorname{curl}\operatorname{curl})u = f,$$

with boundary conditions $u \cdot \nu = 0$, $(\operatorname{curl} u) \times \nu = 0$.

Finally, we interpret the case $k = 0$. Here, in view of the L^2 de Rham sequence, we interpret $H\Lambda^{-1}(\Omega)$ as $\mathbb{R}$ with the operator $H\Lambda^{-1}(\Omega) \to H\Lambda^0(\Omega)$ just the inclusion of $\mathbb{R}$ in $H^1(\Omega)$. Thus, the unknowns are $\sigma \in \mathbb{R}$ and $u \in H^1(\Omega)$, (6.1) just gives the equation $\sigma = \int_\Omega u\,dx/\operatorname{meas}(\Omega)$, while (6.2) is

$$\int_\Omega \operatorname{grad} u \cdot \operatorname{grad} v\,dx + \sigma \int_\Omega v\,dx = \int_\Omega fv\,dx \quad \forall v \in H^1(\Omega).$$

Thus we just have the usual weak formulation of the Neumann problem for the Poisson equation (if $\int_\Omega f\,dx = 0$, then $\sigma = 0$).

Returning now to the case of general n, suppose we are given a triangulation, and let

$$\mathbb{R} \hookrightarrow \Lambda_h^0 \xrightarrow{d_h} \Lambda_h^1 \xrightarrow{d_h} \cdots \xrightarrow{d_h} \Lambda_h^n \to 0 \tag{6.6}$$

denote any of the 2^{n-1} piecewise polynomial de Rham complexes discussed in Section 5, e.g., (5.3) or (5.4). Here we use d_h to denote the restriction of the exterior differential d, and we shall denote by d_h^* its adjoint. We further suppose we have a commuting diagram of the form

$$\begin{array}{ccccccccc}
\mathbb{R} \hookrightarrow & \Lambda^0(\Omega) & \xrightarrow{d} & \Lambda^1(\Omega) & \xrightarrow{d} & \cdots & \xrightarrow{d} & \Lambda^n(\Omega) & \longrightarrow 0 \\
& \Pi_h^0\big\downarrow & & \Pi_h^1\big\downarrow & & & & \Pi_h^n\big\downarrow & \\
\mathbb{R} \hookrightarrow & \Lambda_h^0 & \xrightarrow{d_h} & \Lambda_h^1 & \xrightarrow{d_h} & \cdots & \xrightarrow{d_h} & \Lambda_h^n & \longrightarrow 0.
\end{array} \tag{6.7}$$

We shall demonstrate stability of the finite element method: find $\sigma \in \Lambda_h^{k-1}$, $u \in \Lambda_h^k$ such that

$$\int_\Omega \langle \sigma, \tau \rangle\,dx = \int_\Omega \langle d\tau, u \rangle\,dx \quad \forall \tau \in \Lambda_h^{k-1}, \tag{6.8}$$

$$\int_\Omega \langle d\sigma, v \rangle\,dx + \int_\Omega \langle du, dv \rangle\,dx = \int_\Omega \langle f, v \rangle\,dx \quad \forall v \in \Lambda_h^k. \tag{6.9}$$

Let $B : [H\Lambda^{k-1}(\Omega) \times H\Lambda^k(\Omega)] \times [H\Lambda^{k-1}(\Omega) \times H\Lambda^k(\Omega)] \to \mathbb{R}$ denote the bounded bilinear form

$$B(\sigma, u; \tau, v) = \int_\Omega (\langle \sigma, \tau \rangle - \langle d\tau, u \rangle + \langle d\sigma, v \rangle + \langle du, dv \rangle)\, dx.$$

Stability of the method (6.8), (6.9) is equivalent to the inf-sup condition for B restricted to the finite element spaces [4]. That is, we must establish the existence of constants $\gamma > 0$, $C < \infty$ such that for any $(\sigma, u) \in \Lambda_h^{k-1} \times \Lambda_h^k$ there exists $(\tau, v) \in \Lambda_h^{k-1} \times \Lambda_h^k$ with

$$B(\sigma, u; \tau, v) \geq \gamma(\|\sigma\|^2_{H\Lambda^{k-1}} + \|u\|^2_{H\Lambda^k}), \tag{6.10}$$

$$\|\tau\|_{H\Lambda^{k-1}} + \|v\|_{H\Lambda^k} \leq C(\|\sigma\|_{H\Lambda^{k-1}} + \|u\|_{H\Lambda^k}). \tag{6.11}$$

We shall do so by proving the existence of a discrete Hodge decomposition (Lemma 6.1) and some estimates associated with it (Lemma 6.2). Such discrete Hodge decompositions have been used to establish the stability of mixed methods in specific cases going back at least as far as [15]. See also [6] for a more recent exposition.

LEMMA 6.1. *Given $u \in \Lambda_h^k$, there exist unique forms $\rho \in d_h^*(\Lambda_h^k) \subset \Lambda_h^{k-1}$ and $\phi \in d_h(\Lambda_h^k) \subset \Lambda_h^{k+1}$ with*

$$u = d_h\rho + d_h^*\phi, \quad d_h^*\rho = 0, \quad d_h\phi = 0, \tag{6.12}$$

$$\|u\|^2 = \|d_h\rho\|^2 + \|d_h^*\phi\|^2. \tag{6.13}$$

Proof. This is a special case of a more general result. Let

$$0 \to X \xrightarrow{f} Y \xrightarrow{g} Z \to 0$$

be a short exact sequence where X, Y, and Z are finite-dimensional Hilbert spaces and f and g linear maps. Then Y decomposes into orthogonal summands $A := \mathcal{R}(f) = \mathcal{N}(g)$ and $B := \mathcal{N}(f^*) = \mathcal{R}(g^*)$. Thus any $y \in Y$ may be decomposed as $y = fx + g^*z$ for some unique $x \in X$, $z \in Z$, and we have $\|y\|_Y^2 = \|fx\|_Y^2 + \|g^*z\|_Y^2$. We apply these results with $Y = \Lambda_h^k$, $Z = d_h(\Lambda_h^k) \subset \Lambda_h^{k+1}$, and $X = d_h^*(\Lambda_h^k) \subset \Lambda_h^{k-1}$. □

LEMMA 6.2. *Suppose that for any $u \in \Lambda_h^k$ of the form* (6.12),

$$\|d_h^*\phi\| \leq K\|d_h u\|, \qquad \|\rho\| \leq K'\|d_h\rho\|,$$

where K and K' are constants independent of ρ, ϕ, and h. Then the stability conditions (6.10) *and* (6.11) *are satisfied.*

Proof. Let $\tau = \sigma - t\rho \in \Lambda_h^{k-1}$ and $v = u + d_h\sigma \in \Lambda_h^k$ with $t = 1/(K')^2$. Using (6.13), the hypotheses of the lemma, and a simple computation,

$$
\begin{aligned}
&= \|\sigma\|^2 + \|d_h\sigma\|^2 + \|d_h u\|^2 + t\|d_h\rho\|^2 - t\int_\Omega \langle \sigma, \rho\rangle\, dx \\
&\geq \frac{1}{2}\|\sigma\|^2 + \|d_h\sigma\|^2 + \|d_h u\|^2 + t\|d_h\rho\|^2 - \frac{t^2}{2}\|\rho\|^2 \\
&\geq \frac{1}{2}\|\sigma\|^2 + \|d_h\sigma\|^2 + \|d_h u\|^2 + \|d_h\rho\|^2(t - t^2(K')^2/2) \\
&\geq \frac{1}{2}\|\sigma\|^2 + \|d_h\sigma\|^2 + \frac{1}{2}\|d_h u\|^2 + \frac{1}{2(K')^2}\|d_h\rho\|^2 + \frac{1}{2K^2}\|d_h^*\phi\|^2 \\
&\geq \frac{1}{2}\|\sigma\|^2 + \|d_h\sigma\|^2 + \frac{1}{2}\|d_h u\|^2 + \frac{1}{2(K'')^2}\|u\|^2,
\end{aligned}
$$

where $K'' = \max(K', K)$. Hence, we obtain (6.10) with $\gamma > 0$ depending only on K and K'. The upper bound (6.11) follows from the fact that

$$\|\rho\| \leq K'\|d_h\rho\| \leq K'\|u\|.$$

□

The hypotheses of the lemma are easily seen to be valid if we allow the constants K and K' to depend on h, with K the norm of the inverse of d_h restricted to the orthogonal complement of its kernel in Λ_h^k and K' is the norm of the inverse of d_h restricted to the orthogonal complement of its kernel in Λ_h^{k-1}. To show that the constants K and K' can be taken independent of h, we need to make use of approximation properties of the interpolation operators Π_h^k and elliptic regularity of appropriately chosen boundary value problems. We shall assume that for $u \in H^1\Lambda^{k-1}(\Omega)$ with $du \in \Lambda_h^k$,

$$\|u - \Pi_h^{k-1}u\| \leq Ch\|u\|_1.$$

We note that the condition $du \in \Lambda_h^k$ is needed in some cases for the interpolant $\Pi_h^{k-1}u$ to be defined. We next consider boundary value problems of the form: Given $\sigma_h \in \Lambda_h^{k-1}$, find $(\sigma, u) \in H\Lambda^{k-1}(\Omega) \times dH\Lambda^{k-1}(\Omega)$ determined by the equations

$$\int_\Omega \langle \sigma, \tau\rangle\, dx = \int_\Omega \langle d\tau, u\rangle\, dx \quad \forall \tau \in H\Lambda^{k-1}(\Omega), \tag{6.14}$$

$$\int_\Omega \langle d\sigma, v\rangle\, dx = \int_\Omega \langle d_h\sigma_h, v\rangle\, dx \quad \forall v \in dH\Lambda^{k-1}(\Omega). \tag{6.15}$$

This is a weak formulation of the equations

$$\sigma = d^*u, \quad d\sigma = d_h\sigma_h, \quad du = 0,$$

together with the natural boundary condition that the trace of u as a k-form on $\partial\Omega$ vanishes. We shall assume that the solution satisfies the regularity estimate:

$$\|\sigma\|_1 \leq C\|d_h\sigma_h\|.$$

We apply this result first in the case when $\sigma_h = \rho$. Since $d_h^*\rho = 0$, there exists $u_h \in d_h\Lambda_h^{k-1}$ such that $\rho = d_h^* u_h$, i.e.,

$$\int_\Omega \langle \rho, \tau \rangle \, dx = \int_\Omega \langle d\tau, u_h \rangle \, dx \quad \forall \tau \in \Lambda_h^{k-1}. \tag{6.16}$$

Since $d\sigma = d_h\sigma_h \in \Lambda_h^k$, we have by the commuting diagram (6.7) that $d\Pi_h^{k-1}\sigma = \Pi_h^k d\sigma = d\sigma = d_h\sigma_h = d_h\rho$. Choosing $\tau = \rho - \Pi_h^{k-1}\sigma$, we get

$$\|\rho\| \le \|\Pi_h^{k-1}\sigma\| \le \|\sigma\| + \|\sigma - \Pi_h^{k-1}\sigma\| \le C\|\sigma\|_1 \le \|d_h\rho\|.$$

To establish the first inequality of the lemma, it is enough to show that $\|\phi\| \le C\|d_h^*\phi\|$, since

$$\|d_h^*\phi\|^2 = (\phi, d_h d_h^*\phi) = (\phi, d_h u) \le \|\phi\|\|d_h u\|.$$

Because $d_h\phi = 0$, we can write $\phi = d_h w$, $w \in \Lambda_h^k$. We then apply our regularity result in the case when $\sigma_h = w$ and k is replaced by $k+1$. Hence, $\|\sigma\|_1 \le C\|d_h w\| \le C\|\phi\|$. Since $d\sigma = d_h\sigma_h \in \Lambda_h^{k+1}$, we again use the commuting diagram (6.7) to write

$$d\Pi_h^k\sigma = \Pi_h^{k+1}d\sigma = d\sigma = d_h\sigma_h = d_h w = \phi.$$

Now

$$\|\phi\|^2 = (d\Pi_h^k\sigma, \phi) = (\Pi_h^k\sigma, d_h^*\phi) \le \|\Pi_h^k\sigma\|\|d_h^*\phi\|.$$

But

$$\|\Pi_h^k\sigma\| \le \|\sigma\| + \|\sigma - \Pi_h^k\sigma\| \le C\|\sigma\|_1 \le C\|\phi\|.$$

Combining these results establishes the first inequality.

REFERENCES

[1] D.N. Arnold, *Differential complexes and numerical stability*, in Proceedings of the International Congress of Mathematicians, Vol. I: Plenary Lectures and Ceremonies, L. Tatsien, ed., Higher Education Press, Beijing, 2002, pp. 137–157.

[2] D.N. Arnold and R. Winther, *Mixed finite elements for elasticity*, Numer. Math., 92 (2002), pp. 401–419.

[3] V.I. Arnold, *Mathematical Methods of Classical Physics*, Springer–Verlag, New York, 1978.

[4] I. Babuška and A.K. Aziz, *Survey lectures on the mathematical foundations of the finite element method*, in The Mathematical Foundations of the Finite Element Method with Applications to Partial Differential Equations, Academic Press, New York, 1972.

[5] I. Babuška and R. Narasimhan, *The Babuška-Brezzi condition and the patch test: an example*, Comput. Methods Appl. Mech. Engrg., 140 (1997), pp. 183–199.

[6] P. Bochev and M. Gunzburger, *On least-squares finite element methods for the Poisson equation and their connection to the Dirichlet and Kelvin principles*, SIAM J. Numer. Anal. (2005). to appear.

[7] A. Bossavit, *Whitney forms: a class of finite elements for three-dimensional computations in electromagnetism*, IEEE Proc. A, 135 (1988), pp. 493–500.

[8] F. Brezzi, *On the existence, uniqueness and approximation of saddle-point problems arising from Lagrangian multipliers*, Rev. Française Automat. Informat. Recherche Opérationnelle Sér. Rouge, 8 (1974), pp. 129–151.

[9] F. Brezzi and K.-J. Bathe, *A discourse on the stability conditions for mixed finite element formulations*, Comput. Methods Appl. Mech. Engrg., 82 (1990), pp. 27–57. Reliability in computational mechanics (Austin, TX, 1989).

[10] F. Brezzi, J. Douglas, Jr., R. Durán, and M. Fortin, *Mixed finite elements for second order elliptic problems in three variables*, Numer. Math., 51 (1987).

[11] F. Brezzi, J. Douglas, Jr., and L.D. Marini, *Two families of mixed finite elements for second order elliptic problems*, Numer. Math., 47 (1985), pp. 217–235.

[12] F. Brezzi and M. Fortin, *Mixed and hybrid finite element methods*, Springer-Verlag, New York, 1991.

[13] P.G. Ciarlet, *The finite element method for elliptic problems*, North-Holland, Amsterdam, 1978.

[14] L. Demkowicz, P. Monk, W. Rachowicz, and L. Vardapetyan, *De Rham diagram for hp finite element spaces*, Computers & Mathematics with Applications, 39 (2000), pp. 29–38.

[15] G. Fix, M. Gunzburger, and R. Nicolaides, *On mixed finite element methods for first-order elliptic systems*, Numer. Math., 37 (1981), pp. 29–48.

[16] V. Girault and P.-A. Raviart, *Finite element methods for Navier–Stokes equations*, Springer-Verlag, New York, 1986.

[17] R. Hiptmair, *Canonical construction of finite elements*, Math. Comp., 68 (1999), pp. 1325–1346.

[18] ———, *Finite elements in computational electromagnetism*, Acta Numerica, 11 (2002), pp. 237–340.

[19] J.-C. Nédélec, *Mixed finite elements in* $\mathbf{R}^3$, Numer. Math., 35 (1980), pp. 315–341.

[20] ———, *A new family of mixed finite elements in* $\mathbf{R}^3$, Numer. Math., 50 (1986), pp. 57–81.

[21] P.-A. Raviart and J.M. Thomas, *A mixed finite element method for 2nd order elliptic problems*, in Mathematical aspects of finite element methods (Proc. Conf., Consiglio Naz. delle Ricerche (C.N.R.), Rome, 1975), Springer, Berlin, 1977, pp. 292–315. Lecture Notes in Math., Vol. 606.

[22] H. Whitney, *Geometric Integration Theory*, Princeton University Press, Princeton, 1957.

DIFFERENTIAL COMPLEXES AND STABILITY OF FINITE ELEMENT METHODS II: THE ELASTICITY COMPLEX

DOUGLAS N. ARNOLD*, RICHARD S. FALK†, AND RAGNAR WINTHER‡

Abstract. A close connection between the ordinary de Rham complex and a corresponding elasticity complex is utilized to derive new mixed finite element methods for linear elasticity. For a formulation with weakly imposed symmetry, this approach leads to methods which are simpler than those previously obtained. For example, we construct stable discretizations which use only piecewise linear elements to approximate the stress field and piecewise constant functions to approximate the displacement field. We also discuss how the strongly symmetric methods proposed in [8] can be derived in the present framework. The method of construction works in both two and three space dimensions, but for simplicity the discussion here is limited to the two dimensional case.

Key words. Mixed finite element method, Hellinger–Reissner principle, elasticity.

AMS(MOS) subject classifications. Primary: 65N30, Secondary: 74S05.

1. Introduction. In this paper we discuss finite element methods for the equations of linear elasticity derived from the Hellinger–Reissner variational principle. The equations can be written as a system of the form

$$A\sigma = \epsilon\, u, \qquad \operatorname{div} \sigma = f \quad \text{in } \Omega. \tag{1.1}$$

The unknowns σ and u denote the stress and displacement fields engendered by a body force f acting on a linearly elastic body that occupies a region $\Omega \subset \mathbb{R}^n$, where $n = 2$ or 3. Then σ takes values in the space $\mathbb{S} = \mathbb{R}^{n\times n}_{\text{sym}}$ of symmetric matrices and u takes values in $\mathbb{R}^n$. The differential operator ϵ is the symmetric part of the gradient, the div operator is applied row-wise to a matrix, and the compliance tensor $A = A(x) : \mathbb{S} \to \mathbb{S}$ is a bounded and symmetric, uniformly positive definite operator reflecting the properties of the body. We shall assume that the body is clamped on the boundary $\partial\Omega$ of Ω, so that the proper boundary condition for the system (1.1) is $u = 0$ on $\partial\Omega$.

Alternatively, the pair (σ, u) can be characterized as the unique critical point of the Hellinger–Reissner functional

$$\mathcal{J}(\tau, v) = \int_\Omega \big(\frac{1}{2} A\tau : \tau + \operatorname{div} \tau \cdot v - f \cdot v\big)\, dx. \tag{1.2}$$

*Institute for Mathematics and its Applications, University of Minnesota, Minneapolis, MN 55455. The work of the first author was supported in part by NSF grant DMS-0411388.

†Department of Mathematics, Rutgers University, Hill Center, 110 Frelinghuysen Rd, Rutgers University, Piscataway, NJ 08854-8019. The work of the second author was supported in part by NSF grant DMS03-08347.

‡Centre of Mathematics for Applications and Department of Informatics, University of Oslo, P.O. Box 1053, 0316 Oslo, NORWAY. The work of the third author was supported by the Norwegian Research Council.

The critical point is sought among all $\tau \in H(\operatorname{div}, \Omega; \mathbb{S})$, the space of square-integrable symmetric matrix fields with square-integrable divergence, and all $v \in L^2(\Omega; \mathbb{R}^n)$, the space of square-integrable vector fields. Equivalently, $(\sigma, u) \in H(\operatorname{div}, \Omega; \mathbb{S}) \times L^2(\Omega; \mathbb{R}^n)$ is the unique solution to the following weak formulation of the system (1.1)

$$\begin{aligned} \int_\Omega (A\sigma : \tau + \operatorname{div} \tau \cdot u)\, dx &= 0, \quad \tau \in H(\operatorname{div}, \Omega; \mathbb{S}), \\ \int_\Omega \operatorname{div} \sigma \cdot v\, dx &= \int_\Omega f v\, dx, \quad v \in L^2(\Omega; \mathbb{R}^n). \end{aligned} \tag{1.3}$$

A mixed finite element method determines an approximate stress field σ_h and an approximate displacement field u_h as the critical point of $\mathcal{J}$ over $\Sigma_h \times V_h$ where $\Sigma_h \subset H(\operatorname{div}, \Omega; \mathbb{S})$ and $V_h \subset L^2(\Omega; \mathbb{R}^n)$ are suitable piecewise polynomial subspaces. To ensure that a unique critical point exists and that it provides a good approximation of the true solution, the subspaces Σ_h and V_h must satisfy the stability conditions from Brezzi's theory of mixed methods [11, 12]. However, the construction of such elements has proved to be surprisingly hard, and most of the known results are limited to two space dimensions. In this case, a family of stable finite elements was presented in [8]. For the lowest order element, the space Σ_h is composed of piecewise cubic functions, while the space V_h consists of piecewise linear functions. Another approach that has proved successful in finding stable elements is the use of composite elements, in which V_h consists of piecewise polynomials with respect to one triangulation of the domain, while Σ_h consists of piecewise polynomials with respect to a different, more refined, triangulation [3, 15, 17, 23].

In the search for low order stable elements, several authors have resorted to the use of Lagrangian functionals that are modifications of the Hellinger–Reissner functional given above [1, 2, 4, 19, 20, 21, 22], in which the symmetry of the stress tensor is enforced only weakly or abandoned altogether. In order to discuss these methods, we extend the compliance tensor $A(x)$ to a symmetric and positive definite operator mapping $\mathbb{M}$ into $\mathbb{M}$, where $\mathbb{M}$ is the space of $n \times n$ matrices. In the isotropic case, the mapping $\sigma \mapsto A\sigma$ has the form

$$A\sigma = \frac{1}{2\mu}\Big(\sigma - \frac{\lambda}{2\mu + n\lambda} \operatorname{tr}(\sigma) I\Big),$$

where $\lambda(x), \mu(x)$ are positive scalar coefficients, the Lamé coefficients. A modification of the variational principle discussed above is obtained if we consider the extended Hellinger–Reissner functional

$$\mathcal{J}_e(\tau, v, q) = \mathcal{J}(\tau, v) + \int_\Omega \tau : q\, dx \tag{1.4}$$

over the space $H(\operatorname{div}, \Omega; \mathbb{M}) \times L^2(\Omega; \mathbb{R}^n) \times L^2(\Omega; \mathbb{K})$, where $\mathbb{K}$ denotes the space of skew symmetric matrices. We note that the symmetry condition for the space of matrix fields is now enforced through the introduction of a Lagrange multiplier. A critical point (σ, u, p) of the functional $\mathcal{J}_e$ is characterized as the unique solution of the system

$$\begin{gathered}\int_\Omega (A\sigma : \tau + \operatorname{div}\tau \cdot u + \tau : p)\,dx = 0, \quad \tau \in H(\operatorname{div}, \Omega; \mathbb{M}),\\ \int_\Omega \operatorname{div}\sigma \cdot v\,dx = \int_\Omega fv\,dx, \quad v \in L^2(\Omega; \mathbb{R}^n),\\ \int_\Omega \sigma : q\,dx = 0, \quad q \in L^2(\Omega; \mathbb{K}).\end{gathered} \tag{1.5}$$

In fact, it is clear that if (σ, u, p) is a solution of this system, then σ is symmetric, i.e., $\sigma \in H(\operatorname{div}, \Omega; \mathbb{S})$, and the pair $(\sigma, u) \in H(\operatorname{div}, \Omega; \mathbb{S}) \times L^2(\Omega; \mathbb{R}^n)$ solves the corresponding system (1.3). In this respect, the two systems (1.3) and (1.5) are equivalent. However, the extended system (1.5) leads to new possibilities for discretization. Assume that we choose finite element spaces $\Sigma_h \times V_h \times Q_h \subset H(\operatorname{div}, \Omega; \mathbb{M}) \times L^2(\Omega; \mathbb{R}^n) \times L^2(\Omega; \mathbb{K})$ and consider a discrete system corresponding to (1.5). If $(\sigma_h, u_h, p_h) \in \Sigma_h \times V_h \times Q_h$ is a discrete solution, then σ_h will not necessary inherit the symmetry property of σ. Instead, σ_h will satisfy the weak symmetry condition

$$\int_\Omega \sigma_h : q\,dx = 0, \qquad \text{for all } q \in Q_h.$$

Therefore, these solutions will in general not correspond to solutions of the discrete system obtained from (1.3).

Discretizations based on the system (1.5) will be referred to as mixed finite element methods with weakly imposed symmetry. For two space dimensions, such discretizations were already introduced by Fraejis de Veubeke in [15] and further developed in [2]. In particular, the so-called PEERS element proposed in [2] used an augmented Cartesian product of the Raviart–Thomas finite element space to approximate the stress σ, piecewise constants to approximate the displacements, and continuous piecewise linear functions to approximate the Lagrange multiplier p, as suggested in the element diagram depicted in Fig. 1. In this paper we use homological

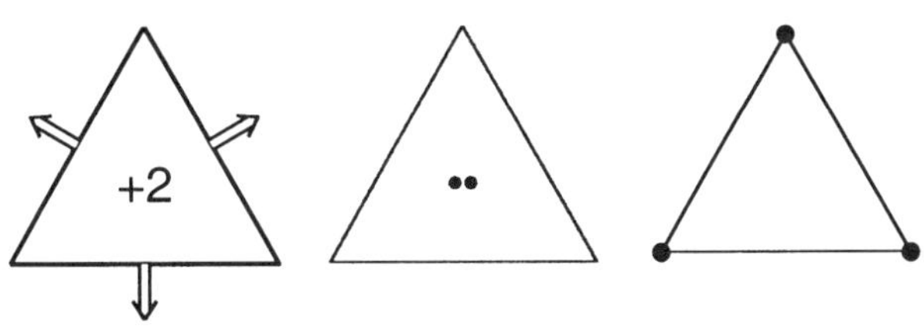

FIG. 1. *Approximation of stress, displacement, and multiplier for PEERS.*

techniques to construct a new family of stable mixed finite elements for elasticity with weakly imposed symmetry, the lowest order case of which is depicted in Fig. 2. The stresses are approximated by the Cartesian product of two copies of the Brezzi–Douglas–Marini finite element space, which means that the shape functions are simply all linear matrix fields and that there are four degrees of freedom per edge. The displacements are approximated by piecewise constants, as for PEERS, but the multipliers are as well, which means that, in contrast to PEERS, the multipliers can be eliminated by static condensation. We will also introduce a reduced version of the element with the same displacement and multiplier spaces, but only three degrees of freedom per edge for the stress. Let us also mention that there exist other mixed elements for elasticity with weakly imposed symmetry, although perhaps none as simple as those presented here. Prior to the PEERS paper, Amara and Thomas [1] developed methods with weakly imposed symmetry using a dual hybrid approach. Other elements based on the formulation in [2], including rectangular elements and elements in three space dimensions, have been developed in [20], [21], [22] and [18].

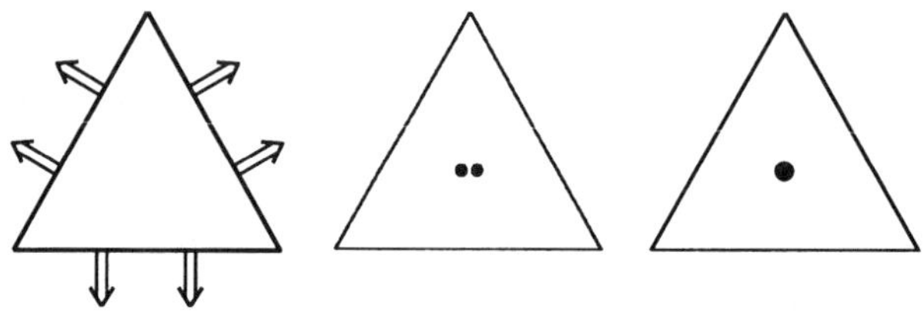

FIG. 2. *Approximation of stress, displacement, and multiplier for an element introduced below.*

Just as there is a close connection between mixed finite elements for Poisson's problem and discretization of the de Rham complex, there is also a close connection between mixed finite elements for elasticity and discretization of another differential complex, the elasticity complex. The importance of this complex was already recognized in [8], where mixed methods for elasticity in two space dimensions were discussed. However, the new ingredient here is that we utilize a close connection between the elasticity complex and the ordinary de Rham complex. This connection is described in Eastwood [13] and is based on a general construction given in [10], the Bernstein–Gelfand–Gelfand resolution. By mimicking this construction in the discrete case, we will be able to derive new mixed finite elements for elasticity in a systematic manner from known discretizations of the de Rham complex. The discussion here will be limited to two space dimensions. However, in a forthcoming paper [7], we will carry out the analogous construction and so obtain mixed finite element methods in three space dimensions.

An outline of the paper is as follows. In Section 2, we describe the notation to be used and recall some standard results about the stability of mixed finite element methods. In Section 3, we give two complexes related to the two mixed formulations of elasticity given by (1.3) and (1.5). In Section 4, we use the framework of differential forms to show how the elasticity complex can be derived from the de Rham complex (basically following the work of Eastwood [13]). In Section 5, we derive discrete analogues of these elasticity complexes beginning from discrete analogues of the de Rham complex, identify the required properties of the discrete spaces necessary for this construction, and explain how a discrete elasticity complex leads to stable finite element methods. In Section 6, we provide examples of finite element spaces that satisfy these conditions. The PEERS element is also discussed in this context. Finally, in Section 7, we show how an element with strongly imposed symmetry, previously obtained in [8], can be derived from discrete de Rham complexes using the methodology introduced in this paper.

2. Notation and preliminaries. We begin with some basic notation and hypotheses. We denote by $\mathbb{M}$ the space of all 2×2 real matrices and by $\mathbb{S}$ and $\mathbb{K}$ the subspaces of symmetric and skew symmetric matrices, respectively. The operators $\operatorname{sym} : \mathbb{M} \to \mathbb{S}$ and $\operatorname{skw} : \mathbb{M} \to \mathbb{K}$ denote the symmetric and skew symmetric parts, respectively. We assume that Ω is a simply connected domain in $\mathbb{R}^2$ with boundary Γ. We shall use the standard function spaces, like the Lebesgue space $L^2(\Omega)$ and the Sobolev space $H^s(\Omega)$. For vector-valued functions, we include the range space in the notation following a semicolon, so $L^2(\Omega; \mathbb{V})$ denotes the space of square integrable functions mapping Ω into a normed vector space $\mathbb{V}$. The space $H(\operatorname{div}, \Omega; \mathbb{R}^2)$ denotes the subspace of (vector-valued) functions in $L^2(\Omega; \mathbb{R}^2)$ whose divergence belongs to $L^2(\Omega)$. Similarly, $H(\operatorname{div}, \Omega; \mathbb{M})$ denotes the subspace of (matrix-valued) functions in $L^2(\Omega; \mathbb{M})$ whose divergence (by rows) belongs to $L^2(\Omega; \mathbb{R}^2)$.

Assuming that $\mathbb{V}$ is an inner product space, then $L^2(\Omega; \mathbb{V})$ has a natural norm and inner product, which will be denoted by $\| \cdot \|$ and $(\cdot , \cdot)$, respectively. For a Sobolev space $H^s(\Omega; \mathbb{V})$, we denote the norm by $\| \cdot \|_s$ and for $H(\operatorname{div}, \Omega; \mathbb{V})$, the norm is denoted by $\|v\|_{\operatorname{div}} := (\|v\|^2 + \| \operatorname{div} v\|^2)^{1/2}$. The space $\mathcal{P}_k(\Omega)$ denotes the space of polynomial functions on Ω of total degree $\le k$. Usually we abbreviate this to just $\mathcal{P}_k$.

We recall that the mixed finite element approximation derived from (1.5) takes the form:

Find $(\sigma_h, u_h, p_h) \in \Sigma_h \times V_h \times Q_h$ such that

$$
\begin{aligned}
(A\sigma_h, \tau) + (\operatorname{div} \tau, u_h) + (\tau, p_h) = 0, \quad & \tau \in \Sigma_h, \\
(\operatorname{div} \sigma_h, v) = (f, v) \quad & v \in V_h, \\
(\sigma_h, q) = 0, \quad & q \in Q_h,
\end{aligned}
\tag{2.1}
$$

where $\Sigma_h \subset H(\operatorname{div}, \Omega; \mathbb{M})$, $V_h \subset L^2(\Omega; \mathbb{R}^2)$, and $Q_h \in L^2(\Omega; \mathbb{K})$ are finite element spaces with h a mesh size parameter. Following the general theory of mixed methods, cf. [11, 12], the stability of the saddle–point system (2.1) is ensured by the following conditions:

(A1) $\|\tau\|_{\text{div}}^2 \le c_1(A\tau,\tau)$ whenever $\tau \in \Sigma_h$ satisfies $(\operatorname{div}\tau, v) = 0\ \forall v \in V_h$ and $(\tau, q) = 0\ \forall q \in Q_h$,

(A2) for all nonzero $(v,q) \in V_h \times Q_h$, there exists nonzero $\tau \in \Sigma_h$ with $(\operatorname{div}\tau, v) + (\tau, q) \ge c_2\|\tau\|_{\text{div}}(\|v\| + \|q\|)$,

where c_1 and c_2 are positive constants independent of h.

If we instead derive the mixed finite element method from the weak formulation (1.3), we need to construct finite element subspaces $\Sigma_h \subset H(\text{div},\Omega;\mathbb{S})$, i.e., with the symmetry condition strongly imposed, and $V_h \subset L^2(\Omega;\mathbb{R}^2)$. The discrete system then determines $(\sigma_h, u_h) \in \Sigma_h \times V_h$ by the equations

$$\begin{aligned} (A\sigma_h,\tau) + (\operatorname{div}\tau, u_h) &= 0, \quad \tau \in \Sigma_h, \\ (\operatorname{div}\sigma_h, v) &= (f,v) \quad v \in V_h. \end{aligned} \tag{2.2}$$

In this case, the stability condition is that Σ_h and V_h must satisfy (A1) and (A2) with $Q_h = 0$. As we shall see below, it is much harder to construct stable elements for elasticity with strongly imposed symmetry than it is with weakly imposed symmetry.

In the preceding paper [6], we have seen the close connection between the construction of stable mixed finite element methods for the approximation of the Poisson problem

$$\Delta p = f \quad \text{in } \Omega, \qquad p = 0 \quad \text{on } \partial\Omega, \tag{2.3}$$

and discrete versions of the de Rham complex. In this paper, we pursue an analogous approach for the elasticity problem.

3. The elasticity complex. We now proceed to a description of two elasticity complexes, corresponding to strongly and weakly imposed symmetry of the stress tensor. For the case of strongly imposed symmetry, corresponding to the mixed elasticity system (1.3), we require a characterization of the divergence-free symmetric matrix fields. In order to give such a characterization, define $J : C^\infty(\Omega) \to C^\infty(\Omega;\mathbb{S})$ by

$$Jq = \begin{pmatrix} \partial^2 q/\partial x_2^2 & -\partial^2 q/\partial x_1 \partial x_2 \\ -\partial^2 q/\partial x_1\partial x_2 & \partial^2 q/\partial x_1^2 \end{pmatrix}.$$

It is easy to check that $\operatorname{div} \circ J = 0$. In other words,

$$\mathcal{P}_1 \hookrightarrow C^\infty \xrightarrow{J} C^\infty(\mathbb{S}) \xrightarrow{\text{div}} C^\infty(\mathbb{R}^2) \to 0, \tag{3.1}$$

is a complex. Here, and frequently in the sequel, the dependence of the domain Ω is suppressed, i.e., $C^\infty(\mathbb{S})$ is short for $C^\infty(\Omega;\mathbb{S})$. When Ω is simply connected, then (3.1) is an exact sequence, a fact which will follow from the discussion below. The complex (3.1) will be referred to as the elasticity complex. If we followed the program that has been outlined in [6] for mixed methods for scalar second order elliptic equations, the construction of stable mixed finite elements for elasticity would be based on extending the sequence (3.1) to a complete commuting diagram of the form

$$\begin{array}{ccccccccc}
\mathcal{P}_1 & \hookrightarrow & C^\infty & \xrightarrow{J} & C^\infty(\mathbb{S}) & \xrightarrow{\operatorname{div}} & C^\infty(\mathbb{R}^2) & \to & 0 \\
 & & \downarrow \Pi_h^2 & & \downarrow \Pi_h^d & & \downarrow \Pi_h^0 & & \\
\mathcal{P}_1 & \hookrightarrow & W_h & \xrightarrow{J} & \Sigma_h & \xrightarrow{\operatorname{div}} & V_h & \to & 0
\end{array}$$

where $W_h \subset H^2(\Omega)$, $\Sigma_h \subset H(\operatorname{div}, \Omega; \mathbb{S})$ and $V_h \subset L^2(\Omega; \mathbb{R}^2)$ are suitable finite element spaces and Π_h^2, Π_h^d, and Π_h^0 are corresponding interpolation operators. This is exactly the construction performed in [8]. In particular, since the finite element space W_h is required to be a subspace of $H^2(\Omega)$, we can conclude that the piecewise polynomial space W_h must contain quintic polynomials, and therefore the lowest order space Σ_h will at least involve piecewise cubics. In fact, for the lowest order elements discussed in [8], W_h is the classical Argyris space, while Σ_h consists of piecewise cubic symmetric matrix fields with a linear divergence. In Section 7 we shall show how the element proposed in [8] arises naturally from the general construction outlined below.

If instead we consider methods with weakly imposed symmetry, i.e., finite element methods based on the mixed formulation (1.5), we are led to study the complex

$$\mathcal{P}_1 \hookrightarrow C^\infty \xrightarrow{J} C^\infty(\mathbb{M}) \xrightarrow{(\operatorname{skw},\operatorname{div})} C^\infty(\mathbb{K} \times \mathbb{R}^2) \to 0. \tag{3.2}$$

Observe that there is a close connection between (3.1) and (3.2). In fact, (3.1) can be derived from (3.2) by performing a projection step. To see this, consider the diagram

$$\begin{array}{ccccccccc}
\mathcal{P}_1 & \hookrightarrow & C^\infty & \xrightarrow{J} & C^\infty(\mathbb{M}) & \xrightarrow{(\operatorname{skw},\operatorname{div})} & C^\infty(\mathbb{K} \times \mathbb{R}^2) & \to & 0 \\
 & & \downarrow id & & \downarrow \operatorname{sym} & & \downarrow \pi & & \\
\mathcal{P}_1 & \hookrightarrow & C^\infty & \xrightarrow{J} & C^\infty(\mathbb{S}) & \xrightarrow{\operatorname{div}} & C^\infty(\mathbb{R}^2) & \to & 0,
\end{array} \tag{3.3}$$

where $\pi(q, u) = u - \operatorname{div} q$. The vertical maps are projections onto subspaces and the diagram commutes. It follows by a simple diagram chase that if the first row is exact, so is the second.

As we shall see below, the complexes (3.1) and (3.2) are closely connected to the standard de Rham complex. In two space dimensions, the de Rham complex is equivalent to the complex

$$\mathbb{R} \hookrightarrow C^\infty \xrightarrow{\operatorname{grad}} C^\infty(\mathbb{R}^2) \xrightarrow{\operatorname{rot}} C^\infty \to 0, \tag{3.4}$$

which is exact when Ω is simply connected. Here $\operatorname{rot} v$, where v is a vector field, is defined as the scalar field $\operatorname{rot} v = \partial v_1/\partial x_2 - \partial v_2/\partial x_1$.

An alternative identification of the de Rham complex in two space dimensions, that we shall use below, is the sequence

$$\mathbb{R} \hookrightarrow C^\infty \xrightarrow{\operatorname{curl}} C^\infty(\mathbb{R}^2) \xrightarrow{\operatorname{div}} C^\infty \to 0, \tag{3.5}$$

where $\operatorname{curl} \phi$ is the vector field defined by $\operatorname{curl} \phi = (-\partial\phi/\partial x_2, \partial\phi/\partial x_1)^T$. The two complexes (3.4) and (3.5) are equivalent. To see this just note that $\operatorname{curl} \phi = (\operatorname{grad} \phi)^\perp$ and $\operatorname{rot} v = \operatorname{div}(v^\perp)$, where $v^\perp$ denotes the vector perpendicular to v given by $v^\perp = (-v_2, v_1)^T$.

4. From the de Rham complex to linear elasticity. In this section we demonstrate the connection between the de Rham complex (3.4) and the elasticity complexes (3.1) and (3.2). Later, we will give an analogous construction to derive discrete elasticity complexes from corresponding discrete de Rham complexes.

We follow the notations of [6] for differential forms. Thus for Ω a domain in $\mathbb{R}^n$, $\Lambda^k = \Lambda^k(\Omega) = C^\infty(\Omega; \operatorname{Alt}^k(\mathbb{R}^n))$ denotes the space of smooth differential k-forms on Ω. Any $\omega \in \Lambda^k$ can be represented as

$$\omega_x = \sum_{i_1 < i_2 < \ldots < i_k} f_{i_1 \ldots i_k}(x) dx^{i_1} \wedge \ldots \wedge dx^{i_k} =: \sum_I f_I(x) dx^I \tag{4.1}$$

with coefficients $f_I \in C^\infty(\Omega)$. In particular, 0–forms can be identified with scalar functions, 1–forms with vector fields under the identification $f_i dx^i \leftrightarrow f_i e_i$, and n–forms can be identified with the scalar function $f_{12\ldots n}$. The spaces $L^2\Lambda^k(\Omega)$, $H^1\Lambda^k(\Omega)$, ..., consist of those ω which can be represented as in (4.1) with the $f_I \in L^2(\Omega)$, $H^1(\Omega)$, ...

The exterior derivative $d : \Lambda^k \to \Lambda^{k+1}$ satisfies

$$d\omega = \sum_{j,I} \frac{\partial f_I}{\partial x_j} dx^j \wedge dx^I,$$

and the de Rham complex is simply

$$\mathbb{R} \hookrightarrow \Lambda^0 \xrightarrow{d} \Lambda^1 \xrightarrow{d} \ldots \xrightarrow{d} \Lambda^n \to 0. \tag{4.2}$$

When $n = 2$, (4.2) becomes (3.4) under the identifications mentioned above. If we instead identify the 1–form $\omega = f_1 dx^1 + f_2 dx^2$ with the vector field $(-f_2, f_1)^T$, we obtain (3.5).

A differential k-form ω on Ω, admits a natural trace, $\operatorname{Tr} \omega$, which is a differential k-form on $\Gamma = \partial\Omega$. Namely, given k vectors $v_1, \cdots, v_k$ tangent to Γ at a point x, we have

$$(\operatorname{Tr} \omega)_x(v_1, \cdots, v_k) = \omega_x(v_1, \cdots, v_k).$$

Denoting by $d_\Gamma : \Lambda^k(\Gamma) \to \Lambda^{k+1}(\Gamma)$ the exterior derivative operator associated with Γ, we have a commuting diagram relating the de Rham complexes on Ω and Γ

$$\begin{array}{ccccccccccc} \mathbb{R} & \hookrightarrow & \Lambda^0(\Omega) & \xrightarrow{d} & \Lambda^1(\Omega) & \xrightarrow{d} & \cdots & \xrightarrow{d} & \Lambda^{n-1}(\Omega) & \xrightarrow{d} & \Lambda^n(\Omega) \to 0 \\ & & \big\downarrow \mathrm{Tr} & & \big\downarrow \mathrm{Tr} & & & & \big\downarrow \mathrm{Tr} & & \\ \mathbb{R} & \hookrightarrow & \Lambda^0(\Gamma) & \xrightarrow{d_\Gamma} & \Lambda^1(\Gamma) & \xrightarrow{d_\Gamma} & \cdots & \xrightarrow{d_\Gamma} & \Lambda^{n-1}(\Gamma) & \to & 0. \end{array} \tag{4.3}$$

The extension to vector-valued differential forms will be important in the sequel. If $\mathbb{V}$ is a vector space, then $\Lambda^k(\mathbb{V}) = \Lambda^k(\Omega; \mathbb{V})$ refers to the k–forms with values in $\mathbb{V}$, i.e., all elements of the form (4.1), but where $f_I \in C^\infty(\Omega; \mathbb{V})$, i.e., $\Lambda^k(\mathbb{V}) = C^\infty(\Omega; \mathrm{Alt}^k(\mathbb{V}))$, where $\mathrm{Alt}^k(\mathbb{V})$ are alternating k-linear forms $\mathbb{R}^n \times \cdots \times \mathbb{R}^n \to \mathbb{V}$.

The exactness of the $\mathbb{V}$-valued de Rham complex

$$\mathbb{V} \hookrightarrow \Lambda^0(\mathbb{V}) \xrightarrow{d} \Lambda^1(\mathbb{V}) \xrightarrow{d} \ldots \xrightarrow{d} \Lambda^n(\mathbb{V}) \to 0, \tag{4.4}$$

for Ω contractible is an obvious consequence of the exactness of (4.2).

We now specialize to the case $n = 2$ and $\Omega \subset \mathbb{R}^2$, and derive the elasticity complex from the de Rham complex with values in the three-dimensional vector space $\mathbb{V} = \mathbb{R} \times \mathbb{R}^2$. Define a map K from $\Lambda^k(\mathbb{R}^2)$ to $\Lambda^k(\mathbb{R})$ by

$$\sum_I f_I(x) dx^I \mapsto \sum_I [f_I(x) \cdot x^\perp] dx^I.$$

If $(\omega, \mu) \in \Lambda^k(\mathbb{R}) \times \Lambda^k(\mathbb{R}^2) = \Lambda^k(\mathbb{V})$, then the map $\Phi(\omega, \mu) := (\omega + K\mu, \mu)$ is an automorphism of $\Lambda^k(\mathbb{V})$, with inverse $\Phi^{-1}(\omega, \mu) = (\omega - K\mu, \mu)$. Define the operator $\mathcal{A} : \Lambda^k(\mathbb{V}) \to \Lambda^{k+1}(\mathbb{V})$ by $\mathcal{A} = \Phi d \Phi^{-1}$. Then the complex

$$\Phi(\mathbb{V}) \hookrightarrow \Lambda^0(\mathbb{V}) \xrightarrow{\mathcal{A}} \Lambda^1(\mathbb{V}) \xrightarrow{\mathcal{A}} \Lambda^2(\mathbb{V}) \to 0 \tag{4.5}$$

is exact when Ω is simply connected, since (4.4) is. The operator $\mathcal{A}$ has the simple form $\mathcal{A}(\omega, \mu) = (d\omega - S\mu, d\mu)$, where $S = dK - Kd : \Lambda^k(\mathbb{R}^2) \to \Lambda^{k+1}(\mathbb{R})$. Since $d \circ d = 0$,

$$dS = d^2K - dKd = -(dK - Kd)d = -Sd. \tag{4.6}$$

Furthermore, S is purely algebraic. In fact, an easy calculation shows that if ω is represented as in (4.1) then

$$S\omega = \sum_I (f_I \cdot e_2 dx^1 \wedge dx^I - f_I \cdot e_1 dx^2 \wedge dx^I).$$

More specifically the action of $S = S_k : \Lambda^k(\mathbb{R}^2) \to \Lambda^{k+1}(\mathbb{R})$, $k = 0, 1$, is given by

$$\begin{pmatrix} f_1 \\ f_2 \end{pmatrix} \overset{S_0}{\longmapsto} f_2 dx^1 - f_1 dx^2,$$

$$\begin{pmatrix} f_{12} \\ f_{22} \end{pmatrix} dx^1 - \begin{pmatrix} f_{11} \\ f_{21} \end{pmatrix} dx^2 \overset{S_1}{\longmapsto} (f_{12} - f_{21}) dx^1 \wedge dx^2.$$

It is important to note that S_0 is invertible (with $S_0^{-1}(f_1 dx^1 + f_2 dx^2) = (-f_2, f_1)^T$). The map S_1 is surjective but not invertible. If we identify $\Lambda^1(\mathbb{R}^2)$ with $C^\infty(\Omega, \mathbb{M})$ by

$$\begin{pmatrix} f_{12} \\ f_{22} \end{pmatrix} dx^1 - \begin{pmatrix} f_{11} \\ f_{21} \end{pmatrix} dx^2 \leftrightarrow (f_{ij}), \tag{4.7}$$

then the kernel of S_1 corresponds to the symmetric matrices.

Note that

$$\Phi(\mathbb{V}) = \{\, (\omega + \mu \cdot x^\perp, \mu) \,|\, \omega \in \mathbb{R}, \mu \in \mathbb{R}^2 \,\} = \{\, (p, S^{-1} dp) \,|\, p \in \mathcal{P}_1 \,\} \cong \mathcal{P}_1,$$

so (4.5) may be viewed as a resolution of $\mathcal{P}_1$.

We now consider a projection of (4.5) onto a subcomplex. Let

$$\Gamma^0 = \{\, (\omega, \mu) \in \Lambda^0(\mathbb{V}) : d\omega = S_0 \mu \,\}, \quad \Gamma^1 = \{\, (\omega, \mu) \in \Lambda^1(\mathbb{V}) : \omega = 0 \,\}$$

and define projections $\pi^0 : \Lambda^0(\mathbb{V}) \to \Gamma^0$, $\pi^1 : \Lambda^1(\mathbb{V}) \to \Gamma^1$ by

$$\pi^0(\omega, \mu) = (\omega, S_0^{-1} d\omega), \quad \pi^1(\omega, \mu) = (0, \mu + d S_0^{-1} \omega).$$

Then the diagram

$$\begin{array}{ccccccccc} \Phi(\mathbb{V}) & \hookrightarrow & \Lambda^0(\mathbb{V}) & \xrightarrow{A} & \Lambda^1(\mathbb{V}) & \xrightarrow{A} & \Lambda^2(\mathbb{V}) & \to & 0 \\ & & \downarrow{\scriptstyle \pi^0} & & \downarrow{\scriptstyle \pi^1} & & \downarrow{\scriptstyle id} & & \\ \Phi(\mathbb{V}) & \hookrightarrow & \Gamma^0 & \xrightarrow{A} & \Gamma^1 & \xrightarrow{A} & \Lambda^2(\mathbb{V}) & \to & 0, \end{array} \tag{4.8}$$

commutes, and so when the first row is exact, the second is as well. Making the obvious correspondences $(\omega, S_0^{-1} d\omega) \leftrightarrow \omega$ and $(0, \mu) \leftrightarrow \mu$, we may identify Γ^0 and Γ^1 with $\Lambda^0(\mathbb{R})$ and $\Lambda^1(\mathbb{R}^2)$, respectively. Thus the bottom row of (4.8) is equivalent to

$$\mathcal{P}_1 \hookrightarrow \Lambda^0(\mathbb{R}) \xrightarrow{d \circ S_0^{-1} \circ d} \Lambda^1(\mathbb{R}^2) \xrightarrow{(-S_1, d)} \Lambda^2(\mathbb{V}) \to 0. \tag{4.9}$$

But this is just another way to write (3.2). In fact, $\Lambda^0(\mathbb{R}) = C^\infty$ and we may identify $\Lambda^1(\mathbb{R}^2)$ with $C^\infty(\mathbb{M})$ as in (4.7). Also, we may identify $\Lambda^2(\mathbb{V})$ with $C^\infty(\mathbb{K} \times \mathbb{R}^2)$ by

$$\left(f, \begin{pmatrix} f_1 \\ f_2 \end{pmatrix} \right) dx^1 \wedge dx^2 \leftrightarrow -\left(\begin{pmatrix} 0 & f/2 \\ f/2 & 0 \end{pmatrix}, \begin{pmatrix} f_1 \\ f_2 \end{pmatrix} \right). \tag{4.10}$$

It is easy to check that, modulo these identifications, (4.9) coincides with (3.2).

Let us summarize the above construction. We began with the $\mathbb{V}$-valued de Rham complex (4.4) and introduced the automorphisms $\mathcal{A}$ to get (4.5). We then projected onto a subcomplex in (4.8) and made some simple identifications to obtain the elasticity complex with weakly imposed symmetry, (3.2). (Of course, we can make the further projection in (3.3) to obtain the elasticity complex with strongly imposed symmetry.)

5. The construction of a discrete elasticity complex. In this section we mimic the above construction on a discrete level to derive discretizations of the elasticity complex from discretizations of the de Rham complex, and use these to derive stable mixed finite elements for elasticity with weakly imposed symmetry.

As explained in [6], there exist a number of discrete de Rham complexes, i.e., complexes of the form

$$\mathbb{R} \hookrightarrow \Lambda_h^0 \xrightarrow{d} \Lambda_h^1 \xrightarrow{d} \Lambda_h^2 \to 0. \tag{5.1}$$

Here the spaces Λ_h^k are spaces of piecewise polynomial differential forms and there exist projections $\Pi_h = \Pi_h^k : \Lambda^k \to \Lambda_h^k$ such that the diagram

$$\begin{array}{ccccccccc} \mathbb{R} & \hookrightarrow & \Lambda^0 & \xrightarrow{d} & \Lambda^1 & \xrightarrow{d} & \Lambda^2 & \to & 0 \\ & & \downarrow \Pi_h & & \downarrow \Pi_h & & \downarrow \Pi_h & & \\ \mathbb{R} & \hookrightarrow & \Lambda_h^0 & \xrightarrow{d} & \Lambda_h^1 & \xrightarrow{d} & \Lambda_h^2 & \to & 0 \end{array} \tag{5.2}$$

commutes.

Our discrete construction begins by taking two discretizations of the de Rham complex, one scalar-valued and one vector-valued. The Cartesian product of these then gives a discretization of the $\mathbb{V}$-valued complex (4.4) which we write

$$\mathbb{V} \hookrightarrow \Lambda_h^0(\mathbb{V}) \xrightarrow{d} \Lambda_h^1(\mathbb{V}) \xrightarrow{d} \Lambda_h^2(\mathbb{V}) \to 0. \tag{5.3}$$

Next we define a discrete analog of the operator K, $K_h : \Lambda_h^k(\mathbb{R}^2) \to \Lambda_h^k(\mathbb{R})$ by $K_h = \Pi_h K$, where Π_h is the projection onto $\Lambda_h^k(\mathbb{R})$ and set $S_h = dK_h - K_h d : \Lambda_h^k(\mathbb{R}^2) \to \Lambda_h^{k+1}(\mathbb{R})$. Observe that the discrete version of (4.6),

$$dS_h = -S_h d \tag{5.4}$$

follows exactly as in the continuous case, and in light of the commutativity (5.2), we find that S_h is simply given by

$$S_h = d\Pi_h K - \Pi_h K d = \Pi_h(dK - Kd) = \Pi_h S.$$

In analogy with the continuous case, we define automorphisms Φ_h on $\Lambda_h^k(\mathbb{V})$ by $\Phi_h(\omega, \mu) = (\omega + K_h\mu, \mu)$ and obtain the exact sequence

$$\Phi_h(\mathbb{V}) \hookrightarrow \Lambda_h^0(\mathbb{V}) \xrightarrow{\mathcal{A}_h} \Lambda_h^1(\mathbb{V}) \xrightarrow{\mathcal{A}_h} \Lambda_h^2(\mathbb{V}) \to 0, \tag{5.5}$$

where $\mathcal{A}_h = \Phi_h d\Phi_h^{-1} : \Lambda_h^k(\mathbb{V}) \to \Lambda_h^{k+1}(\mathbb{V})$, so $\mathcal{A}_h(\omega, \mu) = (d\omega - S_h\mu, d\mu)$.

We now make some requirements on the choice of spaces used in the discrete de Rham complexes. A minor requirement is that the global linear polynomials are contained in the space $\Lambda_h^0(\mathbb{R})$ and the constant forms dx^1 and dx^2 are contained in $\Lambda_h^1(\mathbb{R})$. The *key requirement* is that *the operator* $S_h = S_{0,h} : \Lambda_h^0(\mathbb{R}^2) \to \Lambda_h^1(\mathbb{R})$ *is onto*, and so admits a right inverse $S_h^\dagger : \Lambda_h^1(\mathbb{R}) \to \Lambda_h^0(\mathbb{R}^2)$. We can then define the subspaces Γ_h^k of $\Lambda_h^k(\mathbb{V})$, $k = 0, 1$, by

$$\Gamma_h^0 = \{\, (\omega, \mu) \in \Lambda_h^0(\mathbb{V}) : d\omega = S_h\mu \,\}, \quad \Gamma_h^1 = \{\, (\omega, \mu) \in \Lambda_h^1(\mathbb{V}) : \omega = 0 \,\},$$

and define projections $\pi_h^0 : \Lambda_h^0(\mathbb{V}) \to \Gamma_h^0$, $\pi_h^1 : \Lambda_h^1(\mathbb{V}) \to \Gamma_h^1$ by

$$\pi_h^0(\omega, \mu) = (\omega, \mu - S_h^\dagger S_h\mu + S_h^\dagger d\omega), \quad \pi^1(\omega, \mu) = (0, \mu + dS_h^\dagger\omega).$$

It is easy to check that these are indeed projections onto the relevant subspaces and that the following diagram commutes:

$$\begin{array}{ccccccccc} \Phi(\mathbb{V}) & \hookrightarrow & \Lambda_h^0(\mathbb{V}) & \xrightarrow{\mathcal{A}_h} & \Lambda_h^1(\mathbb{V}) & \xrightarrow{\mathcal{A}_h} & \Lambda_h^2(\mathbb{V}) & \to & 0 \\ & & \Big\downarrow \pi_h^0 & & \Big\downarrow \pi_h^1 & & \Big\downarrow id & & \\ \Phi(\mathbb{V}) & \hookrightarrow & \Gamma_h^0 & \xrightarrow{\mathcal{A}_h} & \Gamma_h^1 & \xrightarrow{\mathcal{A}_h} & \Lambda_h^2(\mathbb{V}) & \to & 0. \end{array} \tag{5.6}$$

Here we have used the fact that $\Lambda_h^0(\mathbb{R})$ contains the linears to see the $\Phi_h(\mathbb{V}) = \Phi(\mathbb{V})$ and the fact that $\Lambda_h^1(\mathbb{R})$ contains the constants to see that $\Phi(\mathbb{V}) \subset \Gamma_h^0$.

The diagram (5.6) is the desired discrete analogue of (4.8), and the bottom row is a discrete analogue of the elasticity complex with weakly imposed symmetry. Under the identification (4.7), $\Gamma_h^1 \cong \Lambda_h^1(\mathbb{R}^2)$ corresponds to a finite element space $\Sigma_h \subset H(\operatorname{div}, \Omega; \mathbb{M})$, while under the identification (4.10), $\Lambda_h^2(\mathbb{V})$ corresponds to a finite element space $Q_h \times V_h \subset L^2(\Omega; \mathbb{K}) \times L^2(\Omega; \mathbb{R}^2)$, and the mapping

$$\Gamma_h^1 \xrightarrow{\mathcal{A}_h} \Lambda_h^2(\mathbb{V})$$

corresponds to

$$\Sigma_h \xrightarrow{(-\Pi_h^Q \operatorname{skw}, \operatorname{div})} Q_h \times V_h,$$

which is the key operator for the stability of a mixed method with weakly imposed symmetry (2.1). The fact that $\operatorname{div}\Sigma_h \subset V_h$, built into our construction, ensures the stability condition (A1), since then we need only show that $\|\tau\|^2 \le c_1(A\tau,\tau)$. It is straightforward to check this condition for fixed λ and μ This condition is also true with c_1 independent of λ for τ satisfying $\operatorname{div}\tau = 0$ and $\int_\Omega \operatorname{tr}(\tau) = 0$. Note this latter condition is implied by the first equation in the mixed method (choosing $\tau = I$), and a simple reformulation of the problem and slight modification of the analysis allows this extra constraint to be easily handled (cf. [3]). The surjectivity of the operator $\mathcal{A}_h$ implies the inequality in (A2), but only for a constant c_2 depending on the mesh size h. Just as in the last section of [6], to obtain a constant independent of h requires a more technical argument, using the properties of the continuous de Rham sequence, the commuting diagram, the approximation properties of an appropriately chosen interpolation operator, and elliptic regularity results. This can be done for all the spaces we consider in the next section. A detailed proof for the three-dimensional case will be provided in a forthcoming paper [7].

Before closing this section, we establish a sufficient condition for the key requirement that $S_h = S_{0,h}$ be surjective which we shall use in the next section. First note that the surjectivity of S_h follows from the commutativity of the diagram

$$\begin{array}{ccc} \Lambda^0(\Omega,\mathbb{R}^2) & \xrightarrow{S} & \Lambda^1(\Omega,\mathbb{R}) \\ \Pi_h^0 \big\downarrow & & \Pi_h^1 \big\downarrow \\ \Lambda_h^0(\mathbb{R}^2) & \xrightarrow{S_h} & \Lambda_h^1(\mathbb{R}). \end{array}$$

Indeed, since Π_h^1 is surjective and S is surjective (even invertible), this certainly implies that S_h is surjective. Recalling that $S_h = \Pi_h^1 S$, the commutativity condition $S_h\Pi_h^0 = \Pi_h^1 S$ may be rewritten

$$\Pi_h^1 S(I - \Pi_h^0) = 0 \text{ on } \Lambda^0(\Omega,\mathbb{R}^2). \tag{5.7}$$

Now $(I - \Pi_h^0)\Lambda^0(\Omega,\mathbb{R}^2)$ is exactly the null space of Π_h^0. Thus we may summarize the condition as follows:

Whenever the projection of $\omega \in \Lambda^0(\Omega,\mathbb{R}^2)$ into $\Lambda_h^0(\mathbb{R}^2)$ vanishes, then the projection of $S\omega = \omega_2 dx^1 - \omega_1 dx^2$ into $\Lambda_h^1(\mathbb{R})$ vanishes.

We close with a summary of the main conclusion of this section. In order to construct stable mixed finite elements for the formulation (2.1), we begin with a discrete de Rham complex

$$\mathbb{R} \hookrightarrow \Lambda_h^0(\mathbb{R}) \xrightarrow{d} \Lambda_h^1(\mathbb{R}) \xrightarrow{d} \Lambda_h^2(\mathbb{R}) \to 0,$$

and a discrete vector-valued de Rham complex

$$\mathbb{R}^2 \hookrightarrow \Lambda_h^0(\mathbb{R}^2) \xrightarrow{d} \Lambda_h^1(\mathbb{R}^2) \xrightarrow{d} \Lambda_h^2(\mathbb{R}^2) \to 0.$$

If these choices satisfy the boxed condition, then the finite element spaces Σ_h corresponding to $\Lambda_h^1(\mathbb{R}^2)$, V_h corresponding to $\Lambda_h^2(\mathbb{R}^2)$, and Q_h corresponding to $\Lambda_h^2(\mathbb{R})$ can be expected to furnish a stable choice of spaces.

6. Examples of stable finite elements. In this section, we apply the construction just presented to derive stable finite element methods for the approximation of the Hellinger-Reissner formulation of linear elasticity with weakly imposed symmetry. The simplest example of such a method will require only piecewise linear functions to approximate stresses and piecewise constants to approximation displacements and multiplier.

Let $\mathcal{T}$ denote a triangular mesh of Ω, one of a shape regular family of meshes with mesh size decreasing to zero. We need to select a scalar-valued and a vector-valued discrete de Rham complex, both of which will be based on piecewise polynomials with respect to $\mathcal{T}$, for which we can verify the boxed condition of the previous section. Starting with the simplest case, we use the Whitney forms for the scalar-valued complex, i.e.,

$$\mathbb{R} \hookrightarrow \mathcal{P}_1\Lambda^0(\mathcal{T};\mathbb{R}) \xrightarrow{d} \mathcal{P}_0^+\Lambda^1(\mathcal{T};\mathbb{R}) \xrightarrow{d} \mathcal{P}_0\Lambda^2(\mathcal{T};\mathbb{R}) \to 0,$$

which is the complex (5.3) of [6] in the case $n = 2$ and $r = 0$. For the vector-valued de Rham complex, we use instead the sequence (5.4) of [6] in the case $n = 2$ and $r = 0$, i.e.,

$$\mathbb{R}^2 \hookrightarrow \mathcal{P}_2\Lambda^0(\mathcal{T};\mathbb{R}^2) \xrightarrow{d} \mathcal{P}_1\Lambda^1(\mathcal{T};\mathbb{R}^2) \xrightarrow{d} \mathcal{P}_0\Lambda^2(\mathcal{T};\mathbb{R}^2) \to 0.$$

These choices lead to the element choice $\Sigma_h \cong \mathcal{P}_1\Lambda^1(\mathcal{T};\mathbb{R}^2)$ for the stress, $V_h \cong \mathcal{P}_0\Lambda^2(\mathcal{T};\mathbb{R}^2)$ for the displacement, and $Q_h \cong \mathcal{P}_0\Lambda^1(\mathcal{T};\mathbb{R})$ for the multiplier, depicted in Fig. 2 above.

The boxed condition requires that whenever ω is a smooth vector field on Ω whose projection into the Lagrange space $\mathcal{P}_2\Lambda^0(\mathcal{T};\mathbb{R}^2)$ of continuous piecewise quadratic vector fields vanishes, then the projection of $\omega_2 dx^1 - \omega_1 dx^2$ into the Raviart–Thomas space $\mathcal{P}_0^+\Lambda^1(\mathcal{T};\mathbb{R})$ vanishes. The vanishing of the projection into the vector-valued quadratic Lagrange space implies that

$$\int_e \omega_i \, de = 0, \quad i = 1, 2, \quad e \in \Delta_1(\mathcal{T}), \tag{6.1}$$

since the edge integrals are among the degrees of freedom ($\Delta_1(\mathcal{T})$ denotes the set of edges of the mesh). We then require that

$$\int_e \mathrm{Tr}_e(\omega_2 dx^1 - \omega_1 dx^2) = 0, \quad e \in \Delta_1(\mathcal{T}),$$

since the quantities $\int_e \mathrm{Tr}_e(\tau)$ determine the projection of a 1-form τ into $\mathcal{P}_0^+\Lambda^1(\mathcal{T};\mathbb{R})$. Now, for any 1-form $\tau = \tau_1 dx^1 + \tau_2 dx^2$,

$$\int_e \mathrm{Tr}_e(\tau) = \int_e (\tau_1 t^1 + \tau_2 t^2) \, de,$$

where (t^1, t^2) is the unit tangent to e. Thus we need to show that

$$\int_e (\omega_2 t^1 - \omega_1 t^2) \, de = 0, \quad e \in \Delta_1(\mathcal{T}),$$

whenever (6.1) holds, which is obvious.

A similar argument can be used to verify the boxed condition for the choice of discrete de Rham sequences

$$\mathbb{R} \hookrightarrow \mathcal{P}_{r+1}\Lambda^0(\mathcal{T};\mathbb{R}) \xrightarrow{d} \mathcal{P}_r^+\Lambda^1(\mathcal{T};\mathbb{R}) \xrightarrow{d} \mathcal{P}_r\Lambda^2(\mathcal{T};\mathbb{R}) \to 0,$$

and

$$\mathbb{R}^2 \hookrightarrow \mathcal{P}_{r+2}\Lambda^0(\mathcal{T};\mathbb{R}^2) \xrightarrow{d} \mathcal{P}_{r+1}\Lambda^1(\mathcal{T};\mathbb{R}^2) \xrightarrow{d} \mathcal{P}_r\Lambda^2(\mathcal{T};\mathbb{R}^2) \to 0,$$

for any $r \geq 0$. Thus we obtain a family of stable finite element methods with $\Sigma_h \cong \mathcal{P}_{r+1}\Lambda^1(\mathcal{T};\mathbb{R}^2)$, $V_h \cong \mathcal{P}_r\Lambda^2(\mathcal{T};\mathbb{R}^2)$, and $Q_h \cong \mathcal{P}_r\Lambda^2(\mathcal{T};\mathbb{R})$.

We also remark that it is possible to reduce the space Σ_h without changing V_h or Q_h and still maintain stability. Returning to the case $r = 0$, we see that we did not use the vanishing of the edge integrals of both components ω_i, but only of the combination $\omega_2 t^1 - \omega_1 t^2$ (the normal component). Hence, instead of the vector-valued quadratic Lagrange space $\mathcal{P}_2\Lambda^0(\mathcal{T};\mathbb{R}^2)$ we can use the reduced space obtained from it by imposing the constraint that the tangential component on each edge vary only linearly on that edge. This space of vector fields, which we denote $\mathcal{P}_2^-\Lambda^0(\mathcal{T};\mathbb{R}^2)$, is well-known as a possible discretization of the velocity field for Stokes flow [9, 14]; see also [16, p. 134 ff., 153 ff.]. An element in it is determined by its vertex values and the integral of its normal component on each edge. In order to complete the construction, we must provide a vector-valued discrete de Rham complex in which the space of 0-forms is $\mathcal{P}_2^-\Lambda^0(\mathcal{T};\mathbb{R}^2)$. This will be the complex

$$\mathbb{R}^2 \hookrightarrow \mathcal{P}_2^-\Lambda^0(\mathcal{T};\mathbb{R}^2) \xrightarrow{d} \mathcal{P}_1^-\Lambda^1(\mathcal{T};\mathbb{R}^2) \xrightarrow{d} \mathcal{P}_0\Lambda^2(\mathcal{T};\mathbb{R}^2) \to 0,$$

where it remains to define $\mathcal{P}_1^-\Lambda^1(\mathcal{T};\mathbb{R}^2)$. This will be the set of $\tau \in \mathcal{P}_1\Lambda^1(\mathcal{T};\mathbb{R}^2)$ for which $\mathrm{Tr}_e(\tau)\cdot t$ is constant on any edge e with unit tangent t and unit normal n. (In more detail: for $\tau \in \mathcal{P}_1\Lambda^1(\mathcal{T};\mathbb{R}^2)$, $\mathrm{Tr}_e(\tau)$ is a vector-valued 1-form on e of the form $g\,ds$ with $\mu : e \to \mathbb{R}^2$ linear and ds the volume form—i.e., length form—on e. If $\mu \cdot t$ is constant, then $\tau \in \mathcal{P}_1^-\Lambda^1(\mathcal{T};\mathbb{R}^2)$.) The natural degrees of freedom for this space are the integral and first moment of $\mathrm{Tr}_e(\tau)\cdot n$ and the integral of $\mathrm{Tr}_e(\tau)\cdot t$. It is straightforward to verify the commutativity of the diagram

$$\begin{array}{ccccccccc}
\mathbb{R}^2 & \hookrightarrow & \Lambda^0(\Omega;\mathbb{R}^2) & \xrightarrow{d} & \Lambda^1(\Omega;\mathbb{R}^2) & \xrightarrow{d} & \Lambda^2(\Omega;\mathbb{R}^2) & \to & 0 \\
 & & \big\downarrow \Pi_h & & \big\downarrow \Pi_h & & \big\downarrow \Pi_h & & \\
\mathbb{R}^2 & \hookrightarrow & \mathcal{P}_2^-\Lambda^0(\mathcal{T};\mathbb{R}^2) & \xrightarrow{d} & \mathcal{P}_1^-\Lambda^1(\mathcal{T};\mathbb{R}^2) & \xrightarrow{d} & \mathcal{P}_0\Lambda^2(\mathcal{T};\mathbb{R}^2) & \to & 0
\end{array}$$

and so the construction may precede. If we use (4.7) to identify vector-valued 1-forms and matrix fields, then the condition for a piecewise linear matrix field F to correspond to an element of $\mathcal{P}_1^-\Lambda^1(\mathcal{T};\mathbb{R}^2)$ is that on each edge e with tangent t and normal n, $Fn \cdot t$ must be constant on e. This defines the reduced space Σ_h, with three degrees of freedom per edge. Together with piecewise constant for displacements and multipliers, this furnishes a stable choice of elements.

We end this section by outlining how the original PEERS element, described in Section 1, cf. Fig. 1, can be derived from a slightly modified version of the theory outlined in Section 5. For this element, the scalar sequence is chosen to be a discrete de Rham sequence with reduced smoothness. The subscript in the spaces defined below indicates this reduced smoothness. Consider the sequence

$$\mathbb{R} \hookrightarrow \mathcal{P}_1\Lambda^0_-(\mathcal{T};\mathbb{R}) \xrightarrow{d} \mathcal{P}_0\Lambda^1_-(\mathcal{T};\mathbb{R}) \xrightarrow{d} \mathcal{P}_1\Lambda^0(\mathcal{T};\mathbb{R})^* \to 0. \tag{6.2}$$

Here $\mathcal{P}_1\Lambda^0_-(\mathcal{T};\mathbb{R})$ is the space of piecewise linear 0-forms with continuity requirement only with respect to the zero order moment on each edge, i.e., $\mathcal{P}_1\Lambda^0_-(\mathcal{T};\mathbb{R})$ is the standard nonconforming $\mathcal{P}_1$ space. Similarly, $\mathcal{P}_0\Lambda^1_-(\mathcal{T};\mathbb{R})$ consists of piecewise constant 1-forms, while the space of 2-forms $\mathcal{P}_1\Lambda^0(\mathcal{T};\mathbb{R})^*$ is the dual of $\mathcal{P}_1\Lambda^0(\mathcal{T};\mathbb{R})$ with respect to the pairing $\int_\Omega \omega \wedge \mu$. The operator $d = d_0 : \mathcal{P}_1\Lambda^0_-(\mathcal{T};\mathbb{R}) \to \mathcal{P}_0\Lambda^1_-(\mathcal{T};\mathbb{R})$ is defined locally on each triangle, and $d = d_1 : \mathcal{P}_0\Lambda^1_-(\mathcal{T};\mathbb{R}) \to \mathcal{P}_1\Lambda^0(\mathcal{T};\mathbb{R})^*$ is defined by $\int_\Omega d\omega \wedge \mu = -\int_\Omega \omega \wedge d\mu$ for $\omega \in \mathcal{P}_0\Lambda^1_-(\mathcal{T};\mathbb{R})$ and $\mu \in \mathcal{P}_1\Lambda^0(\mathcal{T};\mathbb{R})$. The orthogonal decomposition implied by the exact sequence (6.2) has been used previously (e.g., see [5]).

The corresponding vector-valued sequence needed for the PEERS element is dictated by the element itself. We consider the sequence

$$\mathbb{R}^2 \hookrightarrow \mathcal{P}_1\Lambda^0(\mathcal{T};\mathbb{R}^2) + B \xrightarrow{d} \mathcal{P}_0^+\Lambda^1(\mathcal{T};\mathbb{R}^2) + dB \xrightarrow{d} \mathcal{P}_0\Lambda^2(\mathcal{T};\mathbb{R}^2) \to 0,$$

which is exact. Here B denotes the space of vector-valued cubic bubbles, i.e., piecewise cubic vector fields which vanish on the element edges. Note the spaces $\mathcal{P}_0^+\Lambda^1(\mathcal{T};\mathbb{R}^2)+dB$, $\mathcal{P}_0\Lambda^2(\mathcal{T};\mathbb{R}^2)$, and $\mathcal{P}_1\Lambda^0(\mathcal{T};\mathbb{R})^*$ can be identified with the finite element spaces used in PEERS. If we choose the interpolation operator Π_h onto $\mathcal{P}_0\Lambda^1_-(\mathcal{T};\mathbb{R})$ to be the L^2 projection, then clearly

$$S_{0,h} = \Pi_h S_0 : \mathcal{P}_1\Lambda^0(\mathcal{T};\mathbb{R}^2) + B \to \mathcal{P}_0\Lambda^1_-(\mathcal{T};\mathbb{R})$$

is onto. Hence, the theory from Section 5 can be applied.

7. An element with strongly imposed symmetry. In this section, we shall discuss finite elements with strongly imposed symmetry, i.e., we consider the system (2.2). A family of stable elements was derived in [8], where, in the lowest degree case, the stress space $\Sigma_h \subset H(\mathrm{div}, \Omega; \mathbb{S})$ consists of piecewise cubics with linear divergence, while the space $V_h \subset L^2(\Omega;\mathbb{R}^2)$ consists of discontinuous linears. The purpose here is to show how this element can be derived from discrete de Rham complexes using the methodology introduced above.

As in the previous section, we start with one scalar-valued and one vector-valued discrete de Rham complex, which we denote here

$$\mathbb{R} \hookrightarrow \mathcal{P}_5\Lambda^0_\sharp(\mathcal{T};\mathbb{R}) \xrightarrow{d} \mathcal{P}_4\Lambda^1_\sharp(\mathcal{T};\mathbb{R}) \xrightarrow{d} \mathcal{P}_3\Lambda^2_\sharp(\mathcal{T};\mathbb{R}) \to 0 \tag{7.1}$$

and

$$\mathbb{R}^2 \hookrightarrow \mathcal{P}_4\Lambda^0_\flat(\mathcal{T};\mathbb{R}^2) \xrightarrow{d} \mathcal{P}_3\Lambda^1_\flat(\mathcal{T};\mathbb{R}^2) \xrightarrow{d} \mathcal{P}_2\Lambda^2_\flat(\mathcal{T};\mathbb{R}^2) \to 0. \tag{7.2}$$

On a single triangle, the scalar-valuded complex will be simply

$$\mathbb{R} \hookrightarrow \mathcal{P}_5\Lambda^0(T) \xrightarrow{d} \mathcal{P}_4\Lambda^1(T) \xrightarrow{d} \mathcal{P}_3\Lambda^2(T) \to 0,$$

but the degrees of freedom we use will impose extra smoothness on the assembled spaces. This extra smoothness appears to be necessary for the final construction.

For the quintic 0–form space, $\mathcal{P}_5\Lambda^0_\sharp(\mathcal{T};\mathbb{R})$, we determine a form on a triangle T by the following 21 values:

$$\phi(x),\ \operatorname{grad}\phi(x),\ \operatorname{grad}^2\phi(x),\quad x\in\Delta_0(T),\quad \int_e \frac{\partial\phi}{\partial n},\quad e\in\Delta_1(T). \tag{7.3}$$

The resulting space, $\mathcal{P}_5\Lambda^0_\sharp(\mathcal{T};\mathbb{R})$, is then the well-known Argyris space, a subspace of $C^1(\Omega)$.

An element $\omega\in\mathcal{P}_4\Lambda^1(T)$ of the form $\omega = -g_2dx^1 + g_1dx^2$ is determined by the 30 degrees of freedom given as

$$g_i(x),\ \operatorname{grad} g_i(x),\quad x\in\Delta_0(T),\quad \int_e g_i,\ \int_e p\operatorname{div} g,\ p\in\mathcal{P}_1(e), e\in\Delta_1(T),$$

and these determine the assembled space $\mathcal{P}_4\Lambda^1_\sharp(\mathcal{T};\mathbb{R})$. Here $\operatorname{div} g$ is the divergence of the vector field $g=(g_1,g_2)$. It is straightforward to check that these conditions determine an element of $\mathcal{P}_4\Lambda^1(T)$ uniquely. For if all of them are zero, then the cubic polynomial $\operatorname{div} g$ is zero on the boundary, and by the divergence theorem, the mean value of $\operatorname{div} g$ over T is zero. Hence, $\operatorname{div} g$, or $d\omega$, is zero, and therefore $\omega = d\phi$, where $\phi\in\mathcal{P}_5(T)$, and where we can assume that ϕ is zero at one of the vertices. However, it now follows that all the degrees of freedom for ϕ given by (7.3) vanish, and hence $\omega = d\phi$ is zero. If $\omega\in\mathcal{P}_4\Lambda^1_0(\mathcal{T};\mathbb{R})$, then ω is continuous, and, moreover, $d\omega = \operatorname{div} g$ is also continuous.

We complete the description of the desired scalar discrete de Rham complex, by letting $\mathcal{P}_3\Lambda^2_\sharp(\mathcal{T};\mathbb{R})$ denote the space of continuous piecewise cubic 2–forms, with standard Lagrange degrees of freedom, i.e., if $\omega = g dx_1\wedge dx_2$, we specify

$$g(x),\quad x\in\Delta_0(T),\quad \int_e gp,\quad p\in\mathcal{P}_1(e),\ e\in\Delta_1(T),\ \text{and}\ \int_T g.$$

It is easy to check that $d[\mathcal{P}_5\Lambda^0_\sharp(\mathcal{T};\mathbb{R})] \subset \mathcal{P}_4\Lambda^1_\sharp(\mathcal{T};\mathbb{R})$ and $d[\mathcal{P}_4\Lambda^1_\sharp(\mathcal{T};\mathbb{R})] = \mathcal{P}_3\Lambda^2_\sharp(\mathcal{T};\mathbb{R})$. Further, the complex (7.1) is exact. To check this, it is enough to show that

$$\dim \mathcal{P}_5\Lambda^0_\sharp(\mathcal{T};\mathbb{R}) + \dim \mathcal{P}_3\Lambda^2_\sharp(\mathcal{T};\mathbb{R}) = \dim \mathcal{P}_4\Lambda^1_\sharp(\mathcal{T};\mathbb{R}) + 1,$$

and this is a direct consequence of Euler's formula.

We now turn to the description of the spaces entering the vector-valued de Rham complex (7.2). The space $\mathcal{P}_4\Lambda^0_\flat(\mathcal{T};\mathbb{R}^2)$ consists of continuous piecewise quartic vector valued 0–forms $\omega = (f_1, f_2)^T$. The degrees of freedom are taken to be

$$f_i(x),\ \operatorname{grad} f_i(x),\quad x \in \Delta_0(\mathcal{T}),\quad \int_e f_i,\ \int_e p \operatorname{div} f,\ p \in \mathcal{P}_1(e),\ e \in \Delta_1(\mathcal{T}).$$

Note that the space $\mathcal{P}_4\Lambda^0_\flat(\mathcal{T};\mathbb{R}^2)$ is not simply the Cartesian product of two copies of a space of scalar-valued of 0–forms. However, the spaces are constructed exactly such that the operator S_0 (defined in Section 4) maps $\mathcal{P}_4\Lambda^1_\sharp(\mathcal{T};\mathbb{R})$ isomorphically onto $\mathcal{P}_4\Lambda^0_\flat(\mathcal{T};\mathbb{R}^2)$. Thus $S_{0,h}$ is simply the restriction of S_0 in this case. It is invertible, and, certainly the key requirement of Section 5, that it is surjective, is satisfied.

The space $\mathcal{P}_3\Lambda^1_\flat(\mathcal{T};\mathbb{R}^2)$ corresponds to a non-symmetric extension of the stress space used in [8]. On each triangle, the elements consist of cubic 1–forms

$$\omega = \begin{pmatrix} f_{12} \\ f_{22} \end{pmatrix} dx^1 - \begin{pmatrix} f_{11} \\ f_{21} \end{pmatrix} dx^2 \tag{7.4}$$

such that $\operatorname{div} F$ is linear, where $F = (f_{ij})$. This space has dimension $40 - 6 = 34$. In fact, 34 unisolvent degrees of freedom are given by $F(x)$ for $x \in \Delta_0(T)$, $\int_T F$ and basis elements for the spaces of moments

$$\int_e (Fn)\cdot p,\quad p \in \mathcal{P}_1(e;\mathbb{R}^2),\quad \int_e p \operatorname{skw}(F),\quad p \in \mathcal{P}_1(e;\mathbb{K}),\ e \in \Delta_1(T).$$

If all these degrees of these degrees of freedom vanish, then $\operatorname{skw}(F) = 0$ on the triangle T, and the corresponding unisolvence argument given in [8] implies $\omega = 0$ on T.

Finally, the space $\mathcal{P}_1\Lambda^2_\flat(\mathcal{T};\mathbb{R}^2) = \mathcal{P}_1\Lambda^2(\mathcal{T};\mathbb{R}^2)$ is the standard space of discontinuous linear vector-valued 2–forms, with degrees of freedom $\int_T \omega\wedge\mu$ for μ in a basis for $\mathcal{P}_1\Lambda^0(T,\mathbb{R}^2)$. By definition, we have the inclusion $d[\mathcal{P}_3\Lambda^1_\flat(\mathcal{T};\mathbb{R}^2)] \subset \mathcal{P}_1\Lambda^2_\flat(\mathcal{T};\mathbb{R}^2)$, and from [8] we know that the symmetric subspace of $\mathcal{P}_3\Lambda^1_\flat(\mathcal{T};\mathbb{R}^2)$ is mapped onto $\mathcal{P}_1\Lambda^2_\flat(\mathcal{T};\mathbb{R}^2)$ by d. Therefore, $d[\mathcal{P}_3\Lambda^1_\flat(\mathcal{T};\mathbb{R}^2)] = \mathcal{P}_1\Lambda^2_\flat(\mathcal{T};\mathbb{R}^2)$. Furthermore, clearly $d[\mathcal{P}_4\Lambda^0_\flat(\mathcal{T};\mathbb{R}^2)] \subset \mathcal{P}_3\Lambda^1_\flat(\mathcal{T};\mathbb{R}^2)$. Hence, as above we can use a dimension count to show that the complex (7.2) is exact.

Since we have already noted that $S_{0,h}$ is surjective, it follows from the general theory of Section 5, that the bottom row of diagram (5.6) is exact. Furthermore, since $S_{0,h}$ is invertible, we can identify the space Γ_h^0 with $\Lambda_h^0(\mathbb{R})$. Now, if ω given by (7.4) belongs to $\mathcal{P}_3\Lambda_\flat^1(\mathcal{T};\mathbb{R}^2)$, then $S_1\omega = (f_{12} - f_{21})dx^1 \wedge dx^2$ belongs to $\mathcal{P}_3\Lambda_\sharp^2(\mathcal{T};\mathbb{R})$. Hence $S_{1,h}$ is just the restriction to $\mathcal{P}_3\Lambda_\flat^1(\mathcal{T};\mathbb{R}^2)$ of S_1 in this case, and the bottom row of (5.6) can be identified with

$$\mathcal{P}_1 \hookrightarrow \Lambda_h^0(\mathbb{R}) \xrightarrow{d\circ S_0^{-1}\circ d} \Lambda_h^1(\mathbb{R}^2) \xrightarrow{(-S_1,d)} \Lambda_h^2(\mathbb{V}) \to 0, \tag{7.5}$$

which, in the present case and notation, takes the form

$$\begin{aligned}\mathcal{P}_1 \hookrightarrow \mathcal{P}_5\Lambda_\sharp^0(\mathcal{T};\mathbb{R}) &\xrightarrow{d\circ S_0^{-1}\circ d} \mathcal{P}_3\Lambda_\flat^1(\mathcal{T};\mathbb{R}^2)\\ &\xrightarrow{(-S_1,d)} \mathcal{P}_3\Lambda_\sharp^2(\mathcal{T};\mathbb{R}) \times \mathcal{P}_1\Lambda^2(\mathcal{T};\mathbb{R}^2) \to 0.\end{aligned} \tag{7.6}$$

Identifying the spaces of differential forms with spaces of piecewise polynomial scalar, vector, and matrix fields as usual, the form space $\mathcal{P}_3\Lambda_\sharp^2(\mathcal{T};\mathbb{R})$ corresponds to the space Q_h of all continuous piecewise cubic skew matrix fields, $\mathcal{P}_1\Lambda^2(\mathcal{T};\mathbb{R})$ corresponds to the space V_h of all piecewise linear vector fields, and $\mathcal{P}_5\Lambda_\sharp^0(\mathcal{T};\mathbb{R})$ corresponds to the Argyris space of piecewise quintic scalar fields. The space $P_3\Lambda_\flat^1(\mathcal{T};\mathbb{R}^2)$ corresponds to a space Ξ_h consisting of all piecewise cubic matrix fields in $H(\mathrm{div},\Omega;\mathbb{M})$ which have piecewise linear divergence, are continuous at the vertices, and for which the skew part is continuous. With these identifications, the sequence (7.6) is equivalent to

$$\mathcal{P}_1 \hookrightarrow W_h \xrightarrow{J} \Xi_h \xrightarrow{(\mathrm{skw},\mathrm{div})} Q_h \times V_h \to 0,$$

which is a discrete version of (3.2).

In order to derive the desired discrete version of (3.1), we develop a discrete analogue of the projection done in (3.3). Observe that of the 34 degrees of freedom determining an element $F \in \Xi_h$ on a given triangle T, there are 10 that only involve $\mathrm{skw}(F)$, i.e., $\mathrm{skw}(F)$ at each vertex, $\int_T \mathrm{skw}(F)$, and $\int_e p\,\mathrm{skw}(F)$ for $p \in \mathcal{P}_1(e;\mathbb{K})$. Moreover, these are exactly the degrees of freedom of $\mathrm{skw}(F)$ in Q_h. Let L_h denote this set of degrees of freedom, and L_h^c the remaining 24 degrees of freedom. Then we can define an injection $i_h : Q_h \to \Xi_h$, determining $i_h q$ on T by

$$l(i_h q) = l(q), \quad l \in L_h, \qquad l(i_h q) = 0, \quad l \in L_h^c.$$

By construction, $\mathrm{skw}\, i_h q = q$ for all $q \in Q_h$. The operator i_h may be considered a discrete analogue of the inclusion of $C^\infty(\Omega;\mathbb{K}) \hookrightarrow C^\infty(\Omega,\mathbb{M})$. (However Q_h is not contained in Ξ_h, and $i_h q$ need not be skew-symmetric.) The operator $\mathrm{sym}_h := I - i_h\,\mathrm{skw}$ is a projection of Ξ_h onto the subspace Σ_h consisting of the symmetric matrix fields in Ξ_h. That is,

$$\Sigma_h := \mathrm{sym}_h(\Xi_h) = \Xi_h \cap H(\mathrm{div},\Omega;\mathbb{S}).$$

A discrete version of the diagram (3.3) is now given by

$$\begin{array}{ccccccccc}
\mathcal{P}_1 & \hookrightarrow & W_h & \xrightarrow{J} & \Xi_h & \xrightarrow{(\mathrm{skw},\mathrm{div})} & Q_h \times V_h & \to & 0 \\
 & & \big\downarrow id & & \big\downarrow \mathrm{sym}_h & & \big\downarrow \Pi_h & & \\
\mathcal{P}_1 & \hookrightarrow & W_h & \xrightarrow{J} & \Sigma_h & \xrightarrow{\mathrm{div}} & V_h & \to & 0
\end{array}$$

where $\Pi_h(q, v) = v - \operatorname{div} i_h q$. It is straightforward to check this diagram commutes and hence the bottom row is exact. This is exactly the discrete sequence utilized in [8].

REFERENCES

[1] MOHAMED AMARA AND JEAN-MARIE THOMAS, *Equilibrium finite elements for the linear elastic problem*, Numer. Math. **33** (1979), no. 4, 367–383.

[2] DOUGLAS N. ARNOLD, FRANCO BREZZI, AND JIM DOUGLAS, JR., *PEERS: a new mixed finite element for plane elasticity*, Japan J. Appl. Math. **1** (1984), no. 2, 347–367.

[3] DOUGLAS N. ARNOLD, JIM DOUGLAS, JR., AND CHAITAN P. GUPTA, *A family of higher order mixed finite element methods for plane elasticity*, Numer. Math. **45** (1984), no. 1, 1–22.

[4] DOUGLAS N. ARNOLD AND RICHARD S. FALK, *A new mixed formulation for elasticity*, Numer. Math. **53** (1988), no. 1-2, 13–30.

[5] DOUGLAS N. ARNOLD AND RICHARD S. FALK, *A uniformly accurate finite element method for the Reissner-Mindlin plate*, SIAM J. on Numer. Anal. **26** (1989), 1276–1290.

[6] DOUGLAS N. ARNOLD, RICHARD S. FALK, AND RAGNAR WINTHER, *Differential complexes and stability of finite element methods. I. The de Rham complex*, this volume.

[7] DOUGLAS N. ARNOLD, RICHARD S. FALK, AND RAGNAR WINTHER, *Mixed finite element methods for linear elasticity with weakly imposed symmetry*, preprint.

[8] DOUGLAS N. ARNOLD AND RAGNAR WINTHER, *Mixed finite elements for elasticity*, Numer. Math. **92** (2002), no. 3, 401–419.

[9] CHRISTINE BERNARDI AND GENEVIÉVE RAUGEL, *Analysis of some finite elements for the Stokes problem*, Math. Comp. **44** (1985), 71–79.

[10] I.N. BERNSTEIN, I.M. GELFAND, AND S.I. GELFAND, *Differential operators on the baseaffine space and a study of g–modules*, Lie groups and their representation, I.M. Gelfand (ed) (1975), 21–64.

[11] FRANCO BREZZI, *On the existence, uniqueness and approximation of saddle-point problems arising from Lagrangian multipliers*, Rev. Française Automat. Informat. Recherche Opérationnelle Sér. Rouge **8** (1974), no. R-2, 129–151.

[12] FRANCO BREZZI AND MICHEL FORTIN, *Mixed and Hybrid Finite Element Methods*, Springer-Verlag, New York, 1991.

[13] MICHAEL EASTWOOD, *A complex from linear elasticity*, Rend. Circ. Mat. Palermo (2) Suppl. (2000), no. 63, 23–29.

[14] MICHEL FORTIN, *Old and new finite elements for incompressible flows*, Int. J. Numer. Methods Fluids **1** (1981), 347–354.

[15] BAUDOIUN M. FRAEJIS DE VEUBEKE, *Stress function approach*, World Congress on the Finite Element Method in Structural Mechanics, Bornemouth, 1975.

[16] V. GIRAULT AND P.-A. RAVIART, *Finite element methods for Navier-Stokes equations. Theory and algorithms*, Springer Series in Computational Mathematics, 5. Springer-Verlag, Berlin, 1986.

[17] CLAES JOHNSON AND BERTRAND MERCIER, *Some equilibrium finite element methods for two-dimensional elasticity problems*, Numer. Math. **30** (1978), no. 1, 103–116.

[18] MARY E. MORLEY, *A family of mixed finite elements for linear elasticity*, Numer. Math. **55** (1989), no. 6, 633–666.

[19] ERWIN STEIN AND RAIMUND ROLFES, *Mechanical conditions for stability and optimal convergence of mixed finite elements for linear plane elasticity*, Comput. Methods Appl. Mech. Engrg. **84** (1990), no. 1, 77–95.

[20] ROLF STENBERG, *On the construction of optimal mixed finite element methods for the linear elasticity problem*, Numer. Math. **48** (1986), no. 4, 447–462.

[21] ———, *A family of mixed finite elements for the elasticity problem*, Numer. Math. **53** (1988), no. 5, 513–538.

[22] ———, *Two low-order mixed methods for the elasticity problem*, The mathematics of finite elements and applications, VI (Uxbridge, 1987), Academic Press, London, 1988, pp. 271–280.

[23] VERNON B. WATWOOD JR. AND B.J. HARTZ, *An equilibrium stress field model for finite element solution of two-dimensional elastostatic problems*, Internat. Jour. Solids and Structures **4** (1968), 857–873.

ON THE ROLE OF INVOLUTIONS IN THE DISCONTINUOUS GALERKIN DISCRETIZATION OF MAXWELL AND MAGNETOHYDRODYNAMIC SYSTEMS

TIMOTHY BARTH*

Abstract. The role of involutions in energy stability of the discontinuous Galerkin (DG) discretization of Maxwell and magnetohydrodynamic (MHD) systems is examined. Important differences are identified in the symmetrization of the Maxwell and MHD systems that impact the construction of energy stable discretizations using the DG method. Specifically, general sufficient conditions to be imposed on the DG numerical flux and approximation space are given so that energy stability is retained. These sufficient conditions reveal the favorable energy consequence of imposing continuity in the normal component of the magnetic induction field at interelement boundaries for MHD discretizations. Counterintuitively, this condition is not required for stability of Maxwell discretizations using the discontinuous Galerkin method.

Key words. Nonlinear conservation laws, energy stability, Maxwell equations, magnetohydrodynamics, symmetrization, discontinuous Galerkin finite element method.

AMS(MOS) subject classifications. 35L02, 65M02, 65K02, 76N02.

1. Overview.

Various mathematical models such the Maxwell equations governing electrodynamics and the magnetohydrodynamic (MHD) equations modeling fluid plasmas have the added complexity of possessing *involutions*. An involution in the sense of conservation law systems is an additional equation that if satisfied at some initial time is satisfied for all future time for both classical and weak solutions [Boi88, Daf86]. Involutions should not be confused with *constraints* that are needed for closure of the system. An example of such a constraint is the continuity equation in incompressible flow. In this note, the role of involutions in obtaining energy stable discretizations using the discontinuous Galerkin method [RH73, LR74, JP86, CLS89, CHS90] is briefly examined. Specifically, the surprisingly different role played by involutions in the discontinuous Galerkin (DG) discretization of Maxwell and ideal compressible MHD systems is contrasted. Although both systems possess solenoidal involutions, it is the interplay between involutions and symmetrization of the Maxwell and MHD systems that enters fundamentally into the construction of stable discretizations. In this regard, the two systems are vastly different. The Maxwell equations are naturally expressed in essentially symmetric form. Consequently, the analysis given in Sects. 2.1 and 3.1 shows that "standard" DG discretizations can then be used. In contrast, symmetrization of the MHD system utilizes the solenoidal involution as a necessary ingredient in the symmetrization process. Details of this symmetrization process are given in Sect. 2.2. Thus, the precise sense in

*NASA Ames Research Center, M.S. T27A-1, Moffett Field, CA 94035-1000, USA (Timothy.J.Barth@nasa.gov).

which involutions are satisfied in element interiors and across interelement boundaries enters prominently into the MHD discrete energy analysis. The analysis of Sect. 3.2 gives general sufficient conditions to be imposed on the DG numerical flux and approximation space in the presence of involutions so that energy stability is retained. These sufficient conditions reveal the favorable consequences of imposing continuity in the normal component of the magnetic induction field at interelement boundaries for MHD discretizations. This is a condition that is not required for stability of Maxwell discretizations using the discontinuous Galerkin method but is often a requirement of other methods that build satisfaction of solenoidal conditions into the discretization. Techniques for achieving this include staggered mesh and specialized differencing techniques [Yee66] as well as edge, face, and volume finite element formulations [Ned80, Bos98, BR02] or the discrete mimetic approximations as given in [HS99]. The present analysis for MHD also provides alternatives to the "divergence cleaning" procedures designed to exactly or approximately satisfy the solenoidal condition, see [BB80, TÓ0, DKK+02, BK04] and references therein. Since the DG method reduces to the simplest finite volume method in the special case of piecewise constant basis approximation, the results given here impact finite volume discretization as well.

2. Symmetrization of conservation laws without involution. Consider the Cauchy initial value problem for a system of m coupled first-order differential equations in d space coordinates and time which represents a conservation law process. Let $\mathbf{u}(x,t) : \mathbb{R}^d \times \mathbb{R}^+ \mapsto \mathbb{R}^m$ denote the dependent solution variables and $\mathbf{f}(\mathbf{u}) : \mathbb{R}^m \mapsto \mathbb{R}^{m\times d}$ the flux vector. The model Cauchy problem is then given by

$$\begin{cases} \mathbf{u}_{,t} + \mathbf{f}_{i,x_i} = 0 \\ \mathbf{u}(x,0) = \mathbf{u}_0(x) \end{cases} \tag{2.1}$$

with implied summation on the index $i = 1, \ldots, d$. Additionally, the system is assumed to possess a convex scalar entropy extension. Let $U(\mathbf{u}) : \mathbb{R}^m \mapsto \mathbb{R}$ and $F(\mathbf{u}) : \mathbb{R}^m \mapsto \mathbb{R}^d$ denote an entropy-entropy flux pair for the system such that in addition to (2.1) the following inequality holds

$$U_{,t} + F_{i,x_i} \le 0 \tag{2.2}$$

with equality for classical (smooth) solutions. In the symmetrization theory for first-order conservation laws without involution [God61, Moc80], one seeks a mapping $\mathbf{u}(\mathbf{v}) : \mathbb{R}^m \mapsto \mathbb{R}^m$ applied to (2.1) so that when transformed

$$\mathbf{u}_{,\mathbf{v}} \mathbf{v}_{,t} + \mathbf{f}_{i,\mathbf{v}}\, \mathbf{v}_{,x_i} = 0 \tag{2.3}$$

the matrix $\mathbf{u}_{,\mathbf{v}}$ is symmetric positive definite (SPD) and the matrices $\mathbf{f}_{i,\mathbf{v}}$ are symmetric. Clearly, if twice differentiable functions $\mathcal{U}(\mathbf{v}) : \mathbb{R}^m \mapsto \mathbb{R}$

and $\mathcal{F}_i(\mathbf{v}) : \mathbb{R}^m \mapsto \mathbb{R}$ can be found so that

$$\mathbf{u} = \mathcal{U}^T_{,\mathbf{v}}, \quad \mathbf{f}_i = \mathcal{F}^T_{i,\mathbf{v}} \tag{2.4}$$

then the matrices

$$\mathbf{u}_{,\mathbf{v}} = \mathcal{U}_{,\mathbf{vv}}, \quad \mathbf{f}_{i,\mathbf{v}} = \mathcal{F}_{i,\mathbf{vv}}$$

are symmetric. Further, we shall require that $\mathcal{U}(\mathbf{v})$ be a convex function such that

$$\lim_{\mathbf{v}\to\infty} \frac{\mathcal{U}(\mathbf{v})}{|\mathbf{v}|} = +\infty \tag{2.5}$$

so that $U(\mathbf{u})$ can be interpreted as a Legendre transform of $\mathcal{U}(\mathbf{v})$

$$U(\mathbf{u}) = \sup_{\mathbf{v}} \{\mathbf{v}\cdot\mathbf{u} - \mathcal{U}(\mathbf{v})\} \ .$$

From (2.5), it follows that $\exists\ \mathbf{v}^* \in \mathbb{R}^m$ such that $\mathbf{v}\cdot\mathbf{u} - \mathcal{U}(\mathbf{v})$ achieves a maximum at $\mathbf{v}^*$

$$U(\mathbf{u}) = \mathbf{v}^* \cdot \mathbf{u} - \mathcal{U}(\mathbf{v}^*) \ . \tag{2.6}$$

At this maximum $\mathbf{u} = \mathcal{U}_{,\mathbf{v}}(\mathbf{v}^*)$ which can be locally inverted to the form $\mathbf{v}^* = \mathbf{v}(\mathbf{u})$. Elimination of $\mathbf{v}^*$ in (2.6) yields the simplified duality relationship

$$U(\mathbf{u}) = \mathbf{v}(\mathbf{u})\cdot\mathbf{u} - \mathcal{U}(\mathbf{v}(\mathbf{u})) \ .$$

Differentiation of this expression

$$U^T_{,\mathbf{u}} = \mathbf{v} + \mathbf{v}_{,\mathbf{u}}\mathbf{u} - \mathbf{v}_{,\mathbf{u}}\mathcal{U}^T_{,\mathbf{v}} = \mathbf{v} \tag{2.7}$$

gives an explicit formula for the entropy variables $\mathbf{v}$ in terms of derivatives of the entropy function $U(\mathbf{u})$. Using the mapping relation $\mathbf{v}(\mathbf{u})$, a duality pairing for entropy flux components is defined

$$F_i(\mathbf{u}) = \mathbf{v}(\mathbf{u})\cdot\mathbf{f}_i(\mathbf{u}) - \mathcal{F}_i(\mathbf{v}(\mathbf{u})) \ .$$

Differentiation then yields the flux relation

$$F_{i,\mathbf{u}} = \mathbf{v}\cdot\mathbf{f}_{i,\mathbf{u}} + \mathbf{v}_{,\mathbf{u}}\mathbf{f}_i - \mathbf{v}_{,\mathbf{u}}\mathcal{F}^T_{i,\mathbf{v}} = \mathbf{v}\cdot\mathbf{f}_{i,\mathbf{u}}$$

and the fundamental relationship for classical solutions

$$\mathbf{v}\cdot(\mathbf{u}_{,t} + \mathbf{f}_{i,x_i}) = U_{,t} + F_{i,x_i} = 0 \ .$$

These relationships are used extensively in the discrete energy analysis of the discontinuous Galerkin method.

2.1. Maxwell equations in symmetric form. The time-dependent Maxwell equations are given by

$$\frac{\partial}{\partial t}\begin{pmatrix}\mathbf{E}\\ \mathbf{B}\end{pmatrix} + \nabla\times\begin{pmatrix}-c^2\mathbf{B}\\ \mathbf{E}\end{pmatrix} = \begin{pmatrix}-\mathbf{j}/\epsilon_0\\ 0\end{pmatrix} \quad \text{(Maxwell equations)}$$

where $\mathbf{E}\in\mathbb{R}^d$, $\mathbf{B}\in\mathbb{R}^d$, $\rho_c\in\mathbb{R}$, and $\mathbf{j}\in\mathbb{R}^d$ denote the electric field, magnetic induction, charge and current density with ϵ_0 and c the free-space permittivity and speed of light, respectively. If the charge conservation equation

$$(\rho_c)_{,t} + \nabla\cdot\mathbf{j} = 0 \tag{2.8}$$

is satisfied for all time then the Maxwell system possesses the following involutions

$$\begin{aligned}\nabla\cdot\mathbf{E} &= \rho_c/\epsilon_0\\ \nabla\cdot\mathbf{B} &= 0\ .\end{aligned}$$

Writing the Maxwell system in matrix coefficient form reveals that the above system is essentially already in symmetric form using the variables $\mathbf{u}\equiv(\mathbf{E},\mathbf{B})^T$

$$\mathbf{u}_{,t} + A_i\,\mathbf{u}_{,x_i} = \mathbf{q}(\mathbf{u})\ ,\quad A_i = \begin{bmatrix}0 & c^2M_i\\ M_i^T & 0\end{bmatrix}$$

where in three space dimensions

$$M_1 = \begin{bmatrix}0&0&0\\0&0&1\\0&-1&0\end{bmatrix},\quad M_2 = \begin{bmatrix}0&0&-1\\0&0&0\\1&0&0\end{bmatrix},\quad M_3 = \begin{bmatrix}0&1&0\\-1&0&0\\0&0&0\end{bmatrix}.$$

Consequently, a suitable entropy-entropy flux pair for the Maxwell system are given by the scaled "square entropy" and square entropy flux

$$U(\mathbf{u}) = \frac{1}{2}(|\mathbf{E}|^2 + c^2\,|\mathbf{B}|^2)\ ,\quad F(\mathbf{u}) = c^2\,(\mathbf{E}\times\mathbf{B})\ .$$

Using this entropy function, the symmetrization variables and right symmetrizer are then obtained

$$\mathbf{v} = U^T_{,\mathbf{u}} = \begin{pmatrix}\mathbf{E}\\ c^2\mathbf{B}\end{pmatrix}\ ,\quad \mathbf{u}_{,\mathbf{v}} = \begin{bmatrix}I_{d\times d} & \\ & c^{-2}I_{d\times d}\end{bmatrix}$$

thus rendering the coefficient matrices symmetric as expected

$$A_i\mathbf{u}_{,\mathbf{v}} = \begin{bmatrix}0 & M_i\\ M_i^T & 0\end{bmatrix}.$$

Observe that the Maxwell system has been successfully symmetrized without utilizing the involutions. Consequently, the energy analysis for Maxwell's equations in a vacuum domain is identical to the energy analysis for conservation law systems without involution as also observed in [CLS04].

2.2. Ideal MHD in symmetric form. The equations of ideal compressible MHD are given by

$$\frac{\partial}{\partial t}\begin{pmatrix}\rho\\ \rho\mathbf{V}\\ E\\ \mathbf{B}\end{pmatrix}+\nabla\cdot\begin{pmatrix}\rho\mathbf{V}\\ \rho\mathbf{V}\mathbf{V}+I_{d\times d}\left(p+|\mathbf{B}|^2/2\right)-\mathbf{B}\,\mathbf{B}\\ \left(E+p+|\mathbf{B}|^2/2\right)\mathbf{V}-(\mathbf{V}\cdot\mathbf{B})\,\mathbf{B}\\ \mathbf{V}\mathbf{B}-\mathbf{B}\mathbf{V}\end{pmatrix}=0 \quad \text{(Ideal MHD)}$$

where $\rho \in \mathbb{R}$, $\mathbf{V} \in \mathbb{R}^d$, $\mathbf{B} \in \mathbb{R}^d$, and $p \in \mathbb{R}$ denote the fluid density, velocity, magnetic induction, and pressure with $E \in \mathbb{R}$ the total specific energy given by

$$E=\frac{p}{\gamma-1}+\rho|\mathbf{V}|^2/2+|\mathbf{B}|^2/2$$

and γ the ratio of specific heats. In addition, the MHD system possesses the solenoidal involution

$$\nabla\cdot\mathbf{B}=0$$

which is consistent with the absence of experimentally observed magnetic monopoles.

It is well known that thermodynamic entropy s is transported along velocity induced particle paths for ideal MHD. Recall that $s=\log(p\rho^{-\gamma})$ for MHD so that a differential of s is given by

$$d\,s=-\frac{\gamma}{\rho}d\,\rho+\frac{1}{p}d\,p\,.$$

Inserting equations derived from the MHD system (2.2) yields

$$s_{,t}+\mathbf{V}\cdot\nabla s+(\gamma-1)\frac{\mathbf{V}\cdot\mathbf{B}}{p}\nabla\cdot\mathbf{B}=0$$

or after combining with the continuity equation

$$(\rho s)_{,t}+\operatorname{div}(\rho\mathbf{V}s)+(\gamma-1)\frac{\rho\mathbf{V}\cdot\mathbf{B}}{p}\nabla\cdot\mathbf{B}=0$$

suggesting that $U(\mathbf{u})=-\rho s$ may be a suitable entropy function only if the involution $\nabla\cdot\mathbf{B}=0$ is satisfied. Indeed, a straightforward calculation for ideal MHD shows that this entropy function does *not* symmetrize the system under the change of variable $\mathbf{u}\mapsto\mathbf{v}$ with $\mathbf{v}=U^T_{,\mathbf{u}}$ (see for example Barth [Bar98])

$$\mathbf{f}_{,\mathbf{v}}\neq\mathbf{f}^T_{,\mathbf{v}}$$

since the involution equation has not been used. Godunov [God72] observed this phenomenon as well which lead to his development of a symmetrization technique for ideal MHD. The basic technique is reviewed here using a modified presentation from that originally given. The model MHD system with solenoidal involution is given by

$$\begin{cases} \mathbf{u}_{,t} + \mathbf{f}_{i,x_i} = 0 \\ \mathbf{B}_{i,x_i} = 0 \\ \mathbf{u}(x,0) = \mathbf{u}_0(x) \end{cases} \tag{2.9}$$

with convex entropy extension

$$U_{,t} + F_{i,x_i} \le 0 \; . \tag{2.10}$$

To analyze this system, Godunov considered augmenting the MHD system by adding multiples of the involution where the multipliers are themselves the gradient of a scalar homogeneous of degree one function $\phi(\mathbf{v}) : \mathbb{R}^m \mapsto \mathbb{R}$ with respect to the symmetrization variables $\mathbf{v}$

$$\mathbf{u}_{,t} + \mathbf{f}_{i,x_i} + \phi^T_{,\mathbf{v}}\, \mathbf{B}_{i,x_i} = 0 \; .$$

Consider the following ansatz for the dependent variables $\mathbf{u}$ and flux components $\mathbf{f}_i$

$$\begin{aligned} \mathbf{u} &= \mathcal{U}^T_{,\mathbf{v}} \\ \mathbf{f}_i &= \mathcal{F}^T_{i,\mathbf{v}} - \mathbf{r}(\mathbf{v})\, \mathbf{B}_i \end{aligned}$$

with $\mathcal{U}$ a convex scalar function and $\mathbf{r}(\mathbf{v}) : \mathbb{R}^m \mapsto \mathbb{R}^m$ an unknown vector-valued function. Observe that the augmented MHD system

$$(\mathcal{U}_{,\mathbf{v}})_{,t} + (\mathcal{F}_{i,\mathbf{v}} - \mathbf{r}(\mathbf{v})\, \mathbf{B}_i)^T_{,x_i} + \phi^T_{,\mathbf{v}}\, \mathbf{B}_{i,x_i} = 0 \tag{2.11}$$

possesses a symmetric quasilinear form in $\mathbf{v}$ variables whenever $\mathbf{r}(\mathbf{v}) = \phi^T_{,\mathbf{v}}$ since the system (2.11) then reduces to

$$\underbrace{\mathcal{U}_{,\mathbf{vv}}}_{\text{SPD}}\, \mathbf{v}_{,t} + \underbrace{(\mathcal{F}_{i,\mathbf{vv}} - \phi_{,\mathbf{vv}}\mathbf{B}_i)}_{\text{SYMM}}\, \mathbf{v}_{,x_i} = 0$$

so that the final flux relationship is obtained

$$\mathbf{f}_i = \mathcal{F}^T_{i,\mathbf{v}} - \phi^T_{,\mathbf{v}}\, \mathbf{B}_i \; .$$

The entropy function $U(\mathbf{u})$ for MHD can be interpreted as a Legendre transform of $\mathcal{U}(\mathbf{v})$

$$U(\mathbf{u}) = \sup_{\mathbf{v}} \{\mathbf{v} \cdot \mathbf{u} - \mathcal{U}(\mathbf{v})\}$$

eventually producing the generalized duality relationships

$$U(\mathbf{u}) = \mathbf{v}(\mathbf{u}) \cdot \mathcal{U}_{,\mathbf{v}}(\mathbf{v}(\mathbf{u})) - \mathcal{U}(\mathbf{v}(\mathbf{u}))$$

$$F_i(\mathbf{u}) = \mathbf{v}(\mathbf{u}) \cdot \mathcal{F}_{i,\mathbf{v}}(\mathbf{v}(\mathbf{u})) - \mathcal{F}_i(\mathbf{v}(\mathbf{u})) \tag{2.12}$$

so that for classical MHD solutions

$$\mathbf{v} \cdot (\mathbf{u}_{,t} + \mathbf{f}_{i,x_i} + \phi^T_{,\mathbf{v}} \mathbf{B}_{i,x_i}) = U_{,t} + F_{i,x_i} = 0 .$$

This relationship will be used heavily in later analysis of the discontinuous Galerkin method.

Choosing the entropy function $U(\mathbf{u}) = -\rho s$ yields $\phi(\mathbf{v}) = (\gamma - 1)\,\rho \mathbf{V} \cdot \mathbf{B}/p$, a homogeneous function of degree one in $\mathbf{v}$ (as required) so that $0 = \mathbf{v} \cdot \phi_{,\mathbf{vv}}$. The resulting involution multipliers $\phi_{,\mathbf{v}}$ are identical to those derived by Powell [Pow94] using a completely different argument motivated by (in part) the lack of Galilean invariance of the original MHD system and the subsequent addition of a divergence wave family into the local Riemann problem solution to restore Galilean invariance.

REMARK 2.1. Observe that MHD provides one particular example of a symmetrizable system with a given entropy-entropy flux pair $\{U, F_i\}$ for which the flux is not expressed as the gradient of a primative function $\mathcal{F}_i$ but rather

$$\mathbf{f}_i = \mathcal{F}^T_{i,\mathbf{v}} - \phi^T_{,\mathbf{v}} \mathbf{B}_i .$$

In fact, for the specific MHD entropy function $U(\mathbf{u}) = -\rho s$, it is possible to show that there cannot exist a function $\widetilde{\mathcal{F}}_i$ such that

$$\mathbf{f}_i = \widetilde{\mathcal{F}}^T_{i,\mathbf{v}} .$$

Thus, the DG energy analysis of MHD systems is fundamentally different from the energy analysis of systems not possessing involutions.

3. The DG finite element method. Let Ω denote a spatial domain composed of stationary nonoverlapping elements K_i, $\Omega = \cup K_i$, $K_i \cap K_j = \emptyset$, $i \neq j$ and time slab intervals $I^n \equiv [t^n_+, t^{n+1}_-], n = 0, \ldots, N-1$. Both continuous in time approximation and full space-time approximation on tensor space-time elements $K_i \times I^n$ will be considered in the analysis. It is useful to also define the element set $\mathcal{T} = \{K_1, K_2, \ldots\}$ and the interface set $\mathcal{E} = \{e_1, e_2, \ldots\}$ with interface members $\overline{K}_i \cap \overline{K}_j, i \neq j$ of measure $d-1$ corresponding to edges in 2-D and faces in 3-D. Let $\mathcal{P}_k(Q)$ denote the set of polynomials of degree at most k in a domain $Q \subset \mathbb{R}^d$. In the discontinuous Galerkin method, the approximating functions are discontinuous polynomials in both space and time

$$\mathcal{V}^h = \left\{ \mathbf{w} \,|\, \mathbf{w}_{|K \times I^n} \in \Big(\mathcal{P}_k(K \times I^n)\Big)^m, \forall K \in \mathcal{T}, n = 0, \ldots, N-1 \right\} .$$

Alternatively, [CLS89, CHS90, Shu99] utilize a semi-discrete formulation of the DG method together with Runge-Kutta time integration. In this case, the set of approximating functions are discontinuous polynomials in space and continuous functions in time denoted by $\mathcal{V}_c^h$.

For ease of exposition, the spatial domain Ω is assumed either periodic in all space dimensions or nonperiodic with compactly supported initial data. In this domain, we first consider the standard first-order Cauchy initial value problem (without involution)

$$\begin{cases} \mathbf{u}_{,t} + \mathbf{f}_{i,x_i} = 0 \\ \mathbf{u}(x, t_-^0) = \mathbf{u}_0(x) \end{cases} \tag{3.1}$$

with convex entropy extension

$$U_{,t} + F_{i,x_i} \le 0 \ . \tag{3.2}$$

The DG method for the time interval $[t_+^0, t_-^N]$ with weakly imposed initial data $\mathbf{v}_h(x, t_-^0)$ obtained from a suitable projection of the initial data $\mathbf{v}(\mathbf{u}_0(x))$ is given by the following statement:

<u>DG FEM:</u> Find $\mathbf{v}_h \in \mathcal{V}^h$ such that

$$B_{\mathrm{DG}}(\mathbf{v}_h, \mathbf{w}_h) = 0 \ , \ \forall \mathbf{w}_h \in \mathcal{V}^h \tag{3.3}$$

with

$$\begin{aligned} B_{\mathrm{DG}}(\mathbf{v}, \mathbf{w}) = \sum_{n=0}^{N-1} \Bigg(&\sum_{K \in \mathcal{T}} \int_{I^n} \int_K -(\mathbf{u}(\mathbf{v}) \cdot \mathbf{w}_{,t} + \mathbf{f}_i(\mathbf{v}) \cdot \mathbf{w}_{,x_i}) \, dx \, dt \\ &+ \sum_{K \in \mathcal{T}} \int_{I^n} \int_{\partial K} \mathbf{w}(x_-) \cdot \mathbf{h}(\mathbf{v}(x_-), \mathbf{v}(x_+); \mathbf{n}) \, ds \, dt \\ &+ \sum_{K \in \mathcal{T}} \int_K \left(\mathbf{w}(t_-^{n+1}) \cdot \mathbf{u}(\mathbf{v}(t_-^{n+1})) - \mathbf{w}(t_+^n) \cdot \mathbf{u}(\mathbf{v}(t_-^n)) \right) dx \Bigg) \end{aligned} \tag{3.4}$$

with suitable modifications when source terms are present. In this statement $\mathbf{h}(\mathbf{v}_-, \mathbf{v}_+; \mathbf{n}) : \mathbb{R}^m \times \mathbb{R}^m \times \mathbb{R}^d \mapsto \mathbb{R}^m$ denotes a numerical flux function, a vector-valued function of two interface states $\mathbf{v}_\pm$ and an oriented interface normal $\mathbf{n}$ with the following consistency and conservation properties:

- Consistency with the true flux, $\mathbf{h}(\mathbf{v}, \mathbf{v}; \mathbf{n}) = \mathbf{f}(\mathbf{v}) \cdot \mathbf{n}$
- Discrete cell conservation, $\mathbf{h}(\mathbf{v}_-, \mathbf{v}_+; \mathbf{n}) = -\mathbf{h}(\mathbf{v}_+, \mathbf{v}_-; -\mathbf{n})$.

For a given symmetrizable system with entropy function $U(\mathbf{u})$, the DG method is uniquely specified once $\mathcal{V}^h$, the entropy function $U(\mathbf{u})$, and the numerical flux function $\mathbf{h}(\mathbf{v}_-, \mathbf{v}_+; \mathbf{n})$ are chosen. In this formulation, the finite-dimensional space of symmetrization variables $\mathbf{v}_h$ are the basic unknowns in the trial space $\mathcal{V}^h$ and the dependent variables are then derived via $\mathbf{u}(\mathbf{v}_h)$. When not needed for clarity, this mapping is sometimes explicitly omitted, e.g. $U(\mathbf{v}_h)$ is written rather than $U(\mathbf{u}(\mathbf{v}_h))$. An important product of the DG energy analysis given below are sufficient conditions to be imposed on the numerical flux so that discrete entropy inequalities and total entropy bounds of the following form are obtained for the discretization of the Cauchy initial value problem (no boundary conditions):

- A local cell entropy inequality assuming continuous in time approximation, $\mathbf{v}_h \in \mathcal{V}_c^h$

$$\frac{d}{dt}\int_K U(\mathbf{v}_h)\, dx + \int_{\partial K} \overline{F}(\mathbf{v}_{-,h}, \mathbf{v}_{+,h}; \mathbf{n})\, ds \le 0 \ , \qquad \text{for each } K \in \mathcal{T} \tag{3.5}$$

where $\overline{F}(\mathbf{v}_{-,h}, \mathbf{v}_{+,h}; \mathbf{n})$ denotes a conservative numerical entropy flux. Summing over all elements then yields the global inequality

$$\frac{d}{dt}\int_\Omega U(\mathbf{v}_h)\, dx \le 0 \ . \tag{3.6}$$

- A total entropy bound assuming full space-time approximation, $\mathbf{v}_h \in \mathcal{V}^h$

$$\begin{aligned}\int_\Omega U(\mathbf{u}^*(t_-^0))\, dx &\le \int_\Omega U(\mathbf{u}(\mathbf{v}_h(x, t_-^N)))\, dx \\ &\le \int_\Omega U(\mathbf{u}(\mathbf{v}_h(x, t_-^0))\, dx\end{aligned} \tag{3.7}$$

where $\mathbf{u}^*(t_-^0)$ denotes the minimum total entropy state of the projected initial data

$$\mathbf{u}^*(t_-^0) \equiv \frac{1}{\text{meas}(\Omega)}\int_\Omega \mathbf{u}(\mathbf{v}_h(x, t_-^0))\, dx \ .$$

Under the assumption that the symmetrizer $\mathbf{u}_{,\mathbf{v}}$ remains spectrally bounded in space-time, i.e. there exist positive constants c_0 and C_0 independent of $\mathbf{v}_h$ such that

$$0 < c_0\, \|\mathbf{z}\|^2 \le \mathbf{z} \cdot \mathbf{u}_{,\mathbf{v}}(\mathbf{v}_h(x,t))\, \mathbf{z} \le C_0\, \|\mathbf{z}\|^2$$

for all $\mathbf{z} \ne 0$, the following L_2 stability result is then readily obtained for the Cauchy problem

$$\|\mathbf{u}(\mathbf{v}_h(\cdot, t_-^N)) - \mathbf{u}^*(t_-^0)\|_{L_2(\Omega)} \le \left(\frac{C_0}{c_0}\right)^{1/2} \|\mathbf{u}(\mathbf{v}_h(\cdot, t_-^0)) - \mathbf{u}^*(t_-^0)\|_{L_2(\Omega)}$$

3.1. DG energy analysis for systems without involution. In this section, the DG energy analysis for systems of conservation laws without involution is reviewed. From Sect. 2.1 it was shown that this analysis is also the relevant analysis for the Maxwell system since this system can be symmetrized without using the Maxwell system involutions. Consequently, consider the DG method applied to the nonlinear system (3.1). For brevity, we avoid the introduction of trace operators and instead use the shorthand notation for interface quantities $f_\pm \equiv f(\mathbf{v}(x_\pm))$, $\langle f\rangle_-^+ \equiv (f_- + f_+)/2$ and $[f]_-^+ = f_+ - f_-$. An energy analysis assuming continuous in time functions, $\mathbf{v}_h \in \mathcal{V}_c^h$, yields the following cell-wise local entropy inequality which build upon previous scalar conservation law analysis for DG by [JJS95, JS94] and further related DG analysis for systems in [CS97] and [Bar98, Bar99].

THEOREM 3.1 (**DG semi-discrete cell entropy inequality**). *Let $\mathbf{v}_h \in \mathcal{V}_c^h$ denote a numerical solution obtained using the discontinuous Galerkin method (3.4) assuming a continuous in time approximation for the Cauchy initial value problem (3.1) with convex entropy extension (3.2). Assume the numerical flux $\mathbf{h}(\mathbf{v}_-, \mathbf{v}_+; \mathbf{n})$ satisfies the system E-flux condition*

$$[\mathbf{v}]_-^+ \cdot (\mathbf{h}(\mathbf{v}_-, \mathbf{v}_+; \mathbf{n}) - \mathbf{f}(\mathbf{v}(\theta)) \cdot \mathbf{n}) \le 0 \ , \quad \forall \theta \in [0,1] \tag{3.8}$$

where $\mathbf{v}(\theta) = \mathbf{v}_- + \theta\, [\mathbf{v}]_-^+$. The numerical solution $\mathbf{v}_h$ then satisfies the local semi-discrete cell entropy inequality

$$\frac{d}{dt}\int_K U(\mathbf{v}_h)\; dx + \int_{\partial K} \overline{F}(\mathbf{v}_{-,h}, \mathbf{v}_{+,h}; \mathbf{n})\, ds \le 0 \ , \quad \textit{for each } K \in \mathcal{T} \tag{3.9}$$

with

$$\overline{F}(\mathbf{v}_-, \mathbf{v}_+; \mathbf{n}) \equiv \langle \mathbf{v}\rangle_-^+ \cdot \mathbf{h}(\mathbf{v}_-, \mathbf{v}_+; \mathbf{n}) - \langle \mathcal{F}\cdot\mathbf{n}\rangle_-^+ \tag{3.10}$$

as well as the global semi-discrete entropy inequality

$$\frac{d}{dt}\int_\Omega U(\mathbf{v}_h)\; dx \le 0 \ . \tag{3.11}$$

Proof. Evaluate the energy, $B_{\rm DG}(\mathbf{v}_h, \mathbf{v}_h)$, for a single stationary element K assuming continuous in time functions

$$\begin{aligned}
\int_K \mathbf{v}\cdot\mathbf{u}_{,t}\; dx &= \frac{d}{dt}\int_K U\, dx \\
&= -\left(\int_K -\mathbf{v}_{,x_i}\cdot \mathbf{f}_i\; dx + \int_{\partial K} \mathbf{v}_-\cdot \mathbf{h}\, ds\right) \\
&= -\left(\int_K -\mathcal{F}_{i,x_i}\; dx + \int_{\partial K} \mathbf{v}_-\cdot \mathbf{h}\; ds\right) \\
&= -\int_{\partial K} (-\mathcal{F}_-\cdot\mathbf{n} + \mathbf{v}_-\cdot\mathbf{h})\; ds \\
&= -\int_{\partial K} \Big(\underbrace{\overline{F}(\mathbf{v}_-, \mathbf{v}_+; \mathbf{n})}_{\text{Conservative Flux}} + \underbrace{D(\mathbf{v}_-, \mathbf{v}_+; \mathbf{n})}_{\text{Entropy Dissipation}}\Big)\; ds
\end{aligned}$$

for carefully chosen conservative entropy flux and entropy dissipation functions

$$\overline{F}(\mathbf{v}_-, \mathbf{v}_+; \mathbf{n}) \equiv \langle \mathbf{v}\rangle_-^+ \cdot \mathbf{h}(\mathbf{v}_-, \mathbf{v}_+; \mathbf{n}) - \langle \mathcal{F}\cdot\mathbf{n}\rangle_-^+$$

$$D(\mathbf{v}_-,\mathbf{v}_+;\mathbf{n}) \equiv -\frac{1}{2}([\mathbf{v}]_-^+ \cdot \mathbf{h}(\mathbf{v}_-,\mathbf{v}_+;\mathbf{n}) - [\mathcal{F}\cdot\mathbf{n}]_-^+) \ .$$

Observe that the chosen form of $\overline{F}(\mathbf{v}_-,\mathbf{v}_+;\mathbf{n})$ is a consistent and conservative approximation to the true entropy flux $F(\mathbf{v})$

- $\overline{F}(\mathbf{v},\mathbf{v};\mathbf{n}) = (\mathbf{v}\cdot\mathbf{f} - \mathcal{F})\cdot\mathbf{n} = F\cdot\mathbf{n}$ (consistency)
- $\overline{F}(\mathbf{v}_-,\mathbf{v}_+;\mathbf{n}) = -\overline{F}(\mathbf{v}_+,\mathbf{v}_-;-\mathbf{n})$ (conservation) .

The only remaining task is to determine sufficient conditions in the design of the numerical flux $\mathbf{h}(\mathbf{v}_-,\mathbf{v}_+;\mathbf{n})$ so that $D(\mathbf{v}_-,\mathbf{v}_+;\mathbf{n}) \geq 0$. Rewriting the jump term appearing in the entropy dissipation term as a path integration in state space

$$\begin{aligned} D(\mathbf{v}_-,\mathbf{v}_+;\mathbf{n}) &= -\frac{1}{2}([\mathbf{v}]_-^+ \cdot \mathbf{h}(\mathbf{v}_-,\mathbf{v}_+;\mathbf{n}) - [\mathcal{F}\cdot\mathbf{n}]_-^+) \\ &= -\frac{1}{2}[\mathbf{v}]_-^+ \cdot \left(\mathbf{h}(\mathbf{v}_-,\mathbf{v}_+;\mathbf{n}) - \int_0^1 \mathcal{F}^T_{,\mathbf{v}}(\mathbf{v}(\theta))\cdot\mathbf{n}\ d\theta\right) \\ &= -\frac{1}{2}[\mathbf{v}]_-^+ \cdot \left(\mathbf{h}(\mathbf{v}_-,\mathbf{v}_+;\mathbf{n}) - \int_0^1 \mathbf{f}(\mathbf{v}(\theta))\cdot\mathbf{n}\ d\theta\right) \\ &= -\frac{1}{2}\int_0^1 [\mathbf{v}]_-^+ \cdot (\mathbf{h}(\mathbf{v}_-,\mathbf{v}_+;\mathbf{n}) - \mathbf{f}(\mathbf{v}(\theta))\cdot\mathbf{n})\ d\theta \ . \end{aligned}$$

A sufficient condition for nonnegativity of $D(\mathbf{v}_-,\mathbf{v}_+;\mathbf{n})$ and the local cell entropy inequality (3.9) when applied to finite-dimensional subspaces is that the integrand be nonpositive. This yields a system generalization of Osher's famous E-flux condition for scalar conservation laws given in [Osh84]

$$[\mathbf{v}]_-^+ \cdot (\mathbf{h}(\mathbf{v}_-,\mathbf{v}_+;\mathbf{n}) - \mathbf{f}(\mathbf{v}(\theta))\cdot\mathbf{n}) \leq 0 \ , \quad \forall\theta\in[0,1] \ . \tag{3.12}$$

Summation of (3.9) over all elements in the mesh together with the conservative telescoping property of $\overline{F}(\mathbf{v}_-,\mathbf{v}_+;\mathbf{n})$ yields the global entropy inequality (3.11). □

Let $\lambda_1 \leq \lambda_2 \leq \cdots \leq \lambda_m$ denote ordered eigenvalues of $\mathbf{f}_{,\mathbf{u}}$. Some specific examples of system E-fluxes (proofs omitted here) include

- Symmetric variable variant of the local Lax-Friedrichs flux

$$\mathbf{h}_{\mathrm{SLF}}(\mathbf{v}_-,\mathbf{v}_+;\mathbf{n}) = \langle\mathbf{f}\cdot\mathbf{n}\rangle_-^+ - \frac{1}{2}\lambda_{\max}\ [\mathbf{u}(\mathbf{v})]_{x_-}^{x_+} \tag{3.13}$$

  with

$$\lambda_{\max} \equiv \sup_{0\leq\xi\leq 1}\ \max_{1\leq i\leq m} |\lambda_i(\mathbf{v}(\xi))|$$

  where $\mathbf{v}(\xi) = \mathbf{v}_- + \xi\ [\mathbf{v}]_-^+$.

- Symmetric variable variant of the Harten-Lax-van Leer-Einfeldt flux [HLvL83, EMRS92]

$$\mathbf{h}_{\mathrm{SHLLE}}(\mathbf{v}_-,\mathbf{v}_+;\mathbf{n}) = \langle\mathbf{f}\cdot\mathbf{n}\rangle_-^+ - \frac{1}{2}\,\mathbf{h}^d_{\mathrm{SHLLE}}(\mathbf{v}_-,\mathbf{v}_+;\mathbf{n}) \tag{3.14}$$

with

$$\mathbf{h}^d_{\mathrm{SHLLE}}(\mathbf{v}_-,\mathbf{v}_+;\mathbf{n}) = \frac{\lambda_{\max}+\lambda_{\min}}{\lambda_{\max}-\lambda_{\min}}[\mathbf{f}(\mathbf{v};\mathbf{n})]_-^+ - \frac{2\lambda_{\max}\lambda_{\min}}{\lambda_{\max}-\lambda_{\min}}[\mathbf{u}(\mathbf{v})]_-^+$$

and

$$\lambda_{\max} \equiv \sup_{0\le\xi\le1} \max(0,\lambda_m(\mathbf{v}(\xi))) \ , \quad \lambda_{\min} \equiv \inf_{0\le\xi\le1} \min(0,\lambda_1(\mathbf{v}(\xi)))$$

where $\mathbf{v}(\xi) = \mathbf{v}_- + \xi\ [\mathbf{v}]_-^+$.

Fully discrete entropy bounds are readily derived assuming DG finite element discretization in time.

THEOREM 3.2 (**DG fully-discrete total entropy bounds**). *Let $\mathbf{v}_h \in \mathcal{V}^h$ denote the space-time numerical solution obtained using the discontinuous Galerkin method (3.4) for the Cauchy initial value problem (3.1) with convex entropy extension (3.2). Assume the numerical flux $\mathbf{h}(\mathbf{v}_-,\mathbf{v}_+;\mathbf{n})$ satisfies the system E-flux condition*

$$[\mathbf{v}]_-^+ \cdot (\mathbf{h}(\mathbf{v}_-,\mathbf{v}_+;\mathbf{n}) - \mathbf{f}(\mathbf{v}(\theta))\cdot\mathbf{n}) \le 0 \ , \quad \forall\theta\in[0,1]$$

where $\mathbf{v}(\theta) = \mathbf{v}_- + \theta\ [\mathbf{v}]_-^+$. The numerical solution $\mathbf{v}_h$ then satisfies the total entropy bound

$$\int_\Omega U(\mathbf{u}^*(t^0_-))\,dx \le \int_\Omega U(\mathbf{u}(\mathbf{v}_h(x,t^N_-)))\,dx \le \int_\Omega U(\mathbf{u}(\mathbf{v}_h(x,t^0_-)))\,dx \qquad (3.15)$$

where $\mathbf{u}^*(t^0_-)$ denotes the minimum total entropy state of the initial projected data

$$\mathbf{u}^*(t^0_-) \equiv \frac{1}{\mathrm{meas}(\Omega)}\int_\Omega \mathbf{u}(\mathbf{v}_h(x,t^0_-))\,dx \ .$$

Proof. Analysis of the spatial terms follows the same path taken in Theorem 3.1 (omitted here) with an additional integration performed in the time coordinate. Consider the energy of the remaining time evolution terms in (3.4) after integration-by-parts for a single time slab interval I^n

$$\begin{aligned}\int_{I^n}\int_\Omega \mathbf{v}\cdot\mathbf{u}_{,t}\,dx\,dt + \int_\Omega \mathbf{v}(t^n_+)\ \cdot\ [\mathbf{u}]_{t^n_-}^{t^n_+}\,dx \\ = \int_\Omega\int_{I^n} U_{,t}\,dt\,dx + \int_\Omega \mathbf{v}(t^n_+)\cdot[\mathbf{u}]_{t^n_-}^{t^n_+}\,dx \\ = \int_\Omega \left([U]_{t^n_+}^{t^{n+1}_-} - [U]_{t^n_-}^{t^n_+} + \mathbf{v}(t^n_+)\cdot[\mathbf{u}]_{t^n_-}^{t^n_+}\right)dx\end{aligned}$$

Taylor series with integral remainder together with the duality relationship (2) yields

$$[U]_{t^n_-}^{t^n_+} - \mathbf{v}(t^n_-)\cdot[\mathbf{u}]_{t^n_-}^{t^n_+} + R^n = 0 \ ,$$

$$R^n \equiv \int_0^1 (1-\theta)\ [\mathbf{v}]_{t^n_-}^{t^n_+}\cdot\mathbf{u}_{,\mathbf{v}}(\mathbf{v}(\theta))\ [\mathbf{v}]_{t^n_-}^{t^n_+}\,d\theta \ge 0$$

where $\mathbf{v}(\theta) = \mathbf{v}(t^n_-) + \theta\, [\mathbf{v}]_{t^n_-}^{t^n_+}$. Inserting into the time evolution terms

$$\int_{I^n}\int_\Omega \mathbf{v}\cdot\mathbf{u}_{,t}\,dx\,dt + \int_\Omega \mathbf{v}(t^n_+)\cdot[\mathbf{u}]_{t^n_-}^{t^n_+}\,dx = \int_\Omega \left([U]_{t^n_-}^{t^{n+1}_-} + R^n\right) dx\ .$$

Summing over all time slabs, the first term on the right-hand side of this equation vanishes except for initial and final time slab contributions. Utilizing nonnegativity of the remainder terms R^n then yields the following inequality for the time evolution terms

$$\sum_{n=0}^{N-1}\left(\int_{I^n}\int_\Omega \mathbf{v}\cdot\mathbf{u}_{,t}\,dx\,dt + \int_\Omega \mathbf{v}(t^n_+)\cdot[\mathbf{u}]_{t^n_-}^{t^n_+}\,dx\right) \geq \int_\Omega \left(U(t^N_-) - U(t^0_-)\right) dx\ .$$

Assume satisfaction of the system E-flux condition, the spatial term analysis used in the proof of Theorem 3.1 reduces to the inequality

$$\sum_{n=0}^{N-1}\sum_{K\in T}\int_{I^n}\left(\int_K -\mathbf{v}_{,x_i}\cdot\mathbf{f}_i\,dx + \int_{\partial K}\mathbf{v}_-\cdot\mathbf{h}\,ds\right) dt \geq 0\ .$$

Combining temporal and spatial results yields

$$0 = B_{\mathrm{DG}}(\mathbf{v},\mathbf{v}) \geq \int_\Omega \left(U(t^N_-) - U(t^0_-)\right) dx\ .$$

Hence, the desired upper bound in (3.15) is established when applied to finite-dimensional subspaces

$$\int_\Omega U(\mathbf{u}(\mathbf{v}_h(x,t^N_-)))\,dx \leq \int_\Omega U(\mathbf{u}(\mathbf{v}_h(x,t^0_-))\,dx\ . \tag{3.16}$$

To obtain the lower bound in (3.15), we exploit the well-known thermodynamic concept of a *minimum total entropy state* (see for example [Mer88]). Define the integral average state $\mathbf{u}^*$ at time slab boundaries

$$\mathbf{u}^*(t^n_-) \equiv \frac{1}{\mathrm{meas}(\Omega)}\int_\Omega \mathbf{u}(\mathbf{v}_h(x,t^n_-))\,dx\ ,\quad n = 0,\ldots,N\ .$$

For the DG space-time discretization of the Cauchy initial value problem, $\mathbf{u}^*$ is invariant when evaluated at time slab boundaries, i.e.

$$\mathbf{u}^*(t^n_-) = \mathbf{u}^*(t^{n-1}_-) = \cdots = \mathbf{u}^*(t^0_-) \tag{3.17}$$

owing to discrete conservation in both space and time. A Taylor series with integral remainder expansion of the entropy function given two states $\mathbf{u}^*(t^n_-)$ and $\mathbf{u}(\mathbf{v}_h(x,t^n_-))$ for a fixed n yields

$$U(\mathbf{u}) = U(\mathbf{u}^*) + \mathbf{v}(\mathbf{u}^*)\cdot(\mathbf{u}-\mathbf{u}^*) + \int_0^1 (1-\theta))(\mathbf{u}-\mathbf{u}^*)\cdot U_{,\mathbf{uu}}(\theta)(\mathbf{u}-\mathbf{u}^*)d\theta\ .$$

When integrated over Ω, the second right-hand side term vanishes identically by the definition of $\mathbf{u}^*$

$$\int_\Omega U(\mathbf{u})dx = \int_\Omega U(\mathbf{u}^*)dx + \int_\Omega \int_0^1 (1-\theta))(\mathbf{u}-\mathbf{u}^*) \cdot U_{,\mathbf{uu}}(\theta)(\mathbf{u}-\mathbf{u}^*)d\theta\, dx \ .$$

From strict convexity of the entropy function, it follows that $\mathbf{u}^*$ is a minimum total entropy state since $\int_\Omega U\, dx$ is minimized when $\mathbf{u} = \mathbf{u}^*$. Finally, since $\mathbf{u}^*(t_-^n)$ is constant for $n = 0, \dots, N$, then

$$\int_\Omega U(\mathbf{u}^*(t_-^0))\, dx = \int_\Omega U(\mathbf{u}^*(t_-^N))\, dx \le \int_\Omega U(\mathbf{u}(\mathbf{v}_h(x, t_-^N)))\, dx \ .$$

This establishes the lower bound in (3.15). □

3.2. DG Stability analysis for systems with solenoidal involution. Our attention shifts to the MHD system with solenoidal involution

$$\begin{cases} \mathbf{u}_{,t} + \mathbf{f}_{i,x_i} = 0 \\ \mathbf{B}_{i,x_i} = 0 \\ \mathbf{u}(x, t_-^0) = \mathbf{u}_0(x) \end{cases} \tag{3.18}$$

with convex entropy extension

$$U_{,t} + F_{i,x_i} \le 0 \ . \tag{3.19}$$

The goal is to derive sufficient conditions for MHD system discretizations so that the cell entropy inequality (3.5), the global semi-discrete bound (3.6), and the global space-time bound (3.7) are obtained. Motivated by the Godunov MHD symmetrization theory, we consider an implementation of the DG method using the Godunov augmented MHD system.

DG FEM for MHD: Find $\mathbf{v}_h \in \mathcal{V}^h$ such that

$$B_{\text{DG–MHD}}(\mathbf{v}_h, \mathbf{w}_h) = 0 \ , \ \forall\, \mathbf{w}_h \in \mathcal{V}^h \tag{3.20}$$

with

$$\begin{aligned} B_{\text{DG–MHD}}&(\mathbf{v}, \mathbf{w}) \\ &= \sum_{n=0}^{N-1} \Bigg(\sum_{K \in \mathcal{T}} \int_{I^n} \int_K -(\mathbf{u}(\mathbf{v}) \cdot \mathbf{w}_{,t} + \mathbf{f}_i(\mathbf{v}) \cdot \mathbf{w}_{,x_i})\, dx\, dt \\ &\quad - \sum_{K \in \mathcal{T}} \int_{I^n} \int_K \sigma_K\, (\mathbf{w} \cdot \phi_{,\mathbf{v}}^T)\, \nabla \cdot \mathbf{B}(\mathbf{v})\, dx\, dt \\ &\quad + \sum_{K \in \mathcal{T}} \int_{I^n} \int_{\partial K} \mathbf{w}(x_-) \cdot \mathbf{h}(\mathbf{v}(x_-), \mathbf{v}(x_+); \mathbf{n})\, ds\, dt \\ &\quad + \sum_{K \in \mathcal{T}} \int_K \left(\mathbf{w}(t_-^{n+1}) \cdot \mathbf{u}(\mathbf{v}(t_-^{n+1})) - \mathbf{w}(t_+^n) \cdot \mathbf{u}(\mathbf{v}(t_-^n))\right) dx \Bigg) \ . \end{aligned} \tag{3.21}$$

Observe the added $\nabla \cdot \mathbf{B}$ term with adjustable coefficient σ_K is motivated by the theory given in Sect. 2.2. The value of σ_K will be determined from the discrete energy analysis. This term is identical to that proposed by Powell [Pow94] using a different motivating argument. Unfortunately, without placing further constraints on the discrete $\mathbf{B}$ field, the Powell term is only valid for classical (smooth) solutions since this term cannot be written in divergence form. Consequently incorrect Rankine-Hugonoit jump conditions are observed for computed weak (discontinuous) solutions [Csi02]. Note that this term vanishes identically and correct weak solutions are obtained when a locally divergence-free basis is employed.

A DG analysis similar to that used in Theorem 3.1 yields the following conditions for a discrete cell entropy inequality for the MHD formulation.

THEOREM 3.3 (**DG semi-discrete MHD cell entropy inequality**). *Let $\mathbf{v}_h \in \mathcal{V}_c^h$ denote a numerical solution obtained using the discontinuous Galerkin method (3.21) assuming continuous in time approximation for the MHD Cauchy initial value problem (3.18) with convex entropy extension (3.19). Assume the following conditions are satisfied:*

1. *Either $\sigma_K = 1$ or the pointwise solenoidal condition*

$$\nabla \cdot \mathbf{B}(\mathbf{v}_h)|_K = 0 \ , \quad \forall K \in \mathcal{T} \ .$$

2. *The MHD system E-flux condition*

$$[\mathbf{v}]_-^+ \cdot (\mathbf{h}(\mathbf{v}_-, \mathbf{v}_+; \mathbf{n}) - \mathbf{f}(\mathbf{v}(\theta)) \cdot \mathbf{n} + \phi(\mathbf{v}(\theta)) \ (\mathbf{B}(\mathbf{v}(\theta)) \cdot \mathbf{n})_{,\mathbf{v}}^T) \leq 0 \ ,$$

$\forall \theta \in [0,1]$ *where* $\mathbf{v}(\theta) = \mathbf{v}_- + \theta \ [\mathbf{v}]_-^+$.

The numerical solution $\mathbf{v}_h$ then satisfies the local semi-discrete cell entropy inequality

$$\frac{d}{dt} \int_K U(\mathbf{v}_h) \ dx + \int_{\partial K} \overline{F}(\mathbf{v}_{-,h}, \mathbf{v}_{+,h}; \mathbf{n}) \, ds \leq 0 \ , \qquad \text{for each } K \in \mathcal{T} \tag{3.22}$$

with

$$\overline{F}(\mathbf{v}_-, \mathbf{v}_+; \mathbf{n}) \equiv \langle \mathbf{v} \rangle_-^+ \cdot \mathbf{h}(\mathbf{v}_-, \mathbf{v}_+; \mathbf{n}) - \langle \mathcal{F} \cdot \mathbf{n} - \phi \ \mathbf{B} \cdot \mathbf{n} \rangle_-^+ \tag{3.23}$$

as well as the global semi-discrete entropy inequality

$$\frac{d}{dt} \int_\Omega U(\mathbf{v}_h) \ dx \leq 0 \ . \tag{3.24}$$

Proof. Evaluate the energy, $B_{\mathrm{DG}}(\mathbf{v}_h, \mathbf{v}_h)$, for a single stationary element K in the DG discretization of the MHD system assuming continuous in time approximation

$$\frac{d}{dt} \int_K U \ dx = \int_K (\ \mathbf{v}_{,x_i} \cdot \mathbf{f}_i) \, dx + \int_{\partial K} \mathbf{v}_- \cdot \mathbf{h} \, ds$$

$$
\begin{aligned}
&= -\int_{\partial K} \left(-\mathcal{F}_- \cdot \mathbf{n} + \phi_- \left(\mathbf{B}_- \cdot \mathbf{n}\right) + \mathbf{v}_- \cdot \mathbf{h}\right) ds \\
&\quad -\int_K (1-\sigma_K)\, \phi \nabla \cdot \mathbf{B}\, dx \\
&= -\int_{\partial K} \left(\overline{F}(\mathbf{v}_-, \mathbf{v}_+; \mathbf{n}) + D(\mathbf{v}_-, \mathbf{v}_+; \mathbf{n})\right) ds \\
&\quad -\int_K (1-\sigma_K)\, \phi \nabla \cdot \mathbf{B}\, dx \ .
\end{aligned}
$$

The remaining element interior term vanishes identically by either imposing $\sigma_K = 1$ or the local solenoidal condition on the magnetic induction field, $\nabla \cdot \mathbf{B}|_K = 0$. Suitable definitions for the conservative entropy flux and entropy dissipation are given by

$$
\begin{aligned}
\overline{F}(\mathbf{v}_-, \mathbf{v}_+; \mathbf{n}) &\equiv \langle \mathbf{v} \rangle_-^+ \cdot \mathbf{h}(\mathbf{v}_-, \mathbf{v}_+; \mathbf{n}) + \langle -\mathcal{F} \cdot \mathbf{n} + \phi\, \mathbf{B} \cdot \mathbf{n} \rangle_-^+ \\
D(\mathbf{v}_-, \mathbf{v}_+; \mathbf{n}) &\equiv -\frac{1}{2}([\mathbf{v}]_-^+ \cdot \mathbf{h} + [-\mathcal{F} \cdot \mathbf{n} + \phi\, \mathbf{B} \cdot \mathbf{n}]_-^+) \ .
\end{aligned}
$$

This choice of numerical entropy flux satisfies conservation and consistency properties

- $\overline{F}(\mathbf{v}_-, \mathbf{v}_+; \mathbf{n}) = -\overline{F}(\mathbf{v}_+, \mathbf{v}_-; -\mathbf{n})$ (conservation)
- $\overline{F}(\mathbf{v}, \mathbf{v}; \mathbf{n}) = (\mathbf{v} \cdot \mathbf{f} - \mathcal{F} + \phi\, \mathbf{B}) \cdot \mathbf{n} = F \cdot \mathbf{n}$ (consistency) .

Rewriting the jump term appearing in the entropy dissipation term as a path integration assuming a parameterized state space $\mathbf{v}(\theta) = \mathbf{v}_- + \theta\, [\mathbf{v}]_-^+$

$$
\begin{aligned}
D(\mathbf{v}_-, \mathbf{v}_+; \mathbf{n}) &= -\frac{1}{2}([\mathbf{v}]_-^+ \cdot \mathbf{h} + [-\mathcal{F} \cdot \mathbf{n} + \phi\, \mathbf{B} \cdot \mathbf{n}]_-^+) \\
&= -\frac{1}{2}[\mathbf{v}]_-^+ \cdot \left(\mathbf{h} - \int_0^1 (\mathcal{F}_{,\mathbf{v}}^T(\mathbf{v}(\theta)) \cdot \mathbf{n} - (\phi\, \mathbf{B} \cdot \mathbf{n})_{,\mathbf{v}}^T(\mathbf{v}(\theta)))\, d\theta\right) \\
&= -\frac{1}{2}\int_0^1 [\mathbf{v}]_-^+ \cdot (\mathbf{h} - \mathbf{f}(\mathbf{v}(\theta)) \cdot \mathbf{n} + \phi(\mathbf{v}(\theta))\, (\mathbf{B}(\mathbf{v}(\theta)) \cdot \mathbf{n})_{,\mathbf{v}}^T)\, d\theta.
\end{aligned}
$$

A sufficient condition for nonnegativity of $D(\mathbf{v}_-, \mathbf{v}_+; \mathbf{n})$ is that the integrand be nonpositive. This yields the MHD E-flux condition

$$
[\mathbf{v}]_-^+ \cdot (\mathbf{h}(\mathbf{v}_-, \mathbf{v}_+; \mathbf{n}) - \mathbf{f}(\mathbf{v}(\theta)) \cdot \mathbf{n} + \phi(\mathbf{v}(\theta))\, (\mathbf{B}(\mathbf{v}(\theta)) \cdot \mathbf{n})_{,\mathbf{v}}^T) \le 0 \ , \quad \forall \theta \in [0,1] \ .
$$

This establishes the semi-discrete cell entropy inequality for MHD. Summation of (3.22) over all elements in the mesh together with the conservative telescoping property of $\overline{F}(\mathbf{v}_-, \mathbf{v}_+; \mathbf{n})$ yields the global semi-discrete entropy inequality (3.24). □

The conditions set forth in Theorem 3.3 are also sufficient to establish two-sided bounds on the total entropy.

THEOREM 3.4 (**DG fully-discrete MHD total entropy bounds**). *Let $\mathbf{v}_h \in \mathcal{V}^h$ denote the space-time numerical solution obtained using the discontinuous Galerkin method (3.21) for the MHD Cauchy initial value problem (3.18) with convex entropy extension (3.19). Assume the following conditions are satisfied:*

1. *Either* $\sigma_K = 1$ *and the cellwise condition*

$$\int_K \phi^T_{,\mathbf{v}} \nabla \cdot \mathbf{B}(\mathbf{v}_h)\, dx = 0 \ , \quad \forall K \in \mathcal{T} \ .$$

or $\sigma_K \neq 1$ *and the pointwise condition*

$$\nabla \cdot \mathbf{B}(\mathbf{v}_h)|_K = 0, \quad \forall K \in \mathcal{T} \ .$$

2. *The MHD system E-flux condition*

$$[\mathbf{v}]^+_- \cdot (\mathbf{h}(\mathbf{v}_-, \mathbf{v}_+; \mathbf{n}) - \mathbf{f}(\mathbf{v}(\theta)) \cdot \mathbf{n} + \phi(\mathbf{v}(\theta))\ (\mathbf{B}(\mathbf{v}(\theta)) \cdot \mathbf{n})^T_{,\mathbf{v}}) \leq 0 \ ,$$

$\forall \theta \in [0,1]$ *where* $\mathbf{v}(\theta) = \mathbf{v}_- + \theta\ [\mathbf{v}]^+_-$.

The numerical solution $\mathbf{v}_h$ *then satisfies the total entropy bound*

$$\begin{aligned} \int_\Omega U(\mathbf{u}^*(t^0_-))\ dx &\leq \int_\Omega U(\mathbf{u}(\mathbf{v}_h(x, t^N_-)))\, dx \\ &\leq \int_\Omega U(\mathbf{u}(\mathbf{v}_h(x, t^0_-))\, dx \end{aligned} \tag{3.25}$$

where $\mathbf{u}^*(t^0_-)$ *denotes the minimum total entropy state of the initial projected data*

$$\mathbf{u}^*(t^0_-) \equiv \frac{1}{\text{meas}(\Omega)} \int_\Omega \mathbf{u}(\mathbf{v}_h(x, t^0_-))\, dx \ .$$

Proof. Omitted, see Theorem 3.2. The cellwise condition arises from the establishment of the minimum entropy state, $\mathbf{u}^*$. □

3.2.1. A compatible B field representation. Unfortunately, conventional system E-fluxes do not satisfy the MHD system E-flux condition. Furthermore, calculation of the actual symmetrization variables for the MHD system (2.2) associated with the entropy function, $U(\mathbf{u}) = -\rho\, s$, reveals that $\mathbf{B}$ is not a vector component of $\mathbf{v}$, viz.

$$\mathbf{v} = U^T_{\mathbf{u}} = (\gamma - 1) \begin{pmatrix} \frac{\gamma - s}{\gamma - 1} + \frac{\rho \mathbf{V}^2}{2p} \\ \frac{\rho \mathbf{V}}{p} \\ -\frac{\rho}{p} \\ \frac{\rho \mathbf{B}}{p} \end{pmatrix} . \tag{3.26}$$

Observe, however, that the last vector component $\rho\mathbf{B}/p$ is a $-\mathbf{B}$ multiple of the preceding component $-\rho/p$. Hence, it is possible to parameterize $\mathbf{v}$ on a line, $\mathbf{v}(\theta) = \mathbf{v}_- + \theta\, [\mathbf{v}]^+_-$, *and* constrain $\mathbf{B} \cdot \mathbf{n}$ independent of θ so that $[\mathbf{B} \cdot \mathbf{n}]^+_- = 0$. The following lemma states that under this constraint, the MHD system E-flux condition reduces to a constrained variant of the system E-flux condition (3.8).

LEMMA 3.1 (**B field compatibility**). *Assume the MHD system E-flux condition as given in Theorems 3.3 and 3.4. In addition, assume that* $\mathbf{B}(\mathbf{v}) \cdot \mathbf{n}$ *is constrained to be continuous at interelement interfaces, i.e.* $[\mathbf{B}(\mathbf{v}) \cdot \mathbf{n}]_-^+ = 0$. *Then, under this assumption, the results of Theorems 3.3 and 3.4 are identically obtained with the MHD system E-flux condition*

$$[\mathbf{v}]_-^+ \cdot (\mathbf{h} - \mathbf{f}(\mathbf{v}(\theta)) \cdot \mathbf{n} + \phi(\mathbf{v}(\theta))(\mathbf{B}(\mathbf{v}(\theta)) \cdot \mathbf{n})_{,\mathbf{v}}^T) \leq 0 \quad , \quad \forall \theta \in [0,1]$$

replaced by the constrained system E-flux condition

$$[\mathbf{v}]_-^+ \cdot (\mathbf{h} - \mathbf{f}(\mathbf{v}(\theta)) \cdot \mathbf{n})|_{\mathbf{B}\cdot\mathbf{n}\,\mathrm{const}} \leq 0 \quad , \quad \forall \theta \in [0,1] \; .$$

Proof. The result follows immediately since

$$[\mathbf{v}]_-^+ \cdot (\mathbf{B}(\mathbf{v}(\theta)) \cdot \mathbf{n})_{,\mathbf{v}}^T = \frac{d\mathbf{B}(\mathbf{v}(\theta)) \cdot \mathbf{n}}{d\theta} = 0 \qquad (3.27)$$

due to the θ independence of $\mathbf{B} \cdot \mathbf{n}$ at element interfaces. □

This result indicates the underlying intrinsic compatibility requirement of continuity in the normal component of the magnetic induction field for DG discretizations of MHD. Precise implementational details are given in a separate work [Bar04]. In that same work, several other DG discretization formulations and simplified flux functions are given which satisfy the sufficient conditions given in Theorems 3.3 and 3.4

- Transformed variable formulations
- Constrained formulations
- Penalty formulations

4. Conclusions. The energy analysis presented herein reveals the subtle interplay of involutions in the nonlinear stability of the DG method. Sufficient conditions for energy stability of DG discretizations of Maxwell and MHD systems have been obtained. From the viewpoint of discrete energy stability, analysis indicates that "standard" DG discretization Maxwell's equations are energy stable without modification. Surprisingly, sufficient conditions for MHD discretization stability place more demanding requirements as set forth in Theorems 3.3 and 3.4. More complete details and DG formulations for MHD can be found in [Bar04].

REFERENCES

[Bar98] T.J. BARTH. Numerical methods for gasdynamic systems on unstructured meshes. In Kröner, Ohlberger, and Rohde, editors, *An Introduction to Recent Developments in Theory and Numerics for Conservation Laws*, volume 5 of *Lecture Notes in Computational Science and Engineering*, pages 195–285. Springer-Verlag, Heidelberg, 1998.

[Bar99] T.J. BARTH. Simplified discontinuous Galerkin methods for systems of conservation laws with convex extension. In Cockburn, Karniadakis, and

Shu, editors, *Discontinuous Galerkin Methods*, Volume 11 of *Lecture Notes in Computational Science and Engineering*. Springer-Verlag, Heidelberg, 1999.

[Bar04] T.J. BARTH. On the discontinuous Galerkin approximation of compressible ideal magnetohydrodynamics I: Energy stable discretizations. *In preparation*, 2005.

[BB80] J.U. BRACKBILL AND D.C. BARNES. The effect of nonzero $\nabla \cdot \mathbf{B}$ on the numerical solution of the magnetohydrodynamic equations. *J. Comp. Phys.*, 35:426–430, 1980.

[BK04] N. BESSE AND D. KRÖNER. Convergence of locally divergence-free discontinuous Galerkin methods for the induction equations of the MHD system. Technical Report Submitted to M2AN, Wolfgang Pauli Institute, Austria, 2004.

[Boi88] G. BOILLAT. Involutions des systéms conservatifs. *C. R. Acad. Sci. Paris, Sére I*, 307:891–894, 1988.

[Bos98] A. BOSSAVIT. *Computational Electromagnetism, Variational Formulations, Complementarity, Edge Elements.* Academic Press, San Diego, 1998.

[BR02] P.B. BOCHEV AND A.C. ROBINSON. Matching algorithms and physics: Exact sequences of finite element spaces. In D. Estep and S. Tavener, editors, **Collected Lectures on the Preservation of Stability Under Discretization**, Philadephia, 2002. SIAM.

[CHS90] B. COCKBURN, S. HOU, AND C.W. SHU. TVB Runge-Kutta local projection discontinuous Galerkin finite element method for conservation laws IV: The multidimensional case. *Math. Comp.*, 54:545–581, 1990.

[CLS89] B. COCKBURN, S.Y. LIN, AND C.W. SHU. TVB Runge-Kutta local projection discontinuous Galerkin finite element method for conservation laws III: One dimensional systems. *J. Comp. Phys.*, 84:90–113, 1989.

[CLS04] B. COCKBURN, F. LI, AND C.W. SHU. Locally divergence-free discontinuous Galerkin methods for Maxwell equations. *J. Comp. Phys.*, 194:588–610, 2004.

[CS97] B. COCKBURN AND C.W. SHU. The Runge-Kutta discontinuous Galerkin method for conservation laws V: Multidimensional systems. Technical Report 201737, Institite for Computer Applications in Science and Engineering (ICASE), NASA Langley R.C., 1997.

[Csi02] A. CSIK. *Upwind Residual Distribution Schemes for General Hyperbolic Conservation Laws with Application to Ideal Magnetohydrodynamics.* PhD thesis, University of Leuven, Belgium, 2002.

[Daf86] C. DAFERMOS. Quasilinear hyperbolic systems with involutions. *Arch. Rational Mech. Anal.*, 106:373–389, 1986.

[DKK+02] A. DEDNER, F. KEMM, D. KRÖNER, C.-D. MUNZ, T. SCHNITNER, AND M. WESENBERG. Hyperbolic divergence cleaning for the MHD equations. *J. Comp. Phys.*, 175:645–673, 2002.

[EMRS92] B. EINFELDT, C. MUNZ, P. ROE, AND B. SJÖGREEN. On Godunov-type methods near low densities. *J. Comp. Phys.*, 92:273–295, 1992.

[God61] S.K. GODUNOV. An interesting class of quasilinear systems. *Dokl. Akad. Nauk. SSSR*, 139:521–523, 1961.

[God72] S.K. GODUNOV. The symmetric form of magnetohydrodynamics equation. *Num. Meth. Mech. Cont. Media*, 1:26–34, 1972.

[HLvL83] A. HARTEN, P.D. LAX, AND B. VAN LEER. On upstream differencing and Godunov-type schemes for hyperbolic conservation laws. *SIAM Rev.*, 25:35–61, 1983.

[HS99] J.M. HYMAN AND M. SHASHKOV. Mimetic discretizations for Maxwell's equations. *J. Comp. Phys.*, 151:881–909, 1999.

[JJS95] J. JAFFRE, C. JOHNSON, AND A. SZEPESSY. Convergence of the discontinuous Galerkin finite element method for hyperbolic conservation laws. *Math. Models and Methods in Appl. Sci.*, 5(3):367–386, 1995.

[JP86] C. Johnson and J. Pitkäranta. An analysis of the discontinuous Galerkin method for a scalar hyperbolic equation. *Math. Comp.*, 46:1–26, 1986.

[JS94] G. Jiang and C.-W. Shu. On a cell entropy inequality for discontinuous galerkin methods. *Math. Comp.*, 62:531–538, 1994.

[LR74] P. LeSaint and P.A. Raviart. On a finite element method for solving the neuton transport equation. In C. de Boor, editor, *Mathematical Aspects of Finite Elements in Partial Differential Equations*, pages 89–145. Academic Press, 1974.

[Mer88] M.L. Merriam. *An Entropy-Based Approach to Nonlinear Stability*. PhD thesis, Stanford University, 1988.

[Moc80] M.S. Mock. Systems of conservation laws of mixed type. *J. Diff. Eqns.*, 37:70–88, 1980.

[Ned80] J.C. Nedelec. Mixed finite elements in $\mathbb{R}^3$. *Numer. Math.*, 35:315–341, 1980.

[Osh84] S. Osher. Riemann solvers, the entropy condition, and difference approximations. *SIAM J. Numer. Anal.*, 21(2):217–235, 1984.

[Pow94] K.G. Powell. An approximate Riemann solver for magnetohydrodynamics (that works in more than one dimension). Technical Report 94-24, Institite for Computer Applications in Science and Engineering (ICASE), NASA Langley R.C., 1994.

[RH73] W.H. Reed and T.R. Hill. Triangular mesh methods for the neutron transport equation. Technical Report LA-UR-73-479, Los Alamos National Laboratory, Los Alamos, New Mexico, 1973.

[Shu99] C.-W. Shu. Discontinuous Galerkin methods for convection-dominated problems. In Barth and Deconinck, editors, *High-Order Discretization Methods in Computational Physics*, Volume 9 of *Lecture Notes in Computational Science and Engineering*. Springer-Verlag, Heidelberg, 1999.

[Tó0] G. Tóth. The $\nabla \cdot \mathbf{B} = 0$ constraint in shock-capturing magnetohydrodynamics codes. *J. Comp. Phys.*, 161:605–652, 2000.

[Yee66] K.S. Yee. Numerical solution of initial boundary value problems involving Maxwell's equations in isotropic media. *IEEE Trans. Ant. Prop.*, AP-14:302–307, 1966.

PRINCIPLES OF MIMETIC DISCRETIZATIONS OF DIFFERENTIAL OPERATORS

PAVEL B. BOCHEV* AND JAMES M. HYMAN†

Abstract. Compatible discretizations transform partial differential equations to discrete algebraic problems that mimic fundamental properties of the continuum equations. We provide a common framework for mimetic discretizations using algebraic topology to guide our analysis. The framework and all attendant discrete structures are put together by using two basic mappings between differential forms and cochains. The key concept of the framework is a natural inner product on cochains which induces a combinatorial Hodge theory on the cochain complex. The framework supports mutually consistent operations of differentiation and integration, has a combinatorial Stokes theorem, and preserves the invariants of the De Rham cohomology groups. This allows, among other things, for an elementary calculation of the kernel of the discrete Laplacian. Our framework provides an abstraction that includes examples of compatible finite element, finite volume, and finite difference methods. We describe how these methods result from a choice of the reconstruction operator and explain when they are equivalent. We demonstrate how to apply the framework for compatible discretization for two scalar versions of the Hodge Laplacian.

Key words. Mimetic discretizations, compatible spatial discretizations, finite element methods, support operator methods, algebraic topology, De Rham complex, Hodge operator, Stokes theorem.

AMS(MOS) subject classifications. 65N06, 65N12, 65N30.

1. Introduction.

Partial differential equations (PDEs) are ubiquitous in science and engineering. A key step in their numerical solution is the *discretization* that replaces the PDEs by a system of algebraic equations. Like any other model reduction, discretization is accompanied by losses of information about the original problem and its structure. One of the principal tasks in numerical analysis is to develop *compatible*, or *mimetic*, algebraic models that yield stable, accurate, and physically consistent approximate solutions. Historically, finite element (FE), finite volume (FV), and finite difference (FD) methods have achieved compatibility by following different paths that reflected their specific approaches to discretization.

Finite element methods begin by converting the PDEs into an equivalent variational equation and then restrict that equation to finite dimensional subspaces. Compatibility of the discrete problem is governed by variational inf-sup conditions, which imply existence of uniformly bounded discrete solution operators; see [6, 18, 46]. In finite volume methods the PDEs are first replaced by equivalent integral equations that express balance of global quantities valid on all subdomains of the problem domain.

*Computational Mathematics and Algorithms, Mail Stop 1110, Sandia National Laboratories, Albuquerque, NM 87185 (pbboche@sandia.gov).

†Mathematical Modeling and Analysis, T-7 Mail Stop B284, Los Alamos National Laboratory, Los Alamos, NM 87545 (hyman@lanl.gov).

The algebraic equations are derived by sampling balance equations on a finite set of admissible subdomains (the finite volumes). Their compatibility is achieved by using the Stokes theorem to define the discrete differential operators [32, 42, 44, 58]. Finite difference methods approximate vector and scalar functions by discrete values on a grid and compatibility is realized by choosing the locations of these variables on the grid [28, 33, 34, 51, 61].

In spite of their differences, compatible FE, FV, and FD methods can result in discrete problems with remarkably similar properties. The observation that their compatibility is tantamount to having discrete structures that mimic vector calculus identities and theorems emerged independently and at about the same time in the FE, FV, and FD literature. For instance in [14, 15, 16, 37] Bossavit and Kotiuga demonstrated connections between stable finite elements for the Maxwell's equations and Whitney forms. In finite volume methods the idea of discrete field theory guided development of covolume methods [42, 43, 44], while support operator and mimetic methods [48, 50, 33, 34, 35, 36] combined the Stokes theorem with variational Green's identities to derive compatible finite differences. Algebraic topology was used to analyze mimetic discretizations by Hyman and Scovel in [31] and more recently by Mattiussi [39], Schwalm et al. [47] and Teixeira [53, 54]. Further research also revealed connections between some compatible methods. For instance, mimetic FD for the Poisson equation can be obtained from mixed FE by quadrature choice [12, 13, 19]. Another example is the equivalence between a covolume method and the classical Marker-and-Cell (MAC) scheme on uniform grids [43] and the analysis of [39] that relates finite volume and finite elements by using the concept of a "spread cell".

This research helped to evolve and clarify the notion of spatial compatibility to its present meaning of a discrete setting that provides mutually consistent operations for discrete integration and differentiation that obey the standard vector identities and theorems, such as the Stokes theorem. It also highlighted the role of differential forms and algebraic topology in the design and analysis of compatible discretizations. The recent work in [2, 8, 9, 10, 22, 29, 30, 39, 44, 47, 52, 53, 58] and the papers in this volume further affirm that these tools are gaining wider acceptance among mathematicians and engineers. For instance, FE methods that have traditionally relied upon nonconstructive variational [6, 18] stability criteria[1] now are being derived by topological approaches that reveal physically relevant degrees of freedom and their proper encoding. Of particular note are the papers by Arnold et al. [4, 2] which develop stable finite elements for mixed elasticity, and by Hiptmair [29], Demkowicz et al. [22] and Arnold et

[1]One exception in FEM was the Grid Decomposition Property (GDP), formulated by Fix et al. [26], that gives a topological rather than variational stability condition for mixed discretizations of the Kelvin principle derived from the Hodge decomposition. The GDP is essentially equivalent to an inf-sup condition; see Bochev and Gunzburger [7]

al. [3] which define canonical procedures for building piecewise polynomial differential complexes.

The key role played by differential forms and algebraic topology in compatible discretizations is not accidental. Exterior calculus provides powerful tools and concise formalism to encode the structure of many PDEs and to expose their local and global invariants. For instance, integration of differential forms is an abstraction of the measurement process, while the Stokes theorem connects differentiation and integration to reveal global equilibrium relations. Algebraic topology, on the other hand, supplies structures that mimic exterior calculus on finite grids and so is a natural discretization tool for differential forms. The application of algebraic topology in modeling dates back to 1923 when H. Weyl [59] used it to describe electrical networks. Other early works of note are Branin [17] and in particular Dodziuk [24] whose combinatorial Hodge theory has great similarity with mixed FE on simplices. However, these papers contained few applications to numerical analysis. The first deliberate application of algebraic topology to solve PDEs numerically is due to Tishkin et al. [55] and Hyman and Scovel [31] who, drawing upon some of the ideas in [24], used it to develop mimetic finite difference methods.

The present paper extends the approach originated in [31] to create a general framework for compatible discretizations that includes FE, FV, and FD methods as special cases. We first translate scalar and vector functions to their differential form equivalents and consider the computational grid to be an algebraic topological complex. The grid consists of 0-cells (nodes), 1-cells (edges), 2-cells (faces), and 3-cells (volumes) which combine to form k-chains; $k = 0, 1, 2, 3$. For simplicity we focus on simplicial grids; however, most of the developments easily carry over to general polyhedral domain partitions.

All necessary discrete structures in our framework are put together by two basic operations: a reduction map $\mathcal{R}$ and a reconstruction map $\mathcal{I}$, such that $\mathcal{I}$ is a right inverse of $\mathcal{R}$. We take $\mathcal{R}$ to be the De Rham map that reduces differential forms to linear functionals on chains, i.e., cochains. Therefore, discrete k-forms are encoded as k-cell quantities. For differential forms, the operators Div, Grad and Curl are generated by the exterior derivative d. Stokes theorem states that d is dual to the boundary operator ∂ with respect to the pairing between forms and chains. To define the discrete operators we mimic this property and use the duality between chains and cochains. Thus, the discrete Div, Grad and Curl are generated by the coboundary δ which is dual to ∂ with respect to this pairing.

The reconstruction map $\mathcal{I}$ translates cochains back to differential forms and induces the *natural* inner product that is central to our approach. This product gives rise to a *derived* adjoint δ^*, a discrete Laplacian $-\triangle = \delta\delta^* + \delta^*\delta$ and hence a combinatorial Hodge theory [25, 24]. By applying a discrete version of Hodge's theorem and De Rham's theorem, we can compute the size of the kernel of this Laplacian in an elementary way.

The global (combinatorial) and the local (metric) properties of the discrete models are determined by $\mathcal{R}$ and $\mathcal{I}$, respectively. The discrete derivative, induced by $\mathcal{R}$, is purely combinatorial and invariant under homeomorphisms. The adjoint δ^* is induced by the inner product and depends on the choice of $\mathcal{I}$.

The present work, based on mappings between differential forms and cochains, differs from other approaches that use differential forms and algebraic topology to provide common frameworks for compatible discretizations. Most notably, we make the inner product on cochains the key concept of our approach because it is sufficient to generate a combinatorial Hodge theory. As a result, distinctions between compatible FE, FV, and FD methods arise from the choice of $\mathcal{I}$ and so equivalence of different models can be established by comparing their reconstruction operators. In contrast, the primary concept in [30, 52, 54] is the discrete $\star$ operator. Different models are distinguished by their choice of the discrete $\star$ and its construction is the central problem.

As an aside, we point out that developments in the FE literature focus primarily on approximation of differential forms by piecewise polynomials of arbitrary degree [1, 3, 22] and less on the equivalence between the discrete models. Except in the lowest-order case, such spaces include degrees of freedom that are not cochains and result in differential operators that are not purely combinatorial. The main advantage of cochain encoding used in this work is seen in the possibility to maintain a clear distinction between the global and the local features in the discrete model. High-order formulations on cochains are also possible by using an appropriate reconstruction operator [32, 58]. Generally, reconstruction stencils for $\mathcal{I}$ grow, which is seen as the principal drawback of this approach. However, the number of degrees of freedom does not increase.

2. Differential forms. We review the basic concepts necessary for the numerical framework. Given an n-dimensional vector space E and an integer $0 \le k \le n$, we denote by Λ^k the vector space of algebraic k-forms, that is, all k-linear, antisymmetric maps[2] $\omega_k : E\times\ldots\times E \mapsto \mathbb{R}$; see [5]. The subscript k in ω_k will be used only when necessary to distinguish between different forms. Dimension of Λ^k is C^n_k and the unique element ω_n of Λ^n is a volume form. We recall the wedge product $\wedge : \Lambda^k \times \Lambda^l \mapsto \Lambda^{k+l}$ for $k+l \le n$ with the property that $\omega_k \wedge \omega_l = (-1)^{kl}\omega_l \wedge \omega_k$. An inner product $(\cdot,\cdot)$ on $E\times E$ induces an inner product $(\cdot,\cdot)$ on $\Lambda^k \times \Lambda^k$. The latter gives rise to a unique metric conjugation operator $\star : \Lambda^k \mapsto \Lambda^{n-k}$, defined by the relation [23, 27]

$$\omega \wedge \star\xi = (\omega,\xi)\omega_n \quad \forall \omega,\xi \in \Lambda^k\,. \tag{2.1}$$

Let $T\Omega$ denote the tangent bundle of a differentiable manifold Ω. A

[2]Equivalently, Λ^k can be defined as the dual of Λ_k - the space of all k-vectors; see [23, 27].

differential k-form on Ω is a map $\Omega \ni x \mapsto \omega(x) \in \Lambda^k(T_x\Omega)$, where $T_x\Omega$ is the tangent space at x. In what follows the set of all smooth k-forms on Ω is denoted by $\Lambda^k(\Omega)$. The exterior derivative $d : \Lambda^k \mapsto \Lambda^{k+1}$; $k = 0, 1, \ldots, n-1$ satisfies

$$d(\omega_k \wedge \omega_l) = (d\omega_k) \wedge \omega_l + (-1)^k \omega_k \wedge (d\omega_l)\,; k + l < n \tag{2.2}$$

and $dd = 0$ and therefore gives rise to an exact sequence

$$\mathbb{R} \hookrightarrow \Lambda^0(\Omega) \overset{d}{\longmapsto} \Lambda^1(\Omega) \overset{d}{\longmapsto} \Lambda^2(\Omega) \overset{d}{\longmapsto} \Lambda^3(\Omega) \longmapsto 0 \tag{2.3}$$

called De Rham complex.

Integration operation for differential k-forms can be defined on k-dimensional manifolds without any reference to a metric structure [5, 23]. The Stokes theorem

$$\int_{\partial\Omega} \omega = \int_\Omega d\omega\,, \tag{2.4}$$

expresses the classical Newton-Leibnitz, Gauss divergence, and Stokes circulation theorems. As a corollary to this theorem and (2.2), we have, for $k + l + 1 = n$, the integration by parts formula

$$\int_{\partial\Omega} \omega_k \wedge \omega_l = \int_\Omega (d\omega_k) \wedge \omega_l + (-1)^k \int_\Omega \omega_k \wedge (d\omega_l)\,. \tag{2.5}$$

On a Riemannian manifold Ω the metric tensor g_{ij} induces Euclidean structure on $T_x\Omega$ and inner product $(\cdot,\cdot)$ on $\Lambda^k(T_x\Omega)$. The latter brings about an L^2 inner product on $\Lambda^k(\Omega)$ defined by

$$(\omega, \xi)_\Omega = \int_\Omega (\omega, \xi)\omega_n\,. \tag{2.6}$$

In view of (2.1), an equivalent definition is

$$(\omega, \xi)_\Omega = \int_\Omega \omega \wedge \star\xi\,. \tag{2.7}$$

The Hilbert spaces obtained by completion of smooth k-forms in the metric induced by (2.6) will be denoted by $\Lambda^k(L^2, \Omega)$.

It is also profitable to introduce the Sobolev spaces [3]

$$\Lambda^k(d, \Omega) = \{\omega \in \Lambda^k(L^2, \Omega) \mid d\omega \in \Lambda^{k+1}(L^2, \Omega)\}\,,$$

of square integrable k-forms whose exterior derivative is also square integrable.

The inner product (2.6) gives rise to an adjoint operator $d^* : \Lambda^k(\Omega) \mapsto \Lambda^{k-1}(\Omega)$. Assuming that Ω is the whole manifold, or that one of the forms has compact support, the adjoint is defined by

$$(d\omega, \xi)_\Omega = (\omega, d^*\xi)_\Omega \quad \text{for all } \omega \in \Lambda^{k-1}(\Omega)\,,\ \xi \in \Lambda^k(\Omega)\,.$$

The adjoint gives rise to the Hodge Laplacian $-\Delta_k = dd^* + d^*d$, which is a mapping $\Lambda^k(\Omega) \mapsto \Lambda^k(\Omega)$.

We assume that the boundary $\partial\Omega$ of domain Ω for the PDEs consists of two disjointed, smooth, possibly empty boundary components Γ_1 and Γ_2. At any boundary point a form can be decomposed into its tangential and normal components, $\omega = \omega_t + \omega_n$. If η is the inward pointing unit covector, then $\omega_n = g \wedge \eta$ where $\star g = \star\omega \wedge \eta$. The Green's formula

$$(d\omega, \xi)_\Omega - (\omega, d^*\xi)_\Omega = \int_{\partial\Omega} \omega \wedge \star\xi = \int_{\partial\Omega} \omega_t \wedge \star\xi_n \tag{2.8}$$

follows from (2.4) and (2.5).

Let $\Lambda_0^k(\Omega)$ be the smooth k-forms ω such that

$$\omega_t = 0 \text{ on } \Gamma_1 \quad \text{and} \quad \omega_n = 0 \text{ on } \Gamma_2 . \tag{2.9}$$

The boundary conditions imposed on $\Lambda_0^k(\Omega)$ imply that $d^* = (-1)^k \star d\star$. Thus, the adjoint has the property that $d^*d^* = 0$. If the metric is the standard Euclidean metric, then the effect of d^* on scalar and vector functions is the same as that of d.

Using (2.8) we see that for $\omega, \xi \in \Lambda_0^k(\Omega)$

$$(-\Delta_k\omega, \xi)_\Omega = (d\omega, d\xi)_\Omega + (d^*\omega, d^*\xi)_\Omega .$$

The right-hand side in the above formula is the Dirichlet integral.

The relation between forms and vector and scalar functions in $\mathbb{R}^3$ is determined as follows. Let $\{x_i\}_{i=1}^3$ and $\{dx_i\}_{i=1}^3$ denote the local coordinates and their conjugates, respectively, that is, $dx_i(x_j) = \delta_{ij}$. A 0-form is dual to zero-dimensional manifolds (points) and so it is a scalar function. A 3-form is dual to three-dimensional manifolds (volumes) and so it has the form

$$\omega = \phi(\mathbf{x}) dx_1 \wedge dx_2 \wedge dx_3 .$$

This defines a relation $\omega \leftrightarrow \phi$ where ϕ is a scalar function. Therefore, 0- and 3-forms can be identified with scalar functions. A 1-form is dual to one-dimensional manifolds and can be written as

$$\omega = \mathbf{u}_1(\mathbf{x})dx_1 + \mathbf{u}_2(\mathbf{x})dx_2 + \mathbf{u}_3(\mathbf{x})dx_3 ,$$

while a 2-form is dual to two-dimensional manifolds and can be written as

$$\omega = \mathbf{u}_1(\mathbf{x})dx_2 \wedge dx_3 + \mathbf{u}_2(\mathbf{x})dx_3 \wedge dx_1 + \mathbf{u}_3(\mathbf{x})dx_1 \wedge dx_2 .$$

This defines a relation $\omega \leftrightarrow \mathbf{u}$, between 1- and 2-forms and vector fields in $\mathbb{R}^3$.

To emphasize correspondences between forms and fields, sometimes we will write $\omega^{\mathbf{u}}$ or ω^ϕ so that

$$d\omega_0^\phi = \omega_1^{\nabla\phi}; \quad d\omega_1^{\mathbf{u}} = \omega_2^{\nabla\times\mathbf{u}}; \quad \text{and} \quad d\omega_2^{\mathbf{u}} = \omega_3^{\nabla\cdot\mathbf{u}} . \tag{2.10}$$

That is, exterior derivative of a $0-, 1-, 2-$form is equivalent to application of Grad, Curl, or Div, respectively, to the corresponding scalar or vector field.

Furthermore, if $\omega^{\mathbf{u}}$ and $\xi^{\mathbf{v}}$ are two 1-forms, then the wedge product $\omega^{\mathbf{u}} \wedge \xi^{\mathbf{v}}$ is a 2-form with corresponding vector function $\mathbf{u} \times \mathbf{v}$. If $\eta^{\mathbf{w}}$ is a 2-form, then the wedge $\omega^{\mathbf{u}} \wedge \eta^{\mathbf{w}}$ is a 3-form with scalar function $\mathbf{u} \cdot \mathbf{w}$.

For the Hilbert spaces $\Lambda^k(d, \Omega)$ boundary conditions are imposed for $k = 0, 1, 2$ either on Γ_1 or Γ_2 but not both at the same time. In this paper we consider the spaces $\Lambda_i^k(d, \Omega)$ with boundary conditions on Γ_i; $i = 1, 2$. The correspondence (2.10) allows us to identify $\Lambda^k(d, \Omega)$, $k = 0, 1, 2$ with the Sobolev spaces $H(\Omega, \mathrm{grad})$, $H(\Omega, \mathbf{curl})$, and $H(\Omega, \mathrm{div})$ of square integrable functions whose gradient, curl, and divergence are also square integrable. With $\Lambda^3(d, \Omega) \simeq L^2(\Omega)$ we have an L^2 version of the De Rham complex (2.3):

$$\mathbb{R} \hookrightarrow H(\Omega, \mathrm{grad}) \overset{\nabla}{\longmapsto} H(\Omega, \mathbf{curl}) \overset{\nabla\times}{\longmapsto} H(\Omega, \mathrm{div}) \overset{\nabla\cdot}{\longmapsto} L^2(\Omega) \longmapsto 0. \quad (2.11)$$

The spaces $\Lambda_i^k(d, \Omega)$ correspond to Sobolev spaces constrained by boundary conditions on Γ_i:

$$\begin{aligned} H_i(\Omega, \mathrm{grad}) &= \{\phi \in H(\Omega, \mathrm{grad}) \mid \phi = 0 \quad \text{on } \Gamma_i\} \\ H_i(\Omega, \mathbf{curl}) &= \{\mathbf{w} \in H(\Omega, \mathbf{curl}) \mid \mathbf{w} \times \mathbf{n} = 0 \quad \text{on } \Gamma_i\} \\ H_i(\Omega, \mathrm{div}) &= \{\mathbf{w} \in H(\Omega, \mathrm{div}) \mid \mathbf{w} \cdot \mathbf{n} = 0 \quad \text{on } \Gamma_i\}. \end{aligned}$$

They form a De Rham complex relative to Γ_i.

3. Algebraic topology. Our goal is to develop discrete structures that support mutually consistent, mimetic notions of integral, derivative, and inner product. The approach adopted in this paper is guided by algebraic topology and draws upon the ideas of [31]. This section reviews the necessary basic concepts. For further details we refer the reader to Cairns [21] or Flanders [27].

For brevity we restrict our attention to computational grids that are triangulations of Ω by a simplicial complex. All discrete structures developed in this paper and their mimetic properties can be extended to general polyhedral partitions of Ω such as considered in [38].

A k-simplex s_k is an ordered collection $[\mathbf{p}_0, \ldots, \mathbf{p}_k]$ of $(k+1)$, $k \le n$ distinct points in $\mathbb{R}^n$ such that they span a k-plane. A k-chain is a formal linear combination

$$c_k = \sum_i a_i s_k^i$$

where a_i are real constants and s_k^i are k-simplices. A set of k-chains is denoted by C_k.

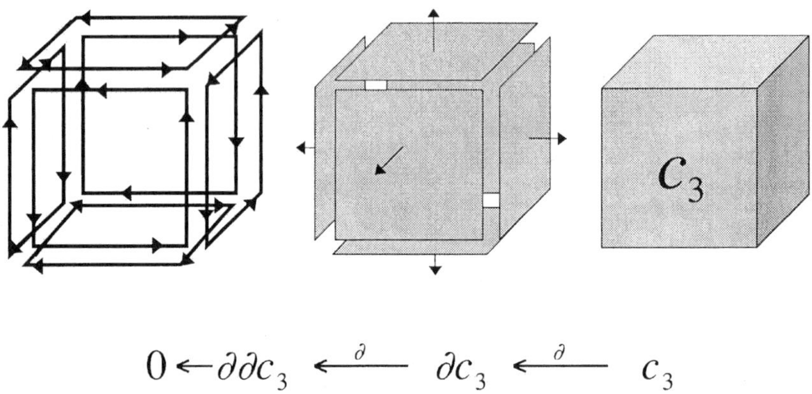

FIG. 1. *The boundary ∂ of a k-simplex is $(k-1)$-chain. The action of ∂ on a 3-cell illustrates that $\partial\partial c_3 = 0$.*

The boundary ∂ of a k-simplex is $(k-1)$-chain is defined by the formula

$$\partial[\mathbf{p}_0, \ldots, \mathbf{p}_k] = \sum_{i=1}^{k} (-1)^i [\mathbf{p}_0, \ldots, \mathbf{p}_{i-1}, \mathbf{p}_{i+1}, \ldots, \mathbf{p}_k] \,. \tag{3.1}$$

A direct calculation shows that $\partial\partial = 0$. Boundary of a chain is defined by linearity; see Fig. 1

$$\partial c = \sum_i a_i \partial c_k^i \,. \tag{3.2}$$

The collection $\{C_0, C_1, C_2, C_3\}$ is called *complex* if for any $c \in C_k$, $\partial c \in C_{k-1}$. This gives rise to an exact sequence

$$0 \longleftarrow C_0 \overset{\partial_0}{\longleftarrow} C_1 \overset{\partial_1}{\longleftarrow} C_2 \overset{\partial_2}{\longleftarrow} C_3 \longleftarrow 0 \tag{3.3}$$

where $\partial_k : C_{k+1} \mapsto C_k$ is the boundary operator on k-chains. The sequence (3.3) is called *exact* since Range $\partial_k \subset$ Ker ∂_{k-1}, which follows from $\partial\partial = 0$.

The geometric realization of a k-simplex $[\mathbf{p}_0, \ldots \mathbf{p}_k]$ is the map

$$t_i \mapsto \sum_{i=0}^{k} t_i \mathbf{p}_i \,, \quad \text{where } t_i \geq 0 \text{ and } \quad \sum_{i=1}^{k} t_i = 1 \,.$$

This map returns the convex hull of the points $[\mathbf{p}_0, \dots \mathbf{p}_k]$. The numbers t_i are called *barycentric* coordinates, and they turn the complex $\{C_0, C_1, C_2, C_3\}$ into a metric space K. A triangulation of Ω is a homeomorphism $K \mapsto \Omega$. Given K, we denote by $L_1 \subset K$ and $L_2 \subset K$ the triangulations of Γ_1 and Γ_2.

The chain C_0 is a collection of zero simplices, i.e., points. We require that these points be given an ordering. This ordering determines an orientation for each k-simplex in K. A simplex $[\mathbf{p}_{i_0}, \dots, \mathbf{p}_{i_k}]$ has positive orientation if $\pi = \{i_0, \dots, i_k\}$ is an even permutation of the symbols $\{0, \dots, k\}$ and negative orientation otherwise. The subsets

$$
\begin{aligned}
Z_k &= \{c_k \in C_k \,|\, \partial_{k-1} c_k = 0\} \quad \text{and} \\
B_k &= \{b_k \in C_k \,|\, b_k = \partial_k c_{k+1} \text{ for } c_{k+1} \in C_{k+1}\}
\end{aligned}
$$

of C_k are called k-cycles and k-boundaries, respectively. Because $\partial\partial = 0$, B_k is a subgroup of Z_k. The kth homology group of K over $\mathbb{R}$, $\mathcal{H}_k(K, \mathbb{R}) = Z_k / B_k$ contains all cycles that are not boundary chains.

The dual C^k is the collection of all linear functionals on C_k. The elements of C^k are called k-cochains. We use the bracket notation $\langle \cdot, \cdot \rangle$ to denote the duality pairing of chains and cochains. The adjoint of ∂, $\delta : C^k \mapsto C^{k+1}$, defined by

$$
\langle a, \partial c \rangle = \langle \delta a, c \rangle \tag{3.4}
$$

satisfies $\delta\delta = 0$ and forms an exact sequence

$$
0 \longrightarrow C^0 \xrightarrow{\delta_0} C^1 \xrightarrow{\delta_1} C^2 \xrightarrow{\delta_2} C^3 \longrightarrow 0\,, \tag{3.5}
$$

dual to (3.3). As before, we define the k-cocycles Z^k, the k-coboundaries B^k of C^k, and the kth cohomology group $\mathcal{H}^k(K, \mathbb{R}) = Z^k / B^k$.

The collection $\{\sigma_k^i\}$, $i = 1, 2, \dots$ of positively oriented k-chains forms a basis for the chain complex. Since K is finite, C_k is finite dimensional and isomorphic to C^k. The isomorphism $J : C^k \mapsto C_k$ is given by

$$
Ja = \sum_i \langle a, \sigma_k^i \rangle \sigma_k^i\,. \tag{3.6}
$$

We identify σ_k^i with its dual so that $\langle \sigma_k^i, \sigma_k^j \rangle = \delta_{ij}$. Then a cochain can be written as $a = \sum a_i \sigma_k^i$ and its action on a chain $c = \sum c_i \sigma_k^i$ is given by

$$
\langle a, c \rangle = \sum_i a_i c_i\,.
$$

From this, the coboundary operator is computed to be

$$
\delta[\mathbf{p}_0, \dots, \mathbf{p}_k] = \sum_{\mathbf{p}} [\mathbf{p}, \mathbf{p}_0, \dots, \mathbf{p}_k]
$$

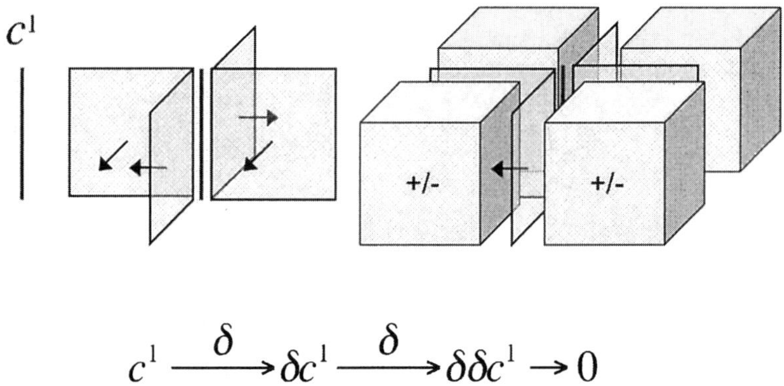

FIG. 2. *The coboundary operator is defined by* $\delta[\mathbf{p}_0, \dots, \mathbf{p}_k] = \sum_{\mathbf{p}}[\mathbf{p}, \mathbf{p}_0, \dots, \mathbf{p}_k]$, *where the points* $[\mathbf{p}, \mathbf{p}_0, \dots, \mathbf{p}_k]$ *form a* $(k+1)$*-simplex, returns a cochain that contains all* $(k+1)$ *simplices that have* $[\mathbf{p}_0, \dots, \mathbf{p}_k]$ *as part of their boundary. The action of* δ *on a 1-cell illustrates that* $\delta\delta c^1 = 0$.

where the sum is over all points $\mathbf{p}$ such that $[\mathbf{p}, \mathbf{p}_0, \dots, \mathbf{p}_k]$ is a $(k+1)$-simplex. In other words, the coboundary returns a cochain that contains all $(k+1)$ simplices that have $[\mathbf{p}_0, \dots, \mathbf{p}_k]$ as part of their boundary; see Fig. 2.

To accommodate boundary conditions, define the subspace $C_i^k \subset C^k$ to be the set of all k-cochains that vanish on L_i, the triangulation of Γ_i:

$$C_i^k = \{a \in C^k \mid \langle a, c_k \rangle = 0 \;\forall c_k \in L_i\},$$

and C_0^k to be the cochains that vanish on $L_1 \cup L_2$. In a similar way we construct the groups Z_i^k, B_i^k and the kth relative cohomology group $\mathcal{H}_i^k = \mathcal{H}^k(K, L_i, \mathbb{R})$.

We stress that geometrically C^k and C_k are distinct despite the isomporphism J. An element of C_k is a formal sum of k-simplices, whereas an element of C^k is a linear function that maps elements of C_k into real numbers. This distinction also extends to the role of chains and cochains in the discretization. The k-chains represent subsets of the nodes, edges, faces, and cells in the grid. The k-cochains are the collections of real numbers $\{a_i\}$ associated with these subsets. Therefore, the chains are the physical objects that make the computational grid, while the cochains are the discrete functions that live on that grid. In particular, the proper way to store scalar functions on the grid is as 0- or 3-cochains, while the proper way to store vector fields is as 1- or 2-cochains.

4. Framework for mimetic discretizations. This section develops structures for mimetic discretization of PDEs by using algebraic topology and two basic operations. A reduction operator maps forms to cochains and gives rise to combinatorial operations of differentiation and integration that satisfy a Stokes theorem. A reconstruction operator translates cochains to differential forms and is used to obtain the *natural* inner and wedge product operations. The natural operations provide the *derived* analogues of the adjoint d^* and the Hodge Laplacian.

4.1. Basic operations.

Reduction. Information about physical quantities is obtained by measuring. Integration of differential forms is an abstraction of this process and motivates our choice of the De Rham map $\Lambda^k(\Omega) \mapsto C^k$ for the reduction operation. This map is defined by

$$\langle \mathcal{R}\omega, c \rangle = \int_c \omega \tag{4.1}$$

where $c \in C_k$ is a k-chain and $\omega \in \Lambda^k(\Omega)$ is a k-form. The mapping $\omega \mapsto \mathcal{R}\omega$ establishes discrete representation of k-forms in terms of global quantities associated with a chain complex. Thus, we encode discrete k-forms as k-cell quantities. The following property of $\mathcal{R}$ will prove useful in the sequel.

LEMMA 4.1. *The De Rham map has the commuting diagram property* $\mathcal{R}d = \delta\mathcal{R}$.

Proof. Using the Stokes formula (2.4) and the duality of ∂ and δ gives

$$\langle \mathcal{R}d\omega, c \rangle = \int_c d\omega = \int_{\partial c} \omega = \langle \mathcal{R}\omega, \partial c \rangle = \langle \delta\mathcal{R}\omega, c \rangle . \tag{4.2}$$

□

In what follows we refer to this property as *CDP1*, the first commuting diagram.

Reconstruction. Central to our approach is the notion of an inner product on cochains. Its *natural* definition requires an operation $\mathcal{I}$ that serves as an approximate inverse to $\mathcal{R}$ and translates the global information stored in C^k back to local representations. In contrast to $\mathcal{R}$, where the De Rham map (4.1) is the obvious candidate, the choice of $\mathcal{I}$ is flexible because of the many possible ways in which global data from C^k can be combined in a local field representation.

The operator $\mathcal{I}$ must satisfy two basic conditions. We will call a bounded linear mapping $\mathcal{I} : C^k \mapsto \Lambda^k(L^2, \Omega)$ an L^2 *mimetic reconstruction operator* if $\mathcal{I}$ is a right inverse of $\mathcal{R}$ (*consistency property*)

$$\mathcal{R}\mathcal{I} = id \tag{4.3}$$

and an approximate left inverse of that operator (*approximation property*)

$$\mathcal{IR} = id + O(h^s), \tag{4.4}$$

where s and h are positive real numbers that give the approximation order and the partition size in K, respectively.

From (4.3) it follows that $\mathcal{I}$ is *unisolvent* in the sense that

$$\operatorname{Ker}\mathcal{I} = \{0\}. \tag{4.5}$$

We require the range of $\mathcal{I}$ to contain square integrable k-forms and (4.3) implies that these forms are continuous on the k-chains of the complex K. However, they may be discontinuities along the $m \neq k$-cells of the complex, or even within the k-cells of K, and so they may not belong to $\Lambda^k(d,\Omega)$. For mimetic reconstruction operators $\mathcal{I}$ whose range is a subspace of the Sobolev space $\Lambda^k(d,\Omega)$ we impose an additional condition that serves to coordinate the action of the exterior derivative and the coboundary operator. This condition takes the form of a second commuting diagram property, *CDP2*,

$$d\mathcal{I} = \mathcal{I}\delta. \tag{4.6}$$

We will call such mappings *conforming* mimetic reconstruction operators. The Whitney map [60, 24, 31] is an example of a regular mimetic reconstruction operator.

4.2. Discrete structures. For a mimetic reconstruction operator $\mathcal{I}$, the range of $\mathcal{IR}$, considered as an operator $\Lambda^k(d,\Omega) \mapsto \Lambda^k(L^2,\Omega)$, is a subspace of $\Lambda^k(L^2,\Omega)$ given by

$$\Lambda^k(L^2,K) = \{\omega_h \in \Lambda^k(L^2,\Omega) \,|\, \omega_h = \mathcal{IR}\omega \;\text{ for some } \omega \in \Lambda^k(d,\Omega)\}. \tag{4.7}$$

When $\mathcal{I}$ is a conforming operator, the range of $\mathcal{IR}$ is a subspace of $\Lambda^k(d,\Omega)$ given by

$$\Lambda^k(d,K) = \{\omega_h \in \Lambda^k(d,\Omega) \,|\, \omega_h = \mathcal{IR}\omega \;\text{ for some } \omega \in \Lambda^k(d,\Omega)\}. \tag{4.8}$$

The spaces $\Lambda^k_i(L^2,K)$ and $\Lambda^k_i(d,K)$ are defined similarly using $\Lambda^k_i(d,\Omega)$.

4.2.1. Combinatorial operations. These operations are induced by the action of $\mathcal{R}$ and are completely independent of any metric structures.

Exterior derivative. Formula (2.10) shows that Grad, Curl and Div are generated by the action of d on 0-, 1-, and 2-forms. Therefore, their discrete versions will be generated by a discrete counterpart of d acting on 0-,1-, and 2-cochains. To find the discrete version of d on K we note that forms are dual to manifolds with respect to the pairing induced by integration and that according to the Stokes theorem (2.4), d is the adjoint of ∂. To define a discrete derivative we mimic this by using the duality of C^k and

C_k and formula (3.4) which states that δ is dual to ∂. Thus, the discrete Grad, Curl and Div are generated by the coboundary. The CDP1 property asserts the consistency of this definition: The action of d on ω followed by a reduction to cochain equals the reduction of ω to cochain followed by the action of δ.

Integration. The integral of $a \in C^k$ is defined on chains C_k by duality:

$$\int_\sigma a = \langle a, \sigma \rangle \quad \forall a \in C^k;\ \sigma \in C_k\,. \tag{4.9}$$

4.2.2. Natural operations. These are defined by composition of $\mathcal{I}$ and the desired analytic operation. Natural operations are the best imitation of the analytic operations on cochains.

Inner product. The L^2 inner product (2.6) on $\Lambda^k(\Omega)$ is the integral of the inner product on $\Lambda^k(T_x\Omega)$. We mimic this relationship by setting up the *local* inner product

$$(a, b) \stackrel{\text{def}}{=} (\mathcal{I}a, \mathcal{I}b) \qquad \forall\, a, b \in C^k\,. \tag{4.10}$$

The discrete L^2 inner product on C^k is the integral of (4.10):

$$(a, b)_\Omega \stackrel{\text{def}}{=} \int_\Omega (a, b)\omega_n \qquad \forall\, a, b \in C^k\,. \tag{4.11}$$

Unisolvency (4.5) of $\mathcal{I}$ guarantees that (4.10) and (4.11) are nondegenerate and are indeed inner products.

Wedge product. The operation $\wedge : C^k \times C^p \mapsto C^{p+k}$ is introduced by using the wedge product of differential forms. Specifically, we set

$$a \wedge b = \mathcal{R}(\mathcal{I}a \wedge \mathcal{I}b) \quad \forall\, a \in C^k\,;\, b \in C^p\,. \tag{4.12}$$

4.2.3. Derived operations. These operations are induced by the existing natural operations.

The discrete adjoint. The inner product on C_0^k induces an adjoint δ^* of δ characterized by the identity

$$(\delta a, b)_\Omega = (a, \delta^* b)_\Omega \qquad \forall\, a \in C_0^k;\ b \in C_0^{k+1}\,. \tag{4.13}$$

The adjoint is a mapping $C_0^{k+1} \mapsto C_0^k$, has the property that $\delta^*\delta^* = 0$, and provides a second set of discrete Grad, Curl, and Div operations. In PDEs modeling physical problems, often a vector function is associated naturally with a 1-form or a 2-form, while a scalar function can be associated with a 0-form or a 3-form. This identification determines whether the vector function should be encoded in C_0^1 or C_0^2 and the scalar function in C_0^0 or C^3. This in turn determines the discrete version of Div, Curl and Grad to use.

Hodge Laplacian. We define the discrete Laplacian $\mathcal{D} : C_0^k \mapsto C_0^k$ with δ and its adjoint δ^* as

$$-\mathcal{D} = \delta^*\delta + \delta\delta^* \tag{4.14}$$

to mimic $-\Delta = d^*d + dd^*$.

REMARK 4.1. Derived operations are needed to avoid internal inconsistencies between the discrete operations. Because $\mathcal{I}$ is only an approximate left inverse of $\mathcal{R}$, some natural definitions with clash with each other. For example, a natural counterpart of (4.13) mimics $d^* = (-1)^k \star d\star$ and defines $\delta^* = (-1)^k \mathcal{R}\star d \star \mathcal{I}$. Besides the fact that this requires $\mathcal{I}$ to be conforming, the real problem is that the *natural* δ^* is not the adjoint of δ with respect to the *natural* inner product (4.11). Indeed, from (4.4) and (4.6)

$$\begin{aligned}(\delta^* a, b)_\Omega &= (-1)^k (\mathcal{I}\mathcal{R} \star d \star \mathcal{I}a, \mathcal{I}b)_\Omega = (-1)^k (\star d \star \mathcal{I}a, \mathcal{I}b)_\Omega + O(h^s) \\ &= (\mathcal{I}a, d\mathcal{I}b)_\Omega + O(h^s) = (\mathcal{I}a, \mathcal{I}\delta b)_\Omega + O(h^s) \\ &= (a, \delta b)_\Omega + O(h^s).\end{aligned}$$

4.3. Mimetic properties. We now establish the mimetic properties of the discrete operations.

Derivative and integral. In addition to $\delta\delta = \delta^*\delta^* = 0$, derivatives have the following mimetic property.

LEMMA 4.2. *Assume that $\mathcal{I}$ is conforming and let $a_h = \mathcal{I}a$, $b_h = \mathcal{I}b$ for $a \in C^k$, $b \in C^{k+1}$. Then*

$$(da_h, b_h)_\Omega = (\delta a, b)_\Omega \quad \textit{and} \quad (a_h, d^* b_h)_\Omega = (a, \delta^* b)_\Omega . \tag{4.15}$$

Proof. The first identity follows directly from CDP2 (4.6) and the definition of the mimetic inner product. To prove the second identity we use (4.6), (4.11), and that d^* is the adjoint of d:

$$(a_h, d^* b_h)_\Omega = (da_h, b_h)_\Omega = (d\mathcal{I}a, \mathcal{I}b)_\Omega = (\mathcal{I}\delta a, \mathcal{I}b)_\Omega = (\delta a, b)_\Omega = (a, \delta^* b)_\Omega .$$

□

The discrete Stokes theorem is a consequence of the identity

$$\langle \delta a, \sigma \rangle = \langle a, \partial\sigma \rangle \quad \forall a \in C^k;\ \sigma \in C_{k+1} .$$

From (4.1), (4.3), and (4.9) we have the property

$$\int_\sigma a = \langle a, \sigma \rangle = \langle \mathcal{R}\mathcal{I}a, \sigma \rangle = \int_\sigma \mathcal{I}a . \tag{4.16}$$

Combinatorial Hodge theory. We recall the relative singular cohomology of Ω over $\mathbb{R}$:

$$\bar{\mathcal{H}}_0^k = \operatorname{Ker}\delta/\operatorname{Range}\delta \quad \text{on singular } k\text{-cochains that vanish on } \Gamma_1\,,$$

the De Rham cohomology:

$$\bar{\mathcal{H}}^k = \operatorname{Ker} d/\operatorname{Range} d \quad \text{on } \Lambda_1^k$$

and the De Rham theorem

$$\bar{\mathcal{H}}^k \simeq \bar{\mathcal{H}}_0^k\,.$$

Let $H^k(\Omega) = \{h \in \Lambda_0^k(\Omega) \,|\, \triangle h = 0\}$, the space of smooth harmonic k-forms. The Hodge decomposition[3] theorem [23] states that $\dim(\operatorname{Ker}\triangle_k) = \dim(\bar{\mathcal{H}}^k)$ and every $\omega \in \Lambda_0^k(\Omega)$ has a decomposition

$$\omega = df + h + d^*g \tag{4.17}$$

where $f \in \Lambda_0^{k-1}(\Omega)$, $g \in \Lambda_0^{k+1}(\Omega)$, and $h \in H^k(\Omega)$. In the vector calculus this theorem implies that any vector function $\mathbf{u}$ has a decomposition $\mathbf{u} = \nabla \times \mathbf{w} + \nabla\phi + h$ where h is harmonic and ϕ is a scalar. It also implies that any real function has the decomposition $f = g + \nabla \cdot \mathbf{v}$, where g is harmonic.

The kernel of the discrete Laplacian $H^k(K) = \{h \in C_0^k \,|\, \mathcal{D}h = 0\}$ is the set of all harmonic cochains in C_0^k. Its characterization mimics that of $H^k(\Omega)$:

$$H^k(K) = \{c \in C_0^k \,|\, \delta c = \delta^* c = 0\}\,. \tag{4.18}$$

THEOREM 4.1. *Every $a \in C_0^k$ has a decomposition*

$$a = \delta b + h + \delta^* c\,, \tag{4.19}$$

where $b \in C_0^{k-1}$, $c \in C_0^{k+1}$ and $h \in H^k(K)$.
Theorem 4.1 is a consequence of $\delta\delta = 0$ and the definition of δ^* as the adjoint to δ. This is another important reason to choose the derived definition (4.13) of δ^* instead of the natural one in Remark 4.1.

To compute $\dim(\operatorname{Ker}\mathcal{D})$ we need the following result.

LEMMA 4.3. *The kernel of $\mathcal{D}$ is isomorphic to the kth relative cohomology group $\mathcal{H}_0^k$.*

---

[3]This theorem is primarily a consequence of the fact that if $T : V \mapsto V$ is a bounded linear operator on a Hilbert space V such that $T^2 = 0$, then

$$V = \operatorname{Range} T \oplus \operatorname{Range} T^* \oplus H\,,$$

where $H = \{x \in V \,|\, Tx = T^*x = 0\}$. A simple proof is as follows. Define $V' = (\operatorname{Range} T \oplus \operatorname{Range} T^*)^{\perp}$ and let $x \in V'$. Then $\langle Ty, x\rangle = 0$ and $\langle T^*y, x\rangle = 0$ for all $y \in V$ imply that $Tx = T^*x = 0$ and $x \in H$. For $T = d$ the proof is complicated by the fact that d is an unbounded operator on a domain in L^2.

Proof. Note that if $a = \delta b + h + \delta^* c$ is in $\mathrm{Ker}(\delta_k)$, then from $\delta\delta = 0$ and (4.18)

$$0 = \delta a = \delta\delta b + \delta h + \delta\delta^* c = \delta\delta^* c .$$

This identity implies that $(\delta^* c, \delta^* c) = 0$ and hence $\delta^* c = 0$. Thus, if $\delta a = 0$, then $a = h + \delta b$, and the correspondence $a \leftrightarrow h$ provides an isomorphism $\mathrm{Ker}\,\delta/\mathrm{Range}\,\delta \mapsto \mathrm{Ker}\,\mathcal{D}$. □

COROLLARY 4.1. *The size of the kernels of the analytic and discrete Laplacians is the same.*

Proof. From Lemma 4.3 it follows that

$$\dim\left(\mathrm{Ker}\,\mathcal{D}_k\right) = \dim \mathcal{H}_0^k .$$

Furthermore, $\dim(\mathcal{H}_0^k) = \dim(\bar{\mathcal{H}}_0^k)$ (Cairns [21]) and $\dim(\bar{\mathcal{H}}_0^k) = \dim(\bar{\mathcal{H}}^k)$ (De Rham's theorem). The assertion follows from $\dim(\bar{\mathcal{H}}^k) = \dim(\ker \triangle_k)$. □

It is remarkable that the size of the kernel of the analytic and discrete Laplacians depends only upon the topology of the domain and not the specific nature of these Laplacians.

Natural inner product. The definition of the discrete L^2 product (4.11) mimics definition (2.6). Using (4.10) we find that this inner product has the property that

$$(a, b)_\Omega = \int_\Omega (a, b)\omega_n = \int_\Omega (\mathcal{I}a, \mathcal{I}b)\omega_n = \int_\Omega \mathcal{I}a \wedge \star\mathcal{I}b ,$$

which mimics the property (2.7) of the analytic inner product.

Vector calculus. The discrete versions of the vector calculus identities hold exactly for the discrete operators defined by δ and δ^*.

LEMMA 4.4. *The discrete versions of Grad, Curl, and Div satisfy Curl Grad* $\equiv 0$ *and Div Curl* $\equiv 0$

Proof. For the two discrete derivatives the identities are $\delta\delta = 0$ and $\delta^*\delta^* = 0$. The first follows by duality of chains and cochains:

$$\langle \delta\delta a, b\rangle = \langle \delta a, \partial b\rangle = \langle a, \partial\partial b\rangle = 0$$

The second follows by the duality of δ and δ^* with respect to the discrete inner product:

$$(\delta^*\delta^* a, b)_\Omega = (\delta^* a, \delta b)_\Omega = (a, \delta\delta b)_\Omega = 0 .$$

□

As a corollary to this Lemma we also have a discrete version of Poincaré's lemma which states that on a contractable domain every closed form is a differential. The discrete version of this lemma is that every cocycle is a coboundary. Therefore, on contractable domains we have existence of discrete potentials. This mimetic property can be used to transfer solenoidal fields between two different cell complexes [11] and gauge discrete problems [15].

The wedge product. We show that (4.12) has the same commutation property as the true wedge product. If $\mathcal{I}$ is also conforming, then the effect of δ on (4.12) is algebraically the same as that of the exterior derivative on forms, and so properties of the discrete wedge and the discrete derivative are properly coordinated.

LEMMA 4.5. *Let* $\wedge : C^k \times C^p \mapsto C^{k+p}$ *be defined by (4.12). Then*

$$a \wedge b = (-1)^{kp} b \wedge a\,, \tag{4.20}$$

and if $\mathcal{I}$ *is conforming mimetic reconstruction,*

$$\delta(a \wedge b) = \delta a \wedge b + (-1)^k a \wedge \delta b \tag{4.21}$$

for all $a \in C^k$ *and* $b \in C^p$.

Proof. The commutation identity (4.20) follows directly from (4.12) and the like property of forms. The second identity is a consequence of the CDP1 property of $\mathcal{R}$ and the CDP2 property of $\mathcal{I}$:

$$\begin{aligned} \delta(a \wedge b) &= \delta \mathcal{R}(\mathcal{I}a \wedge \mathcal{I}b) \overset{CDP1}{=} \mathcal{R} d(\mathcal{I}a \wedge \mathcal{I}b) \\ &= \mathcal{R}(d\mathcal{I}a \wedge \mathcal{I}b) + (-1)^k \mathcal{R}(\mathcal{I}a \wedge d\mathcal{I}b) \\ &\overset{CDP2}{=} \mathcal{R}(\mathcal{I}\delta a \wedge \mathcal{I}b) + (-1)^k \mathcal{R}(\mathcal{I}a \wedge \mathcal{I}\delta b) \\ &= \delta a \wedge b + (-1)^k a \wedge \delta b\,. \end{aligned}$$

□

The wedge product is nonassociative: $(a \wedge b) \wedge c \neq a \wedge (b \wedge c)$.

4.4. Discrete $\star$. In this section we discuss complications arising in the construction of a discrete $\star$ operation and explain why it is not among the discrete operations that comprise our mimetic framework.

A *natural* discrete $\star$ operation uses $\mathcal{I}$ to translate cochains to forms, applies the analytic $\star$ and then reduces the result back to cochains. Thus, a natural operator $\overset{N}{\star}: C^k \mapsto C^{n-k}$ is defined by

$$\overset{N}{\star} = \mathcal{R} \star \mathcal{I}\,. \tag{4.22}$$

Tarhasaari *et al* [52] proposed this formula for a primal-dual cell complex.

The *derived* discrete $\star$ is defined in terms of the existing natural operations. We use the inner product (4.11) and the wedge product (4.12) to mimic[4] (2.7) and define $\overset{D}{\star}: C^k \mapsto C^{n-k}$ by the formula

$$\int_\Omega a \wedge \overset{D}{\star} b = (a, b)_\Omega \quad \forall\, a, b \in C^k\,. \tag{4.23}$$

In Section 5 we show that the derived $\star$ is related to an algebraic definition proposed by Hiptmair [30].

[4]The discrete $\star$ acts on cochains and is a global operation. Thus, we mimic the global relation (2.7) instead of the local formula (2.1) which defines the analytic $\star$ locally.

LEMMA 4.6. *The operator* $\overset{N}{\star}$ *has a commuting diagram property on the range of* $\mathcal{IR}$, *that is*

$$\overset{N}{\star}\,\mathcal{R}\omega_h = \mathcal{R} \star \omega_h \quad \forall \omega_h \in \Lambda^k(L^2, K). \tag{4.24}$$

Proof. From (4.7) we know that any $\omega_h \in \Lambda^k(L^2, K)$ has the form $\omega_h = \mathcal{IR}\omega$ for some $\omega \in \Lambda^k(d, \Omega)$. Using this characterization and the fact that $\mathcal{RI} = id$ gives

$$(\overset{N}{\star}\,\mathcal{R})\omega_h = (\overset{N}{\star}\,\mathcal{R})(\mathcal{IR}\omega) = (\mathcal{R} \star \mathcal{I})(\mathcal{RI})(\mathcal{R}\omega) = (\mathcal{R}\star)(\mathcal{IR}\omega) = (\mathcal{R}\star)\omega_h .$$

□

LEMMA 4.7. *The operator* $\overset{D}{\star}$ *has a weak commuting diagram property on* C^k:

$$\int_\Omega \mathcal{IR}(\mathcal{I}a \wedge \mathcal{I} \overset{D}{\star} a) = \int_\Omega \mathcal{I}a \wedge \star\mathcal{I}a . \tag{4.25}$$

Proof. Using (4.16) and (4.12)

$$\int_\Omega a \wedge \overset{D}{\star}\, a = \int_\Omega \mathcal{I}(a \wedge \overset{D}{\star}\, a) = \int_\Omega \mathcal{IR}(\mathcal{I}a \wedge \mathcal{I} \overset{D}{\star}\, a),$$

which is the left-hand side in (4.25). Using (4.11) and (2.7)

$$(a, a)_\Omega = (\mathcal{I}a, \mathcal{I}a)_\Omega = \int_\Omega \mathcal{I}a \wedge \star\mathcal{I}a ,$$

which is the right-hand side in (4.25). □

Similar arguments can be used to show that

$$\int_\Omega a \wedge \overset{N}{\star}\, a = (a, a)_\Omega + O(h^s), \tag{4.26}$$

which implies that $\overset{D}{\star} - \overset{N}{\star} = O(h^s)$. Formula (4.26) also means that the natural operator $\overset{N}{\star}$ is not compatible with the natural inner and wedge product definitions, while (4.24) means that it is compatible with the reduction map $\mathcal{R}$. Exactly the opposite is true for the derived operator $\overset{D}{\star}$. By construction this operator is compatible with the natural inner product and the natural wedge product but is incompatible with $\mathcal{R}$ and $\mathcal{I}$. Finally, neither $\overset{N}{\star}$, nor $\overset{D}{\star}$ is compatible with the derived adjoint δ^* defined in (4.13).

The problems with the discrete $\star$ operation arise from the fact that its action must be coordinated with two natural operations. The natural definition fails to accomplish this, while forcing the discrete $\star$ into compliance with the two natural operations leads to other incompatibilities. In contrast, an operation like δ^* requires a single natural operation for its definition and has a "built-in" compatibility with that operation.

These observations show that if a discrete $\star$ operation is required, then it *must* be made the primary object of the discrete framework and then used to define all other necessary structures. However, construction of a good discrete $\star$ is nontrivial and more difficult than the construction of a good inner product. For instance, the analytic $\star$ is local and invertible. To mimic this in finite dimensions the discrete $\star$ must be given by a diagonal matrix with positive entries. This is impossible unless K has a dual complex $\tilde{K}$ such that C^k is isomorphic[5] to $\tilde{C}^{n-k}$. In all other cases, the discrete $\star$ will be a rectangular matrix.

As a rule, the need for a discrete $\star$ arises from discretization of material laws. Because of the difficulties with this operator, we prefer to either incorporate these laws in the inner product or to enforce them in a weak, L^2 sense. In the first case we work with δ^* and in the second we solve a constrained optimization problem. These alternatives to a discrete $\star$ offer several valuable advantages. Besides being sufficient for a combinatorial Hodge theory, the inner product gives rise to a symmetric and positive semidefinite Laplacian. In contrast, direct discretization of material laws by an independently defined discrete $\star$ and the subsequent formation of the Laplacian through this operation may lead to operators that have imaginary and/or negative eigenvalues with the attendant stability problems; see [49] for examples in computational electromagnetism. On the other hand, the weak enforcement of the material laws is justified by their approximate nature as summaries of complex interactions.

In summary, the natural inner product leads to well-behaved discrete structures and is much easier to construct than a good discrete $\star$ operator. Choosing the inner product to be the primary discrete operation will also mimic the analytic case where the $\star$ operator is induced by the inner product, but not vice versa.

5. Algebraic realizations. Let $m_k = \dim C_0^k$. The map

$$a = \sum_{i=1}^{m_k} a_i \sigma_k^i \mapsto \mathbf{a} = (a_1, \dots, a_{m_k})$$

establishes an isomorphism $C_0^k \mapsto \mathbb{R}^{m_k}$. Then $\mathcal{R}$ can be viewed as a map $\Lambda_0^k(\Omega) \mapsto \mathbb{R}^{m_k}$, defined by

$$\mathbf{a} = \mathcal{R}\omega \quad \text{if and only if} \quad a_i = \int_{\sigma_k^i} \omega\,,$$

while $\mathcal{I}$ is an approximate inverse of this map. As a result, all mimetic operations on cochains can be realized by matrices acting on their coefficient

[5] It is worth pointing out that when $\tilde{C}^{n-k}$ and C^k have the same dimension, the *covolume* reconstruction operator gives rise to an inner product that is compatible with a diagonal discrete $\star$; see [44, 45, 58]. Thus, in this case, explicit definition of a discrete $\star$ operation can also be avoided.

vectors. The action of $\delta : C_0^k \mapsto C_0^{k+1}$ is given by a matrix $\mathbb{D}_k \in \mathbb{R}^{m_{k+1} \times m_k}$ with the property that $\mathbb{D}_{k+1}\mathbb{D}_k = 0$. This matrix has elements -1, 0, and 1 which reflect the combinatorial nature of the discrete derivative δ

The local and the L^2 inner products on C_0^k are associated with the symmetric and positive definite matrices $\mathbb{M}_k(x), \mathbb{M}_k \in \mathbb{R}^{m_k \times m_k}$ such that

$$(a,b) = \mathbf{a}^T \mathbb{M}_k(x)\mathbf{b} \quad \text{and} \quad (a,b)_\Omega = \mathbf{a}^T \mathbb{M}_k \mathbf{b}\,, \tag{5.1}$$

respectively.

The action of δ^* is given by a matrix $\mathbb{D}_k^* \in \mathbb{R}^{m_k \times m_{k+1}}$. Since δ^* is derived from δ and the natural inner product, it follows that $\mathbb{D}_k^*$ can be expressed in terms of the matrices that represent these operations. From

$$\mathbf{a}^T (\mathbb{D}_{k+1}^*)^T \mathbb{M}_k \mathbf{b} = (\delta_{k+1}^* a, b)_\Omega = (a, \delta_k b)_\Omega = \mathbf{a}^T \mathbb{M}_{k+1}\mathbb{D}_k \mathbf{b}$$

we see that $\mathbb{D}_{k+1}^* = \mathbb{M}_k^{-1}\mathbb{D}_k^T \mathbb{M}_{k+1}$ and

$$\mathbb{D}_k^*\mathbb{D}_{k+1}^* = \mathbb{M}_{k-1}^{-1}\mathbb{D}_{k-1}^T \mathbb{M}_k \mathbb{M}_k^{-1}\mathbb{D}_k^T \mathbb{M}_{k+1} = \mathbb{M}_{k-1}^{-1}\mathbb{D}_{k-1}^T \mathbb{D}_k^T \mathbb{M}_{k+1} = 0\,,$$

as expected from a derivative.

The discrete Laplacian $\mathcal{D}_k$ is also a derived operation and its action is given by the matrix

$$\mathbb{L}_k = (\mathbb{M}_k^{-1}\mathbb{D}_k^T \mathbb{M}_{k+1}\mathbb{D}_k + \mathbb{D}_{k-1}\mathbb{M}_{k-1}^{-1}\mathbb{D}_{k-1}^T \mathbb{M}_k) \in \mathbb{R}^{m_k \times m_k}\,.$$

We have the formula

$$(\delta_{k+1}^*\delta_k a, b) = \mathbf{a}^T \mathbb{D}_k^T (\mathbb{M}_k^{-1}\mathbb{D}_k^T \mathbb{M}_{k+1})^T \mathbb{M}_k \mathbf{b} = \mathbf{a}^T \mathbb{D}_k^T \mathbb{M}_{k+1}\mathbb{D}_k \mathbf{b} = (\delta_k a, \delta_k b)$$

and a similar formula for $(\delta_{k-1}\delta_k^* a, b)$.

To find a matrix expression for the wedge product $\wedge : C_0^1 \times C_0^1 \mapsto C_0^2$ we use the formula

$$a_1 \wedge b_1 = \mathcal{R}(\mathcal{I}a_1 \wedge \mathcal{I}b_1) = \sum_{i=1}^{m_2} c_i \sigma_2^i$$

and the commutation property (4.20) to conclude that each coefficient c_i is a skew-symmetric bilinear form of the coefficient vectors $\mathbf{a}$ and $\mathbf{b}$. Therefore, c_i is given by a skew-symmetric matrix $\mathbb{W}_{11}^i \in \mathbb{R}^{m_1 \times m_1}$ and

$$a_1 \wedge b_1 = \sum_{i=1}^{m_2} (\mathbf{a}^T \mathbb{W}_{11}^i \mathbf{b})\sigma_2^i\,.$$

For $\wedge : C_0^1 \times C_0^2 \mapsto C_0^3$ and $\wedge : C_0^2 \times C_0^1 \mapsto C_0^3$ we have the formulas

$$a_1 \wedge b_2 = \mathcal{R}(\mathcal{I}a_1 \wedge \mathcal{I}b_2) = \sum_{i=1}^{m_3} c_i^{12}\sigma_3^i \quad \text{and} \quad b_2 \wedge a_1 = \mathcal{R}(\mathcal{I}b_2 \wedge \mathcal{I}a_1) = \sum_{i=1}^{m_3} c_i^{21}\sigma_3^i\,,$$

respectively. The coefficients c_i^{12} and c_i^{21} are bilinear functions of $\mathbf{a}$ and $\mathbf{b}$ and so they are given by matrices $\mathbb{W}_{12}^i \in \mathbb{R}^{m_1 \times m_2}$ and $\mathbb{W}_{21}^i \in \mathbb{R}^{m_2 \times m_1}$, respectively. From (4.20) it follows that $\mathbb{W}_{12}^i = (\mathbb{W}_{12}^i)^T$ and

$$a_1 \wedge b_2 = \sum_{i=1}^{m_3} (\mathbf{a}^T \mathbb{W}_{12}^i \mathbf{b}) \sigma_3^i \quad \text{and} \quad b_2 \wedge a_1 = \sum_{i=1}^{m_3} (\mathbf{b}^T (\mathbb{W}_{12}^i)^T \mathbf{a}) \sigma_3^i . \tag{5.2}$$

Matrix representations for the remaining two wedge products follow in a similar fashion. From (5.1) and (5.2) we can obtain a matrix representation for $\overset{D}{\star}\colon C_0^1 \mapsto C_0^2$. Using (5.2) and definition (4.9) the matrix form of the left hand side in (4.23) is

$$\int_\Omega a \wedge \overset{D}{\star} a = \langle a \wedge \overset{D}{\star} a, \sum_{i=1}^{m_3} \sigma_3^i \rangle = \sum_{i=1}^{m_3} \mathbf{a}^T \mathbb{W}_{12}^i (\overset{D}{\star}\, \mathbf{a}) \langle \sigma_3^i, \sigma_3^i \rangle = \sum_{i=1}^{m_3} \mathbf{a}^T \mathbb{W}_{12}^i (\overset{D}{\star}\, \mathbf{a}) \mu_i$$

where $\overset{D}{\star}\, \mathbf{a} \in \mathbb{R}^{m_2}$ is the coefficient vector of $\overset{D}{\star}\, a$ and $\mu_i = \langle \sigma_3^i, \sigma_3^i \rangle$ is the volume of the ith basis 3-cell. The matrix form of the right hand side in (4.23) is

$$(a, a)_\Omega = \mathbf{a}^T \mathbb{M}_1 \mathbf{a} .$$

Let $\mathbb{W}_{12} = \sum_{i=1}^{m_3} \mu_i \mathbb{W}_{12}^i$. Then, the matrix form of (4.23) is

$$\mathbb{W}_{12} (\overset{D}{\star}\, \mathbf{a}) = \mathbb{M}_1 \mathbf{a} . \tag{5.3}$$

This formula reflects the fact that the derived operator $\overset{D}{\star}$ relies on two natural operations and so is associated with a *pair* of matrices related to these operations. A formula similar to (5.3) was used in [30] for an axiomatic definition of a discrete $\star$ operation.

Algebraic realizations of the mimetic operations are summarized in Table 1.

6. Examples of reconstruction operators. For simplicity we present examples of reconstruction operators in two-dimensions and restrict attention to operators that translate 1-cochains to 1-forms. We will consider three operators $\mathcal{I} : C^1 \mapsto \Lambda^1(L^2, \Omega)$, one of which will be conforming. To explain the action of these operators it suffices to consider a space C_1 consisting of a single 1-chain $c_1 = \sum_{i=1}^3 c_1^i$ which is a boundary of a 2-simplex c_2 that forms the space C_2. In two dimensions c_2 is a triangle and the 1-cells $\{c_1^i\}$ are its edges. Two edges, c_1^i and c_1^j, intersect at a vertex c_0^k, $k \neq i, j$. The set $\{c_0^k\}$ forms the space C_0.

Using the isomorphism (3.6), the elements of C^1 can be written as $c^1 = \sum_{i=1}^3 a_i c_1^i$ where $a_i = \int_{c_1^i} \omega$ for some $\omega \in \Lambda^1(d, \Omega)$. The value of a_i gives the circulation of the vector field $\mathbf{u}$, associated with ω, along the edge c_1^i.

TABLE 1
Algebraic realizations of mimetic operations.

Operation	Matrix form	Type
δ	$\mathbb{D}_k$	incidence matrix
$(\cdot,\cdot)$	$\mathbb{M}_k$	SPD
$a^1 \wedge b^1$	$\sum \mathbb{W}_{11}$	skew symmetric
$a^1 \wedge b^2$	$\sum \mathbb{W}_{12}$	$\mathbb{W}_{12} =$
$b^2 \wedge a^1$	$\sum \mathbb{W}_{21}$	$\mathbb{W}_{21}^T$
δ^*	$\mathbb{M}_k^{-1}\mathbb{D}_k^T\mathbb{M}_{k+1}$	rectangular
$\mathcal{D}$	$\mathbb{M}_k^{-1}\mathbb{D}_k^T\mathbb{M}_{k+1}\mathbb{D}_k + \mathbb{D}_{k-1}\mathbb{M}_{k-1}^{-1}\mathbb{D}_{k-1}^T\mathbb{M}_k$	square
$\overset{D}{\star}: C^1 \mapsto C^2$	$(\mathbb{W}_{12}, \mathbb{M}_1)$	pair

Covolume reconstruction. To define the covolume reconstruction operator [58] the simplex c_2 is divided into three subsimplices c_2^i by connecting the circumcenter of c_2 with its vertices c_0^i as shown in Fig. 3. Each subsimplex is bordered by exactly one of the edges c_1^i; we denote that subsimplex by c_2^i.

The covolume reconstruction operator maps the 1-cochain c^1 into a 1-form $\omega^{\mathbf{u}}$ whose associated vector field $\mathbf{u}$ is piecewise constant on each subsimplex, determined according to the rule

$$\mathbf{u}|_{c_2^i} = a_i c_1^i; \quad i = 1,2,3\,. \tag{6.1}$$

The range of the operator defined in (6.1) is in the Hilbert space $\Lambda^1(L^2,\Omega)$ but not in the Sobolev space $\Lambda^1(d,\Omega)$. Therefore, covolume reconstruction is not conforming. A unique property of covolume reconstruction is that derived operators have local stencils and that there is a discrete $\star$ star operation that is compatible with the natural inner product [58]. As a result, the matrix $\mathbb{M}$ that gives the action of the natural inner product is diagonal

$$\mathbb{M} = \begin{pmatrix} h_1 h_1^{\perp} & 0 & 0 \\ 0 & h_2 h_2^{\perp} & 0 \\ 0 & 0 & h_3 h_3^{\perp} \end{pmatrix}. \tag{6.2}$$

In (6.2) h_i is the length of c_1^i and $h_i^{\perp}$ is the length of the perpendicular from the circumcenter to c_1^i.

These properties follow from the fact that covolume reconstruction can be associated with cochains on a Voronoi-Delaunay grid complex; see

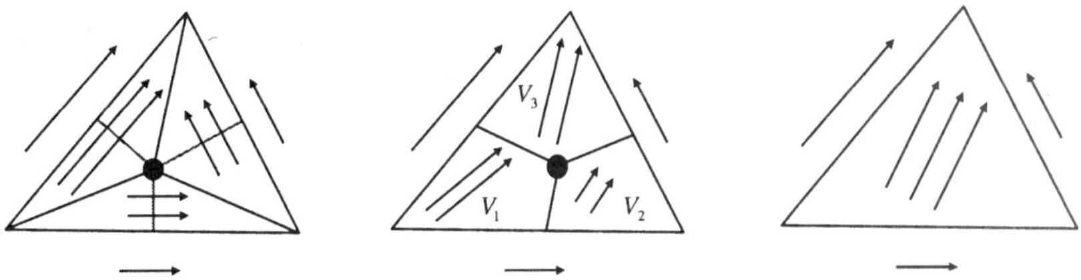

FIG. 3. *The reconstruction operators are shown for the 1-cochains: covolume, mimetic, and Whitney, respectively. In the first figure, the covolume reconstruction operator divides the simplex into three subsimplices by connecting the circumcenter of with its vertices. Each subsimplex is bordered by exactly one of the edges. The covolume reconstruction operator maps the 1-cochain into a 1-form whose associated vector field is piecewise constant on each subsimplex. In the second figure, mimetic reconstruction acts in a similar way to recover a form with a piecewise constant vector field. In the mimetic approach, the subregions are associated with the vertices, have quadrilateral shapes, and are bordered by the edges adjacent to each vertex. The third figure of the Whitney map is an example of a regular mimetic reconstruction operator. In contrast to the previous two reconstruction operators, the Whitney map builds a polynomial 1-form from the cochain using a basis of polynomial 1-forms associated with the edges.*

[42, 44, 58]. This association also implies that existence of the covolume reconstruction is contingent upon the existence of the Voronoi regions and so the simplexes must satisfy an angle condition [44].

Mimetic reconstruction. Mimetic reconstruction [33] acts in a similar way to recover a form $\omega^{\mathbf{u}}$ whose associated vector field $\mathbf{u}$ is a piecewise constant on c_2. As a result, the reconstructed form is in the Hilbert space $\Lambda^1(L^2,\Omega)$ but not in the Sobolev space $\Lambda^1(d,\Omega)$. The main difference between covolume and mimetic reconstruction is in the choice of the subregions. In the mimetic approach, the subregions are associated with the vertices c_0^k of c_2, have quadrilateral shapes, and are bordered by the edges c_1^i and c_1^j, $i,j \neq k$; see Fig. 3. Each subregion is determined by connecting the midpoint of c_1^i with an arbitrary but fixed point inside the triangle. We denote the subregion associated with the vertex c_0^k by q_2^k and its area by V_k. The mimetic reconstruction operator builds on c_2 the following piecewise constant field:

$$\mathbf{u}|_{q_2^k} = a_i c_1^i + a_j c_1^j; \quad k = 1,2,3; \ i,j \neq k\,. \tag{6.3}$$

Mimetic reconstruction is less restrictive than the covolume $\mathcal{I}$ because existence of the subregions is not contingent upon the circumcenter being inside the triangle. However, mimetic reconstruction gives rise to nonlocal derived operators [34]. If ϕ_k is the angle associated with the vertex c_0^k, the inner product matrix on c_2 is given by

$$\mathbb{M} = \begin{pmatrix} \frac{V_2}{\sin^2\phi_2} + \frac{V_3}{\sin^2\phi_3} & \frac{V_3\cos\phi_3}{\sin^2\phi_3} & \frac{V_2\cos\phi_2}{\sin^2\phi_2} \\ \frac{V_3\cos\phi_3}{\sin^2\phi_3} & \frac{V_1}{\sin^2\phi_1} + \frac{V_3}{\sin^2\phi_3} & \frac{V_1\cos\phi_1}{\sin^2\phi_1} \\ \frac{V_2\cos\phi_2}{\sin^2\phi_2} & \frac{V_1\cos\phi_1}{\sin^2\phi_1} & \frac{V_1}{\sin^2\phi_1} + \frac{V_2}{\sin^2\phi_2} \end{pmatrix}. \tag{6.4}$$

Whitney reconstruction. The Whitney map [24, 60] is an example of a conforming reconstruction operator whose range is in the Sobolev space $\Lambda^1(d,\Omega)$. In contrast to the previous two reconstruction operators, the Whitney map builds a polynomial 1-form on c_2 from the cochain c^1 using a *basis* of polynomial 1-forms associated with the edges c_1^i. The basis 1-forms are defined by the formula

$$\omega_k = t_i dt_j - t_j dt_i; \quad i,j \neq k,\ i<j\,, \tag{6.5}$$

where t_i are the barycentric coordinates. The vector field corresponding to the basis 1-form is given by

$$\mathbf{u}_k = t_i \nabla t_j - t_j \nabla t_i; \quad i,j \neq k,\ i<j\,.$$

Therefore, the Whitney reconstruction map translates the cochain c^1 to the 1-form $\omega^1 = \sum_{k=1}^3 a_k\omega_k$ with a vector field

$$\mathbf{u} = \sum_{k=1}^{3} a_k(t_i\nabla t_j - t_j\nabla t_i)\,. \tag{6.6}$$

The reconstructed image of c^1 is in the smooth space $\Lambda^1(\Omega)$. When K consists of more than one 2-simplex, the range of the Whitney map contains piecewise polynomial 1-forms obtained by gluing together the reconstructed images from the individual triangles. It is possible to show [24] that the resulting 1-forms are in the Sobolev space $\Lambda^1(d,\Omega)$.

7. Application to PDEs. We consider mimetic discretizations of the elliptic boundary value problems

$$\begin{cases} -\triangle_0\phi &= f \\ \phi &= 0 \quad \text{on } \Gamma_1 \\ \mathbf{n}\cdot\nabla\phi &= 0 \quad \text{on } \Gamma_2 \end{cases} \quad \text{and} \quad \begin{cases} -\triangle_3\psi &= f \\ \mathbf{n}\cdot\nabla\psi &= 0 \quad \text{on } \Gamma_1 \\ \psi &= 0 \quad \text{on } \Gamma_2 \end{cases} \tag{7.1}$$

respectively. Note that $-\triangle_0 = d^*d$ and $-\triangle_3 = dd^*$. To better illustrate the formation of the discrete mimetic equations we use equivalent first-order formulations of (7.1):

$$\begin{cases} d\phi - \mathbf{u} &= 0 \\ d^*\mathbf{u} &= f \\ \phi &= 0 \quad \text{on } \Gamma_1 \\ \mathbf{n}\cdot\mathbf{u} &= 0 \quad \text{on } \Gamma_2 \end{cases} \quad \text{and} \quad \begin{cases} d^*\psi - \mathbf{v} &= 0 \\ d\mathbf{v} &= f \\ \mathbf{n}\cdot\mathbf{v} &= 0 \quad \text{on } \Gamma_1 \\ \psi &= 0 \quad \text{on } \Gamma_2 \end{cases}. \tag{7.2}$$

In (7.2) the variables acted upon by d, their boundary conditions, and the equations involving d are called *primal.* The other variables, boundary conditions and equations, are called *dual.*

7.1. Direct mimetic discretization. In the direct approach we use that d and $\mathcal{R}$ commute and apply $\mathcal{R}$ to translate the primal equation and boundary condition to combinatorial cochain equations. Reduction of the primal equation fixes the type of cochains in the discrete model. The discrete primal equation is uniquely determined by the mesh topology of the triangulation K and does not require a reconstruction operator $\mathcal{I}$. However, this operator is needed for the discretization of the dual equation. Because $\mathcal{R}$ and d^* do not commute, the discrete dual equation cannot be obtained by an application of $\mathcal{R}$. Instead, we derive it by using the discrete adjoint δ^* to mimic the analytic dual. Therefore, the discrete dual equation depends on the choice of the reconstruction map $\mathcal{I}$ and is not unique. Note that $\mathcal{I}$ is only needed to induce the adjoint and does not have to be a conforming reconstruction operator.

For $\triangle_0$ the primal variable ϕ is 0-form and the dual variable $\mathbf{u}$ is 1-form. We approximate them by $\phi_0 = \mathcal{R}\phi \in C_1^0$ and $\mathbf{u}_1 = \mathcal{R}\mathbf{u} \in C^1$. For $\triangle_3$ the primal variable $\mathbf{v}$ is a 2-form $\mathbf{v}$ and the dual variable is a 3-form ψ. They are approximated by $\mathbf{v}_2 = \mathcal{R}\mathbf{v} \in C_1^2$ and $\psi_3 = \mathcal{R}\psi \in C^3$, respectively. Applying $\mathcal{R}$ to the primal equations in (7.2) and using CDP1 gives

$$
\begin{aligned}
0 &= \mathcal{R}(d\phi - \mathbf{u}) = \delta\mathcal{R}\phi - \mathcal{R}\mathbf{u} = \delta\phi_0 - \mathbf{u}_1 \quad \text{and} \\
0 &= \mathcal{R}(d\mathbf{v} - f) = \delta\mathcal{R}\mathbf{v} - \mathcal{R}f = \delta\mathbf{v}_2 - f_3\,,
\end{aligned}
$$

respectively. Hence, the direct mimetic models for $\triangle_0$ and $\triangle_3$ are

$$
\left\{ \begin{array}{rcl} \delta\phi_0 - \mathbf{u}_1 &=& 0 \\ \delta^*\mathbf{u}_1 &=& f_0 \end{array} \right. \quad \text{and} \quad \left\{ \begin{array}{rcl} \delta^*\psi_3 - \mathbf{v}_2 &=& 0 \\ \delta\mathbf{v}_2 &=& f_3 \end{array} \right. , \tag{7.3}
$$

respectively. In (7.3) the primal boundary conditions on Γ_1 constrain the spaces for the primal variables. The boundary conditions on Γ_2 are enforced weakly through the definition of δ^* as adjoint to δ.

The methods in (7.3) can be realized using any one of the three reconstruction operators (6.1), (6.3), or (6.6). With the covolume reconstruction the derived adjoint δ^* has local stencil and (7.3) is equivalent to a finite volume method on Delaunay-Voronoi grid complex. With the mimetic and Whitney reconstructions the stencil of δ^* is not local. For these two operators (7.3) is a conservative finite difference scheme on an unstructured grid.

If $\mathbf{u}_1$ and $\mathbf{v}_2$ are eliminated from (7.3) we obtain the equations

$$
\delta^*\delta\phi_0 = f_0 \quad \text{and} \quad \delta\delta^*\phi_3 = f_3 \tag{7.4}
$$

that represent direct discretizations of the equations in (7.1) by the discrete Laplace operators $\mathcal{D}_0 = \delta^*\delta$ and $\mathcal{D}_3 = \delta\delta^*$.

7.2. Conforming mimetic discretization. In the conforming approach, the analytic equations are restricted to finite dimensional spaces in the range of $\mathcal{IR}$. In contrast to the direct approach, where only discrete derivatives are used, this requires $\mathcal{I}$ to be conforming. Assuming that such $\mathcal{I}$ is given, we approximate ϕ and $\mathbf{u}$ by $\phi_0^h \in \Lambda_1^0(d,K)$ and $\mathbf{u}_1^h \in \Lambda^1(d,K)$, respectively. For ψ and $\mathbf{v}$ the approximations are $\psi_3^h \in \Lambda^3(K)$ and $\mathbf{v}_2^h \in \Lambda_1^2(d,K)$. The conforming discretizations of (7.2) are given by

$$\left\{\begin{array}{rcl} d\phi_0^h - \mathbf{u}_1^h &=& 0 \\ d^*\mathbf{u}_1^h &=& f_0^h \end{array}\right. \quad \text{and} \quad \left\{\begin{array}{rcl} d^*\psi_3^h - \mathbf{v}_2^h &=& 0 \\ d\mathbf{v}_2^h &=& f_3^h \end{array}\right. , \tag{7.5}$$

respectively, where $f_0^h = \mathcal{IR}f$ and $f_h^3 = \mathcal{IR}f$.

In contrast to the direct methods in (7.3), the methods in (7.5) cannot be realized by the covolume or the mimetic reconstruction operators because they are not conforming. However, for (7.5) we can use the Whitney map (6.6). In this case, the scheme where the scalar is the primal variable reduces to the familiar Galerkin finite element method in which the scalar is approximated by continuous, piecewise linear polynomial finite elements on simplices. The second scheme, where the scalar is the dual variable, reduces to a mixed Galerkin method in which the scalar is approximated by a piecewise constant and the vector is approximated by the lowest order Raviart-Thomas spaces [18, 46]. For this reason we will call the schemes in (7.5) *Galerkin* and *mixed Galerkin*, respectively. The Whitney map has been extensively used in computational electromagnetism where it gives rise to the lowest-order Nedelec edge elements [14, 29, 40, 41].

THEOREM 7.1. *Assume that $\mathcal{I}$ is a conforming reconstruction operator, then the direct and the conforming mimetic models are equivalent.*

Proof. We give the details for $\triangle_0$; the proofs for $\triangle_3$ are very similar. For $\phi_0^h \in \Lambda_1^0(d,K)$ and $\mathbf{u}_1^h \in \Lambda^1(d,K)$ there exist $\phi \in \Lambda_1^0(d,\Omega)$ and $\mathbf{u} \in \Lambda^1(d,\Omega)$, such that $\phi_0^h = \mathcal{IR}\phi$ and $\mathbf{u}_1^h = \mathcal{IR}\mathbf{u}$, respectively. Using (4.6)

$$\begin{aligned} 0 &= d\phi_0^h - \mathbf{u}_1^h = d(\mathcal{IR}\phi) - \mathcal{IR}\mathbf{u} = \mathcal{I}\delta\mathcal{R}\phi - \mathcal{IR}\mathbf{u} \\ &= \mathcal{I}(\delta\mathcal{R}\phi - \mathcal{R}\mathbf{u}) = \mathcal{I}(\delta\phi_0 - \mathbf{u}_1), \end{aligned}$$

where $\phi_0 = \mathcal{R}\phi$ and $\mathbf{u}_1 = \mathcal{R}\mathbf{u}$. From (4.5) we conclude that $\delta\phi_0 - \mathbf{u}_1 = 0$, that is, the degrees of freedom of ϕ_0^h and $\mathbf{u}_1^h$ solve the direct equation. To prove equivalence of the dual equations note that for $\xi_0 \in C_1^0$ - arbitrary, and $\xi_0^h = \mathcal{I}\xi_0$ formula (4.15) implies the identity

$$(d^*\mathbf{u}_1^h, \xi_h^0)_\Omega = (\delta^*\mathbf{u}_1, \xi_0)_\Omega$$

while definition of f_0^h and the L^2 inner product give that

$$(f_h^0, \xi_h^0)_\Omega = (\mathcal{IR}f, \mathcal{I}\xi_0)_\Omega = (\mathcal{R}f, \xi_0)_\Omega = (f_0, \xi_0).$$

Combining the two equations shows that

$$(\delta^*\mathbf{u}_1, \xi_0)_\Omega - (f_0, \xi_0) \quad \forall \xi_0 \in C_1^0 \quad \text{or} \quad \delta^*\mathbf{u}_1 = f_0.$$

Therefore, $\mathbf{u}_1^h$ solves $d^*\mathbf{u}_1^h = f_0^h$ if and only if $\mathbf{u}_1$ solves the direct dual equation $\delta^*\mathbf{u}_1 = f_0$. □

From this theorem we can conclude that realizations of the direct scheme (7.3) and the conforming scheme (7.5) by the Whitney map lead to two completely equivalent discretizations of the PDEs (7.2). Further connections between direct and conforming methods can be established by choosing specific quadrature points to compute the integrals in the conforming method [12, 13, 19]. Note that quadrature selection can be interpreted as yet another choice for the reconstruction operator.

7.3. Mimetic discretization with weak material laws. The first-order systems in (7.2) can be combined into a single problem by keeping the two primal equations and adding the constitutive laws

$$\mathbf{u} = \star\mathbf{v} \quad \text{and} \quad \psi = \star\phi \tag{7.6}$$

that express the dual variables in terms of the primal variables. We write the new system as

$$\begin{cases} d\phi - \mathbf{u} &= 0 \\ d\mathbf{v} + g\psi &= f \end{cases} \quad \text{and} \quad \begin{matrix} \mathbf{u} = \star\mathbf{v} \\ \psi = \star\phi \end{matrix} \tag{7.7}$$

where g is a function that can be identically zero; see [15, 14, 30, 56, 57] for discussions of such *factorization* diagrams. Instead of trying to approximate (7.7), which would require us to deal with the material laws and a discrete $\star$ operation, we first transform this system into an equivalent constrained optimization problem and then discretize that problem. Let

$$\mathcal{J}(\phi, \mathbf{u}; \psi, \mathbf{v}) = \frac{1}{2}\left(\|\psi - \star\phi\|^2 + \|\mathbf{u} - \star\mathbf{v}\|^2\right).$$

The optimization problem: *find* $(\phi, \mathbf{u}) \in \Lambda_1^0(\Omega) \times \Lambda^1(\Omega)$ *and* $(\psi, \mathbf{v}) \in \Lambda^3(\Omega) \times \Lambda_1^2(\Omega)$ *such that for all* $(\hat{\phi}, \hat{\mathbf{u}}) \in \Lambda_1^0(\Omega) \times \Lambda^1(\Omega)$ *and* $(\hat{\psi}, \hat{\mathbf{v}}) \in \Lambda^3(\Omega) \times \Lambda_1^2(\Omega)$

$$\begin{gathered} \mathcal{J}(\phi, \mathbf{u}; \psi, \mathbf{v}) \leq \mathcal{J}(\hat{\phi}, \hat{\mathbf{u}}; \hat{\psi}, \hat{\mathbf{v}}) \\ \textit{subject to } d\hat{\phi} - \hat{\mathbf{u}} = 0 \textit{ and } d\hat{\mathbf{v}} + g\hat{\psi} = f \end{gathered} \tag{7.8}$$

is an equivalent to (7.7). We use this optimization problem to devise direct and conforming mimetic methods in which material laws are enforced weakly and no explicit construction of a discrete $\star$ operation is required.

The idea is to approximate the four variables in (7.8) by the same cochains as in (7.3) or by the same conforming spaces as in (7.5). In the first case we have the constrained optimization problem *find* $(\phi_0, \mathbf{u}_1) \in C_1^0 \times C^1$ *and* $(\psi_3, \mathbf{v}_2) \in C^3 \times C_1^2$ *such that for all* $(\hat{\phi}_0, \hat{\mathbf{u}}_1) \in C_1^0 \times C^1$ *and* $(\hat{\psi}_3, \hat{\mathbf{v}}_2) \in C^3 \times C_1^2$

$$\begin{gathered} \mathcal{J}(\phi_0, \mathbf{u}_1; \psi_3, \mathbf{v}_2) \leq \mathcal{J}(\hat{\phi}_0, \hat{\mathbf{u}}_1; \hat{\psi}_3, \hat{\mathbf{v}}_2) \\ \textit{subject to } \delta\hat{\psi}_0 - \hat{\mathbf{u}}_1 = 0 \textit{ and } \delta\hat{\mathbf{v}}_2 + g\hat{\psi}_3 = f_3 \end{gathered} \tag{7.9}$$

which gives a direct mimetic method. If, instead, we use the conforming spaces, the optimization problem is *find* $(\phi^h, \mathbf{u}^h) \in \Lambda^0_1(d,K) \times \Lambda^1(d,K)$ *and* $(\psi^h, \mathbf{v}^h) \in \Lambda^3(d,K) \times \Lambda^2_1(d,K)$ *such that for all* $(\hat{\phi}^h, \hat{\mathbf{u}}^h) \in \Lambda^0_1(d,K) \times \Lambda^1(d,K)$ *and* $(\hat{\psi}^h, \hat{\mathbf{v}}^h) \in \Lambda^3(d,K) \times \Lambda^2_1(d,K)$

$$\begin{aligned} &\mathcal{J}(\phi^h, \mathbf{u}^h; \psi^h, \mathbf{v}^h) \le \mathcal{J}(\hat{\phi}^h, \hat{\mathbf{u}}^h; \hat{\psi}^h, \hat{\mathbf{v}}^h) \\ &\textit{subject to } d\hat{\phi}^h - \hat{\mathbf{u}}^h = 0 \textit{ and } d\hat{\mathbf{v}}^h + g\hat{\psi}^h = f^h \end{aligned} \tag{7.10}$$

and we have a conforming mimetic method.

Because C^{n-k} and C^k and $\Lambda^k(d,K)$ and $\Lambda^{n-k}(d,K)$ have different dimensions, the primal and the dual variables cannot be related by a one-to-one map. Instead, we minimize their discrepancy in L^2 sense and so the material laws are imposed in a weak sense.

To realize (7.9) we can use any one of the three reconstruction operators (6.1), (6.3), or (6.6) and obtain a finite-difference like scheme. For the conforming method (7.10) we cannot use the covolume or the mimetic reconstruction, but we can use the Whitney map (6.6) to obtain a finite element-like scheme. We note that with the Whitney map realizations of (7.9) and (7.5) are completely equivalent.

For further details on mimetic discretizations with weak constitutive laws and their connection to least-squares minimization principles we refer to [7, 8, 9]. Examples of this idea in magnetostatics can be found in [14] and [20].

8. Conclusions. We described a general framework for mimetic discretizations that uses two basic operators to define all discrete structures. Scalars and vectors are translated to differential forms and then reduced to cochains. *Combinatorial* differentiation and integration operations are induced by the De Rham map which effects the reduction to cochains. The *natural* inner product and wedge product are defined by using a reconstruction operator that translates cochains back to forms. The inner product induces an adjoint derivative and a discrete Laplacian. Together with the combinatorial and natural operations these *derived* operations comprise the core of the mimetic framework.

The choice of the natural and derived operations is determined by the internal consistency of the framework. The natural definitions of the inner product and the wedge product are not compatible with a natural definition of the discrete $\star$. As a result, a consistent discrete framework requires a choice of its primary operation. We choose the primary operation to be the natural inner product on real cochain spaces. It would be equally valid to choose the primary operation to be the discrete $\star$ and its construction to be the principal computational task.

We choose to base our mimetic framework on the natural inner product instead of the $\star$ operation because of the complications that arise in the construction of the latter and because the inner product is sufficient to induce a combinatorial Hodge theory on cochains. For problems that require approximations of material laws we propose to consider constrained optimization formulations that enforce the laws weakly, instead of using their explicit discretization. In all other cases, our framework offers the choice of direct and conforming methods. Direct methods are representative of the type of discretizations that arise in FV and FD methods while conforming methods are typical of FE. We demonstrated that for regular reconstruction operators direct and conforming methods are equivalent. This opens up a possibility to carry out error analysis of direct mimetic methods by using variational tools from FE. Some recent examples are the analyses in [12, 13, 19].

Acknowledgement. We thank Clint Scovel for his insight and guidance in the early stages of this research. This work was funded by the Department of Energy under contracts with the Sandia Corporation (DE-AC04-94-AL85000), Los Alamos National Laboratory (W-7405-ENG-36) and the DOE Office of Science's Advanced Scientific Computing Research (ASCR) Applied Mathematics Research Program (KJ010101). Sandia is a multiprogram laboratory operated by Sandia Corporation, a Lockheed Martin Company, for the US Department of Energy under contract DE-AC04-94-AL85000.

REFERENCES

[1] M. Ainsworth and K. Pinchedez, *hp-Approximation theory for BDFM/RT finite elements and applications*, SIAM Journal on Numerical Analysis, 40(6), pp. 2047–2068, 2003.
[2] D. Arnold, *Differential complexes and numerical stability*, Proceedings of the International Congress of Mathematicians, Beijing 2002, Volume I: Plenary Lectures.
[3] D. Arnold, R. Falk, and R. Winther, *Differential complexes and stability of finite element methods. I. The De Rham complex.* This volume.
[4] D. Arnold and R. Winther, *Mixed finite elements for elasticity*, Numer. Math., 42, pp. 401–419, 2002.
[5] V.I. Arnold, *Mathematical methods of classical mechanics*, Springer-Verlag, 1989.
[6] A. Aziz (Editor), *The Mathematical Foundations of the Finite Element Method with Applications to Partial Differential Equations*, Academic Press, 1972.
[7] P. Bochev and M. Gunzburger, *On least-squares finite elements for the Poisson equation and their connection to the Kelvin and Dirichlet principles*, SIAM J. Num. Anal., Vol. 43/1, pp. 340–362, 2005.
[8] P. Bochev and M. Gunzburger, *Compatible discretizations of second-order elliptic problems*, to appear in Zapiski POMI, St. Petersburg Branch of the Steklov Institute of Mathematics, St. Petersburg, Russia, 2005.
[9] P. Bochev and M. Gunzburger *Locally conservative least-squares methods for Darcy flows*, submitted to Comp. Meth. Mech. Engrg.
[10] P. Bochev and A. Robinson, *Matching algorithms with physics: exact sequences of finite element spaces*, in *Collected Lectures on the Preservation of Stability Under Discretization,* D. Estep and S. Tavener, Eds., SIAM, Philadelphia, 2001, pp. 145–165.
[11] P. Bochev and M. Shashkov, *Constrained Interpolation (remap) of Divergence-free Fields*, Computer Methods in Applied Mechanics and Engineering, 194 (2005), pp. 511–530.

[12] M. Berndt, K. Lipnikov, J. Moulton, and M. Shashkov, *Convergence of mimetic finite difference discretizations of the diffusion equation*, East-West Journal on Numerical Mathematics, Vol. 9/4, pp. 253–316, 2001.

[13] M. Berndt, K. Lipnikov, M. Shashkov, M. Wheeler, and I. Yotov, *Superconvergence of the Velocity in Mimetic Finite Difference Methods on Quadrilaterals*, Los Alamos National Laboratory report LA-UR-03-7904, 2003.

[14] A. Bossavit, *A rationale for "edge-elements" in 3-D fields computations*, IEEE Trans. Mag. 24, pp. 74–79, 1988.

[15] A. Bossavit, *Computational Electromagnetism*, Academic, 1998.

[16] A. Bossavit and J. Verite, *A mixed fem-biem method to solve 3-d eddy current problems*, IEEE Trans. Mag. 18, pp. 431–435, 1982.

[17] F.H. Branin, Jr., *The algebraic topological basis for network analogies and the vector calculus*, IBM Technical Report, TROO, 1495, Poughkeepsie, NY, 1966.

[18] F. Brezzi and M. Fortin, *Mixed and Hybrid Finite Element methods*, Springer-Verlag, 1991.

[19] F. Brezzi, K. Lipnikov, and M. Shashkov, *Convergence of Mimetic Finite Difference Method for Diffusion Problems on Polyhedral Meshes*, Los Alamos National Laboratories Report LA-UR-04-5756, 2004.

[20] F. Brezzi, D. Marini, I. Perugia, P. Di Barba, and A. Savini, *A novel-field-based mixed formulation of magnetostatics*, IEEE Trans. Magnetics, 32/3, May 1996, pp. 635–638.

[21] S.S. Cairns, *Introductory topology*, Ronald Press Co., New York, 1961.

[22] L. Demkowicz, P. Monk, L. Vardapetyan, and W. Rachowicz, *De Rham Diagram for hp-finite element spaces*, TICAM Report 99-06, TICAM, University of Texas, Austin, 1999.

[23] A. Dezin, *Multidimensional analysis and discrete models*, CRC Presss, Boca Raton, 1995.

[24] J. Dodziuk, *Finite-difference approach to the Hodge theory of harmonic forms*, American Journal of Mathematics, 98/1, pp. 79–104, 1973.

[25] B. Eckman, *Harmonische funktionen und randvertanfagaben in einem complex*, Commentarii Math. Helvetici, 17 (1944-45), pp. 240–245.

[26] G. Fix, M. Gunzburger, and R. Nicolaides, *On mixed finite element methods for first-order elliptic systems*, Numer. Math., 37, pp. 29–48, 1981.

[27] H. Flanders, *Differential forms with applications to the physical sciences*, Dover Publications, New York, 1989

[28] G. Forsythe and W. Wasow, *Finite difference methods for partial differential equations*, Wiley, New York, 1960.

[29] R. Hiptmair, *Canonical construction of finite element spaces*, Math. Comp. 68, pp. 1325–1346, 1999.

[30] R. Hiptmair, *Discrete Hodge operators*, Numer. Math. 90, pp. 265–289, 2001.

[31] J.M. Hyman and J.C. Scovel, *Deriving mimetic difference approximations to differential operators using algebraic topology*, Los Alamos National Laboratory, unpublished report, 1988.

[32] J. Hyman, R. Knapp, and J. Scovel, *High-order finite volume approximations of differential operators*, Physica D 60, pp. 112–138, 1992.

[33] J. Hyman and M. Shashkov, *Natural discretizations for the divergence, gradient and curl on logically rectangular grids*, Comput. Math. Appl. 33, pp. 88–104, 1997.

[34] J. Hyman and M. Shashkov, *Adjoint operators for the natural discretizations of the divergence, gradient and curl on logically rectangular grids*, Appl. Num. Math. 25, pp. 413–442, 1997.

[35] J. Hyman and M. Shashkov, *The orthogonal decomposition theorems for mimetic finite difference schemes*, SIAM J. Num. Anal. 36, pp. 788–818, 1999.

[36] J. Hyman and M. Shashkov, *Mimetic Discretizations for Maxwell's Equations*, Journal of Computational Physics, 151, pp. 881–909, 1999.

[37] P.R. Kotiuga, *Hodge Decomposition and Computational Electromagnetics*, Thesis, Department of electrical engineering, McGill University, Montreal, 1984.

[38] Y. Kuznetsov, K. Lipnikiov, and M. Shashkov, *Mimetic Finite-Difference Method on Polygonal Meshes*, Los Alamos National Laboratory report LA-UR-03-7608, 2003.
[39] C. Mattiussi, *An analysis of finite volume, finite element and finite difference methods using some concepts from algebraic topology*, J. Comp. Phys. 133, pp. 289–309, 1997.
[40] J. Nedelec, *Mixed finite elements in* $\mathbb{R}^3$, Numer. Math., 35, pp. 315–341, 1980.
[41] J. Nedelec, *A new family of finite element methods in* $\mathbb{R}^3$, Numer. Math., 50, pp. 57–81, 1986.
[42] R. Nicolaides, *Direct discretization of planar div-curl problems*, SIAM J. Numer. Anal. 29, pp. 32–56, 1992.
[43] R. Nicolaides, *The covolume approach to computing incompressible flows*, in Incompressible fluid dynamics, Trends and advances, M. Gunzburger and R. Nicolaides, Eds., Cambridge University press, pp. 295–334, 1993.
[44] R. Nicolaides and X. Wu, *Covolume solutions of three-dimensional div-curl equations*, SIAM J. Num. Anal. 34, pp. 2195–2203, 1997.
[45] R. Nicolaides and K. Trapp, *Co-volume discretization of differential forms.* This volume.
[46] Raviart P.A. and Thomas J.M., *A mixed finite element method for second order elliptic problems*, Mathematical aspects of the finite element method, I. Galligani, E. Magenes, Eds. Lecture Notes in Math. 606, Springer-Verlag, New York 1977.
[47] W. Schwalm, B. Moritz, M. Giona, and M. Schwalm, *Vector difference calculus for physical lattice models*, Physical Review E, 59/1, pp. 1217–1233, 1999.
[48] A. Samarskii, V. Tishkin, A. Favorskii, and M. Shashkov, *Operational finite difference schemes*, Differential Equations 17, p. 854, 1981.
[49] R. Schumann and T. Weiland, *Stability of the FDTD algorithm on nonorthogonal grids related to the spatial interpolation scheme*, IEEE Trans. Magnetics, 34/5, pp. 2751–2754, September 1998.
[50] S. Steinberg and M. Shashkov, *Support-operators finite difference algorithms for general elliptic problems*, J. Comp. Phys. 118, pp. 131–151, 1995.
[51] M. Shashkov, *Conservative finite difference methods on general grids*, CRC Press, Boca Raton, FL, 1996.
[52] T. Tarhasaari, L. Kettunen, and A. Bossavit, *Some realizations of a discrete Hodge operator: a reinterpretation of finite element techniques*, IEEE Transactions on Magnetics, 35/3, 1999.
[53] F.L. Teixeira (Editor); *Geometric Methods for Computational Electromagnetics*, PIER32, EMW Publishing, Cambridge, MA, 2001.
[54] F.L. Teixeira and W.C. Chew, *Lattice electromagnetic theory from a topological viewpoint*, J. Math. Phys. 40/1, pp. 169–187, 1999.
[55] V.F. Tishkin, A.P. Favorskii, and M.Y. Shashkov, *Using topological methods to construct discrete models* (in Russian), Preprint 96, Institute of Applied Mathematics of the USSR Academy of Sciences, 1983.
[56] E. Tonti, *On the mathematical structure of a large class of physical theories*, Lincei, Rend. Sc. Fis. Mat. e Nat. 52 pp. 51–56, 1972.
[57] E. Tonti, *The algebraic-topological structure of physical theories*, Proc. Conf. on Symmetry, Similarity and Group Theoretic Meth. in Mech., Calgari, Canada, 1974, pp. 441–467.
[58] K. Trapp, *A class of compatible discretizations with applications to Div-Curl systems*, PhD Thesis, Carnegie Mellon University, 2004.
[59] H. Weyl, *Repartition de corriente et uno red conductoru*, Revista Matematica Hispano-Americana 5, pp. 153–164, 1923.
[60] H. Whitney, Geometric Integration Theory, Princeton University Press, 1957.
[61] K.S. Yee; *Numerical solution of initial boundary value problems involving Maxwell's equations in isotropic media*, IEEE Trans. Ant. Propa. 14, 1966, pp. 302–307.

COMPATIBLE DISCRETIZATIONS FOR EIGENVALUE PROBLEMS

DANIELE BOFFI*

Abstract. The choice of discrete spaces for a variationally posed symmetric and compact eigenvalue problem corresponding a source problem is discussed. Any standard Galerkin discretization space that is convergent for the source problem automatically performs well for the eigenvalue problem. On the other hand, mixed discretizations that are convergent (satisfying the classical Brezzi conditions) exhibit spurious low frequency eigenmodes. Examples of discretizations with spurious modes are presented. Moreover, necessary and sufficient conditions on mixed discretization are established for the (ordered) discrete eigenvalues to converge to the corresponding continuous eigenvalues. The theory is applied to the determination of band gaps for photonic crystals and evolution problems.

Key words. Mixed finite element, spurious eigenvalues, Maxwell's eigenvalues, photonic crystals, de Rham complex.

AMS(MOS) subject classifications. Primary 65N25, 65N30, 78M10, 65M60.

Table of contents.

1. Introduction. We present in a general setting different examples of finite element discretizations of eigenvalue problems for partial differential equations (PDEs).

An adequate approximation of the eigensolutions for an elliptic self-adjoint PDE is obtained automatically from any Galerkin scheme that provides a convergent approximation of the corresponding source problem (see Section 2).

On the other hand, the use of *mixed* finite element schemes for the approximation of the eigensolution for an elliptic self-adjoint PDE must

*Dipartimento di Matematica, Università di Pavia, 27100 Pavia, Italy.

be handled with particular care. In particular, in Section 3 we present a discrete scheme which provides a convergent approximation for the source problem (satisfying the classical Brezzi conditions, see [13]), but which exhibits spurious eigensolutions.

In this paper we review the basics of the theory for standard Galerkin and mixed finite element approximations and apply the latter to photonic band gaps computation and evolution problems.

2. Elliptic PDEs, eigenvalue problems, and their Galerkin discretizations. We assume that the reader has a working knowledge of finite element methods on the level of [13]. Issues associated with eigenvalue problems are described in some detail: Rayleigh quotients (2.9), discrete resolvent (2.13), gap (2.22), and the Definition of eigenmode convergence. Proposition 2.1 gives sufficient conditions for eigenmode convergence. The aim of this presentation is to give a short overview of the subject, having in mind the major differences appearing when considering the theory of eigenvalue problems in mixed form, which will be described in the next section. For this reason, we restrict ourselves to symmetric problems, for which the theory is easier to present in general, and more complete in the case of mixed approximations. Let H be a Hilbert space and V a closed subspace of H. Let $a : V \times V \longrightarrow \mathbb{R}$ and $b : H \times H \longrightarrow \mathbb{R}$ be two symmetric and continuous bilinear forms, and consider the problem: find $\lambda \in \mathbb{R}$ such that there exists $u \in V$, with $u \neq 0$ satisfying

$$a(u,v) = \lambda b(u,v) \quad \forall v \in V. \tag{2.1}$$

We suppose that a is V-elliptic, namely there exists $\alpha > 0$ such that

$$a(v,v) \geq \alpha \|v\|_V^2. \tag{2.2}$$

For the sake of simplicity, we also assume that b defines a scalar product in H (indeed, in many applications, b turns out to be the standard inner product of H)

EXAMPLE 1. A basic example the reader should have in mind when reading this section, is the Laplace-Poisson eigenproblem: find $\lambda \in \mathbb{R}$ such that there exists u with $u \neq 0$ satisfying

$$\begin{aligned} -\Delta u &= \lambda u \qquad \text{in } \Omega, \\ u &= 0 \qquad \text{on } \partial\Omega. \end{aligned} \tag{2.3}$$

This problem fits within the framework of (2.1) with the choices $H = L^2(\Omega)$, $V = H_0^1(\Omega)$, and

$$\begin{aligned} a(u,v) &= \int_\Omega \operatorname{grad} u \cdot \operatorname{grad} v \, dx, \\ b(u,v) &= \int_\Omega uv \, dx. \end{aligned} \tag{2.4}$$

From the ellipticity hypothesis (2.2), it follows that the source problem corresponding to (2.1) is uniquely solvable, so that it is possible to define the resolvent operator $T : H \to H$ which, given $f \in H$, satisfies

$$\begin{aligned} &Tf \in V \\ &a(Tf, v) = b(f, v) \quad \forall v \in V. \end{aligned} \tag{2.5}$$

Clearly, the eigenmodes of (2.1) are the same as those associated with the inverse of T. We make the assumption

$$T : H \to H \text{ is a compact operator.} \tag{2.6}$$

In practice, the compactness of T is often the consequence of a compact embedding of V into H.

REMARK 2.1. The ellipticity hypothesis (2.2) can be relaxed by assuming that the form $a(\cdot,\cdot) + \mu b(\cdot,\cdot)$ is elliptic for a suitable positive constant μ. In such case it is possible to define the resolvent operator associated with (2.1) by a standard shift procedure and the results of this section apply to the shifted operator with the natural modifications.

We denote the eigenvalues of (2.1) by λ_i, $i \in \mathbb{N}$, with the natural numbering

$$\lambda_1 \le \lambda_2 \le \cdots \le \lambda_i \le \dots, \tag{2.7}$$

where the same eigenvalue may be repeated several times according to its multiplicity. The corresponding eigenfunctions are denoted by u_i, with the usual normalization $b(u_i, u_i) = 1$, and the associated eigenspaces are $E_i = \operatorname{span}(u_i)$, $i \in \mathbb{N}$, so that we have $V = \oplus_i E_i$. The following orthogonalities are also well known (when $\lambda_i \ne \lambda_j$ they follow from equation (2.1), otherwise we include them in our definition of u_i and u_j)

$$a(u_i, u_j) = b(u_i, u_j) = 0 \quad \text{if } i \ne j. \tag{2.8}$$

A useful representation of eigenvaules and eigenvectors is given in terms of the Rayleigh quotient, namely:

$$R(v) = \frac{a(v,v)}{b(v,v)} : V \setminus \{0\} \to \mathbb{R}. \tag{2.9}$$

We have

$$\lambda_1 = \min_{v \in V} R(v), \qquad u_1 = \arg\min_{v \in V} R(v), \tag{2.10}$$

$$\lambda_i = \min_{v \in \left(\bigoplus_{k=1}^{i-1} E_k\right)^\perp} R(v), \quad u_i = \arg \min_{v \in \left(\bigoplus_{k=1}^{i-1} E_k\right)^\perp} R(v), \quad i > 1. \tag{2.11}$$

A natural way of discretizing problem (2.1) is to introduce a sequence V_h of finite dimensional subspaces of V and to consider the following discrete problem: find $\lambda_h \in \mathbb{R}$ such that there exists $u_h \in V_h$, with $u_h \neq 0$ satisfying

$$a(u_h, v) = \lambda_h(u_h, v) \quad \forall v \in V_h. \tag{2.12}$$

It is beyond the scope of this presentation to discuss the numerical solution of (2.12), which is a generalized algebraic eigenvalue problem of the form $Ax = \nu Mx$. From the ellipticity hypothesis (2.2) it is possible to define a discrete resolvent operator $T_h : H \to H$ in a similar way as for the continuous problem, namely

$$\begin{aligned} &T_h f \in V_h \\ &a(T_h f, v) = b(f, v) \quad \forall v_h \in V_h. \end{aligned} \tag{2.13}$$

Since V_h is finite dimensional, the operator T_h is compact. We shall use the natural notation for the discrete eigenvalues:

$$\lambda_{1,h} \leq \lambda_{2,h} \leq \cdots \leq \lambda_{N(h),h}, \tag{2.14}$$

where $N(h)$ is the dimension of V_h. The corresponding discrete eigenfunctions will be denoted by $u_{i,h}$ (with the normalization $b(u_{i,h}, u_{i,h}) = 1$ and orthogonalities analogous to (2.8)) and the eigenspaces by $E_{i,h} = \operatorname{span}(u_{i,h})$, so that $V_h = \oplus_i E_{i,h}$.

The representations given in (2.10) and (2.11) apply to the discrete eigenmodes as follows

$$\lambda_{1,h} = \min_{v_h \in V_h} R(v_h), \qquad u_{1,h} = \arg\min_{v_h \in V_h} R(v_h)u, \tag{2.15}$$

$$\lambda_{i,h} = \min_{v_h \in \left(\bigoplus_{k=1}^{i-1} E_{k,h}\right)^{\perp}} R(v_h), \quad u_{i,h} = \arg \min_{v_h \in \left(\bigoplus_{k=1}^{i-1} E_{k,h}\right)^{\perp}} R(v_h), \quad i > 1. \tag{2.16}$$

From (2.10) and (2.15), since $V_h \subset V$, we immediately obtain the property

$$\lambda_{1,h} \geq \lambda_1. \tag{2.17}$$

Indeed, the last inequality generalizes to the important result that all eigenvalues are approximated from above. In order to obtain this result, we need a suitable modification of (2.11) and (2.16). The following alternative representation of the eigenvalues holds true

$$\lambda_i = \min_{E \in V_i} \max_{v \in E} R(v), \tag{2.18}$$

where V_i denotes the set of all subspaces V with dimension equal to i. For the reader's convenience, we give a sketch of the proof of this result (taking for granted the better known representation (2.11)). First, we show that $\lambda_i \geq \min_{E\in V_i} \max_{v\in E} R(v)$; let us take $E = \bigoplus_{k=1}^{i} E_k$ and a generic $v = \sum_{k=1}^{i} \alpha_k u_k$. Then it is easily seen that $R(v) \leq \lambda_i$ by using the orthogonalities in (2.8). Since we have to take the minimum among all possible $E \in V_i$, we have the desired inequality. The proof of the opposite inequality gives also the additional information that the minimum in (2.18) is attained for $E = \bigoplus_{k=1}^{i} E_k$ with the choice $v = u_i$. Indeed, if $E = \bigoplus_{k=1}^{i} E_k$ then it is clear that the optimal choice for maximizing $R(v)$ is $v = u_i$. On the other hand, if $E \neq \bigoplus_{k=1}^{i} E_k$ then there exists $v \in E$ with v orthogonal to u_k for all $k \leq i$ and hence $R(v) \geq \lambda_i$, which shows that $E = \bigoplus_{k=1}^{i} E_k$ is an optimal choice for the minimum in (2.18).

Inequality (2.18) has a discrete counterpart which reads

$$\lambda_{i,h} = \min_{E\in V_{i,h}} \max_{v\in E} R(v), \tag{2.19}$$

where $V_{i,h}$ denotes the set of all subspaces of V_h having dimension equal to i. It is now clear that, since the minimum in (2.19) is taken over a smaller set than in (2.18), we have the general result that

$$\lambda_{i,h} \geq \lambda_i \quad \forall i. \tag{2.20}$$

The monotonicity property stated in (2.20), which is an important result by itself, does not answer, however, the question of the convergence of the discrete eigenmodes towards the continuous ones. First, let us define what we mean by convergence. Following [7], we introduce a map $m : \mathbb{N} \to \mathbb{N}$ which associates to every N the dimension of the space generated by the eigenspaces of the first N *distinct* eigenvalues, that is

$$\begin{aligned} m(1) &= \dim\Big\{ \bigoplus_i E_i : \lambda_i = \lambda_1 \Big\}, \\ m(N+1) &= m(N) + \dim\Big\{ \bigoplus_i : \lambda_i = \lambda_{m(N)+1} \Big\} \end{aligned} \tag{2.21}$$

With this notation, $\lambda_{m(1)}, \ldots, \lambda_{m(N)}$ are the first N distinct eigenvalues. A standard notion when dealing with the approximation of eigenspaces, is the following definition of *gap* $\hat{\delta}(E, F)$ between two subspaces E and F of a Hilbert space H

$$\begin{aligned} \delta(E,F) &= \sup_{u\in E,\ \|u\|_H=1} \inf_{v\in F} \|u-v\|_H, \\ \hat{\delta}(E,F) &= \max(\delta(E,F), \delta(F,E)). \end{aligned} \tag{2.22}$$

DEFINITION 2.1. *We say that the discrete eigenmodes* $\{(\lambda_{i,h}, u_{i,h})\}$ converge *to the continuous ones* $\{(\lambda_i, u_i)\}$ *if, given* $\varepsilon > 0$, *for any* $N \in \mathbb{N}$

there exists $h_0 > 0$ such that for all $h < h_0$ we have

$$\max_{i=1,\dots,m(N)} |\lambda_i - \lambda_{i,h}| \le \varepsilon,$$
$$\hat{\delta}\left(\bigoplus_{i=1}^{m(N)} E_i, \bigoplus_{i=1}^{m(N)} E_{i,h}\right) \le \varepsilon. \tag{2.23}$$

We explicitly observe that Definition 2.1, besides the convergence of the eigenmodes, contains also the information that no spurious eigenvalues pollute the spectrum; i.e., 1) each continuous eigenvalue is approximated by a number of discrete eigenvalues (counted with their multiplicity) which corresponds exactly to its multiplicity and 2) each discrete eigenvalue approximates a continuous one.

We also remark that (2.23) does not give any information on the order of convergence for eigenvalues and eigenvectors; this issue will be considered later on in this section.

A sufficient condition for convergence of the discrete eigenmodes (see Chapter IV of [21]) is that the the sequence $\{T_h\}$ converges to T uniformly in the operator norm $\mathcal{L}(H)$, namely

$$\|T - T_h\|_{\mathcal{L}(H)} \to 0, \quad \text{as } h \to 0, \tag{2.24}$$

or, equivalently,

$$\|Tf - T_h f\|_H \le C\rho(h)\|f\|_H \quad \forall f \in H, \tag{2.25}$$

with $\rho(h)$ tending to zero as h goes to zero. Indeed, it turns out that the convergence in norm (2.24) is *equivalent* to the eigenmodes convergence (2.23). The interested reader is referred to [7] for a proof of the necessity of condition (2.24).

A simple way of estimating the norm of the difference $T - T_h$ is the use of standard a priori estimates (when they are available). For instance, considering the model example of Laplace eigenproblem, the choice of continuous piecewise linear functions on a triangular mesh for V_h gives the estimate

$$\|(T - T_h)f\|_H \le Ch^2\|f\|_H \tag{2.26}$$

when the domain is convex or smooth, and (2.26) clearly implies (2.24).

We now present an alternative way of estimating the norm of the difference $T - T_h$, which will give us some interesting hints in order to study eigenvalue problems in mixed form in the next section. We observe, that, by Galerkin orthogonality, we have $T_h = P_h T$, where $P_h : V \to V_h$ is the linear projection with respect to the bilinear form a. Hence, we have $T - T_h = (I - P_h)T$, I being the identity operator, and the following proposition can be used to prove the convergence in norm (2.24).

PROPOSITION 2.1. *Let us suppose that, for any $u \in V$,*

$$\lim_{h\to 0} \|u - P_h u\|_H = 0, \tag{2.27}$$

that is, $I - P_h$ converges pointwise *to zero. Suppose, moreover, that T is compact from H to V. Then it follows that the convergence in norm* (2.24) *holds true.*

Proof. First we show that the sequence $\{\|(I-P_h)\|_{\mathcal{L}(V,H)}\}$ is bounded. Define $c(h,u)$ by $\|(I-P_h)u\|_H = c(h,u)\|u\|_V$. Pointwise convergence means that for each u, there holds $c(h,u) \to 0$. Thus $M(u) = \max_h c(h,u)$ is finite. By the uniform boundedness principle (or Banach–Steinhaus theorem, see [24], p. 196), there exists C such that for all h there holds $\|(I-P_h)\|_{\mathcal{L}(V,H)} \le C$.

Consider some $\{f_h\}$ such that for each h, $\|f_h\|_H = 1$ and $\|T - T_h\|_{\mathcal{L}(H)} = \|Tf_h - T_h f_h\|_H$. Since $\{f_h\}$ is bounded in H and T is compact from H to V, a subsequence $\{Tf_h\}$ (using the same notation as for the sequence) has limit $Tf_h \to w$ in V. We claim that $(I-P_h)Tf_h \to 0$ for the subsequence, and hence for the sequence itself. T is a closed operator: there exists v in H such that $T(v) = w$. By hypothesis $T_h v \to w$. Furthermore $T_h v = P_h T v = P_h w$. The statement of pointwise convergence of both $Tf_h \to w$ and $T_h v \to w$ implies that for any $\varepsilon > 0$, there exists h small enough such that

$$C\|Tf_h - w\| + \|(I-P_h)w\| \le \varepsilon. \tag{2.28}$$

The triangle inequality, the boundedness of $\{\|I-P_h\|\}$ and Equation (2.28) imply that

$$\begin{aligned}\|(I-P_h)Tf_h\| \le &\|(I-P_h)(Tf_h - w)\| + \|(I-P_h)w\| \le \\ &C\|(Tf_h - w)\| + \|(I-P_h)w\| \le \varepsilon.\end{aligned}$$

□

REMARK 2.2. The hypothesis on T can be relaxed by assuming it is compact from V into itself, provided a stronger pointwise convergence than (2.27) is assumed. Namely, the two conditions

$$\begin{aligned}&T \text{ compact from } V \text{ to } V,\\ &\lim_{h\to 0} \|u - P_h u\|_V = 0, \quad \forall u \in V\end{aligned} \tag{2.29}$$

imply (with a similar proof as in Proposition 2.1) the uniform convergence

$$\|T - T_h\|_{\mathcal{L}(V)} \to 0, \quad \text{as } h \to 0. \tag{2.30}$$

It must be noticed that (2.30), which differs from (2.24) for the space in which the norms are evaluated, is equivalent to a type of convergence of eigenvalues/eigenfunctions analogous to (2.23).

We now consider the rate of convergence of eigenvalues/eigenmodes, under the hypothesis that the convergence in norm (2.30) is satisfied. We mainly refer to [3] (where more general estimates can be found for the cases when (2.24) is satisfied and for nonsymmetric problems) and to the references therein. The fundamental estimates (which basically relate the rate of convergence of eigenmodes to the behavior of the convergence of T_h to T) can be stated as follows. Let λ be an eigenvalue of (2.1) and denote by λ_h the average of the discrete eigenvalues converging to it; namely, if $\lambda = \lambda_i$ is the k-th distinct eigenvalue, then $\lambda_h = (1/\mu)\sum_{j=m(k-1)+1}^{m(k)} \lambda_{j,h}$ where $\mu = m(k) - m(k-1)$ is the multiplicity of λ. Moreover, let E denote the μ dimensional eigenspace associated with λ and E_h the direct sum of the μ discrete eigenspaces associated with λ_h. Then there exists C such that

$$
\begin{aligned}
&|\lambda - \lambda_h| \le C\epsilon_h^2 \\
&\hat{\delta}(E, E_h) \le C\epsilon_h,
\end{aligned}
\tag{2.31}
$$

where

$$
\epsilon_h = \sup_{u\in E,\ \|u\|=1} \inf_{v_h\in V_h} \|u - v_h\|_V.
$$

If the eigenfunctions in E all are in $H^\sigma(\Omega)$ and if V_h contains piecewise linear functions, then $\epsilon_h = Ch^t$ for $t = \min(\sigma, 1)$.

3. Mixed discretizations. In this section we generalize the results of the previous section to the case of eigenvalue problems in mixed form. In particular we shall see that the two fundamental properties for the well-posedness of a source mixed problem are *neither sufficient nor necessary* for the eigenmodes convergence.

In order to make things simpler and to emphasize the differences of this case from the setting of previous section, we postpone the presentation of the general abstract results until after we introduce and study the basic example of Laplace eigenproblem in mixed form.

3.1. The eigenvalue problem for the Laplacian. There are two types of finite element element methods for Laplace's equation, $\operatorname{div}\operatorname{grad} u = g$, displacement type methods that explixitly enforce $\boldsymbol{\sigma} = \operatorname{grad} u$ and equilibrium type methods that explicitly enforce $\operatorname{div}\boldsymbol{\sigma} = g$. Equilibrium-type discretizations are applied for example in Darcy flow problems, and take the following form. Given $g \in L^2(\Omega)$, find $(\boldsymbol{\sigma}, u)$ in $\Sigma \times V = H(\operatorname{div};\Omega) \times L^2(\Omega)$ that solves the saddle point problem

$$
\begin{cases}
(\boldsymbol{\sigma}, \boldsymbol{\tau}) + (\operatorname{div}\boldsymbol{\tau}, u) = 0 & \forall \boldsymbol{\tau} \in \Sigma,\\
(\operatorname{div}\boldsymbol{\sigma}, v) = -(g, v) & \forall v \in V.
\end{cases}
\tag{3.1}
$$

The eigenvalue problem associated with (3.1) is: find $\lambda \in \mathbb{R}$ such that there exists $u \in V$, with $u \neq 0$ satisfying

$$
\begin{cases}
(\boldsymbol{\sigma}, \boldsymbol{\tau}) + (\operatorname{div}\boldsymbol{\tau}, u) = 0 & \forall \boldsymbol{\tau} \in \Sigma,\\
(\operatorname{div}\boldsymbol{\sigma}, v) = -\lambda(u, v) & \forall v \in V
\end{cases}
\tag{3.2}
$$

for some $\boldsymbol{\sigma} \in \Sigma$.

Let us suppose that we are given a sequence of pairs of finite element spaces $\{\Sigma_h, V_h\}$ for which (see [13]) the ellipticity in the discrete kernel condition holds

$$\begin{aligned} &(\boldsymbol{\tau}, \boldsymbol{\tau}) \geq \alpha \|\boldsymbol{\tau}\|_V^2 \qquad \forall \boldsymbol{\tau} \in K_h, \\ &K_h = \{\boldsymbol{\tau} \in \Sigma_h : (\operatorname{div} \boldsymbol{\tau}, v) = 0 \ \forall v \in V_h\}, \end{aligned} \tag{3.3}$$

and the inf-sup condition holds

$$\inf_{v \in V_h} \sup_{\boldsymbol{\tau} \in \Sigma_h} \frac{(\operatorname{div} \boldsymbol{\tau}, v)}{\|\boldsymbol{\tau}\|_\Sigma \|v\|_V} \geq \beta. \tag{3.4}$$

This implies the well posedness of the discrete problem: find $(\boldsymbol{\sigma}_h, u_h) \in \Sigma_h \times U_h$ such that

$$\begin{cases} (\boldsymbol{\sigma}_h, \boldsymbol{\tau}) + (\operatorname{div} \boldsymbol{\tau}, u_h) = 0 & \forall \boldsymbol{\tau} \in \Sigma_h, \\ (\operatorname{div} \boldsymbol{\sigma}_h, v) = -(g, v) & \forall v \in V_h \end{cases} \tag{3.5}$$

and the error estimate

$$\|\boldsymbol{\sigma} - \boldsymbol{\sigma}_h\|_{\mathrm{div}} + \|u - u_h\|_0 \leq C \inf_{\substack{\boldsymbol{\tau}_h \in \Sigma_h \\ v_h \in V_h}} \left(\|\boldsymbol{\sigma} - \boldsymbol{\tau}_h\|_{\mathrm{div}} + \|u - v_h\|_0 \right). \tag{3.6}$$

Given discrete spaces Σ_h and V_h, in the discretized eigenvalue problem, the left-most λ_h are sought such that there exists u_h in $V_h \setminus \{0\}$ satisfying

$$\begin{cases} (\boldsymbol{\sigma}_h, \boldsymbol{\tau}) + (\operatorname{div} \boldsymbol{\tau}, u_h) = 0 & \forall \boldsymbol{\tau} \in \Sigma_h, \\ (\operatorname{div} \boldsymbol{\sigma}_h, v) = -\lambda_h(u_h, v) & \forall v \in V_h \end{cases} \tag{3.7}$$

for some $\boldsymbol{\sigma}_h \in \Sigma_h$. From the practical point of view, problem (3.7) is a generalized algebraic eigenvalue problem of the form

$$\begin{pmatrix} A & B^t \\ B & 0 \end{pmatrix} \begin{pmatrix} x \\ y \end{pmatrix} = -\nu \begin{pmatrix} 0 & 0 \\ 0 & M \end{pmatrix} \begin{pmatrix} x \\ y \end{pmatrix} \tag{3.8}$$

where the matrix M is positive definite.

In order to generalize what has been described in the previous section, let us try first to define a suitable *compact* operator (both at continuous and discrete level) so that we can apply Proposition 2.1.

The first naïve attempt is to consider an operator $T_{\Sigma V}$ (where the notation is chosen in order to point out that its definition involves both the spaces of the variational formulation) from $L^2(\Omega) \times L^2(\Omega)$ into itself defined as $T_{\Sigma V}(f_1, f_2) = (\sigma, u)$, where (σ, u) is the solution to (3.1) with datum $g = f_2$. From standard regularity results it turns out that

$$T_{\Sigma V} : L^2(\Omega) \times L^2(\Omega) \to L^2(\Omega) \times L^2(\Omega) \text{ is a compact operator.} \tag{3.9}$$

(For instance, when Ω is for instance convex, we have $\boldsymbol{\sigma} \in H^1(\Omega)$ and $u \in H^2(\Omega)$).

At discrete level, we introduce, in a similar way, a discrete operator $T_{\Sigma V,h}$ from $L^2(\Omega)\times L^2(\Omega)$ into $\Sigma_h \times V_h \subset L^2(\Omega)\times L^2(\Omega)$ as $T_{\Sigma V,h}(f_1, f_2) = (\boldsymbol{\sigma}_h, u_h)$, where $(\boldsymbol{\sigma}_h, u_h)$ is the solution to (3.5) with datum $g = f_2$. Hence, the main question is whether we can prove that $\{T_{\Sigma V,h}\}$ converges to $T_{\Sigma V}$ in a suitable operator norm.

Trying to mimic the theory of the previous section, it is not difficult to see that also in this case we can interpret the discrete operator $T_{\Sigma V,h}$ as the action of a projection Q_h on $T_{\Sigma V}$ and that, thanks to (3.6), we have the pointwise convergence

$$\|(I - Q_h)(\boldsymbol{\sigma}, u)\|_{\Sigma\times V} \to 0 \quad \forall(\boldsymbol{\sigma}, u) \in \Sigma \times U \tag{3.10}$$

whenever the spaces Σ_h and U_h are chosen in a reasonable way (stability properties and approximation capabilities). In order to apply the arguments of Proposition 2.1, we need a stronger compactness than (3.9), namely we need $T_{\Sigma U}$ compact from $L^2(\Omega) \times L^2(\Omega)$ to $\Sigma \times U$. Unfortunately, such compactness is not fulfilled; indeed it can be easily seen that, using the notation $T_{\Sigma U}(f_1, f_2) = (\boldsymbol{\sigma}, u)$ with $(f_1, f_2) \in L^2(\Omega) \times L^2(\Omega)$, we have $\operatorname{div} \boldsymbol{\sigma} = -f_2$ which prevents $\boldsymbol{\sigma}$ from being in a compact subset of Σ.

If we try to relax the required compactness, as explained in Remark 2.2, we need $T_{\Sigma U}$ compact from $\Sigma \times U$ into itself, which is not true either (by the same argument as above).

Since Proposition 2.1 and Remark 2.2 fail to prove the uniform convergence, we can try to use a direct approach using the a priori estimate (3.6) as we did, for instance, in (2.26). We consider $g \in L^2(\Omega)$ in (3.5). Taking advantage of the regularity of u, we have that the term $\|u - v_h\|_0$ can easily be bounded in a uniform way (in terms of powers of h). On the other hand, we cannot bound uniformly the term $\|\boldsymbol{\sigma} - \boldsymbol{\tau}_h\|_{\operatorname{div}}$ since we do not have any extra regularity for $\operatorname{div} \boldsymbol{\sigma}$ which equals the negative of g and hence is only in $L^2(\Omega)$.

In conclusion, it turns out that the techniques of the previous section cannot be applied directly to the setting of mixed eigenproblems.

The previous example presents a major issue which will be made clearer in the next section. Actually, the stability conditions for source mixed problems (3.3) and (3.4) are neither necessary nor sufficient for the good approximation of the corresponding eigenvalue problem. In order to show this result, we start with the presentation of a numerical scheme which is stable and convergent for the source problem (3.5) but which presents spurious eigensolutions when applied to the eigenvalue problem (3.2).

EXAMPLE 2. We follow the presentation and the analysis of [7] (see also [18]). Let Ω be a square and consider a decomposition of Ω into subsquares which are subdivided into four triangles by their diagonals. We consider as Σ_h the space of all vectorfields whose components are continuous piecewise linear functions, and as V_h the space $\operatorname{div} \Sigma_h$, which is made

TABLE 1
Eigenvalue computation on a criss-cross mesh: spurious modes.

exact	computed			
1.00000	1.00428	1.00190	1.00107	1.00068
1.00000	1.00428	1.00190	1.00107	1.00069
2.00000	2.01711	2.00761	2.00428	2.00274
4.00000	4.06804	4.03037	4.01710	4.01095
4.00000	4.06804	4.03037	4.01710	4.01095
5.00000	5.10634	5.04748	5.02674	5.01712
5.00000	5.10634	5.04748	5.02674	5.01712
	5.92293	5.96578	5.98074	5.98767
8.00000	8.27128	8.12151	8.06845	8.04383
9.00000	9.34085	9.15309	9.08640	9.05537
9.00000	9.34085	9.15309	9.08640	9.05537
d.o.f.	254	574	1022	1598

of piecewise constant functions. As a consequence of the definition of V_h, it is trivial to check that (3.3) holds true. In [7] is has been shown that the inf-sup condition (3.4) is satisfied as well. On the other hand, the approximation of the eigenvalue problem (3.2) presents several spurious modes: Table 1 shows the first one (which apparently converges to a value close to six) and in Figure 1 the eigenfunctions corresponding to the first four spurious modes are plotted.

In the next two sections we present the theory for the approximation of eigenvalue problems in mixed form as it has been developed in [7, 6]. The theory, which splits into subsections referring to two different families of mixed eigenproblems, uses a suitable definition of the resolvent operator which will enjoy better compactness property than $T_{\Sigma V}$.

3.2. Equilibrium-type eigenvalue problems. Let V, H, and Q be Hilbert spaces satisfying the relation

$$V \subset H \simeq H' \subset V', \tag{3.11}$$

where H' and V' denote the dual spaces of H and V. Given a bilinear and continuous symmetric form $a : V \times V \to \mathbb{R}$ and a bilinear and continuous form $b : V \times Q \to \mathbb{R}$, we consider the eigenvalue problem: find $\lambda \in \mathbb{R}$ such that there exists $u \in V$ with $u \neq 0$ satisfying

$$\begin{cases} a(u,v) + b(v,p) = \lambda(u,v)_H & \forall v \in V \\ b(u,q) = 0 & \forall q \in Q, \end{cases} \tag{3.12}$$

for some $p \in Q$.

REMARK 3.1. A prototype of equilibrium-type eigenproblem is the eigenvalue problem associated with Stokes system.

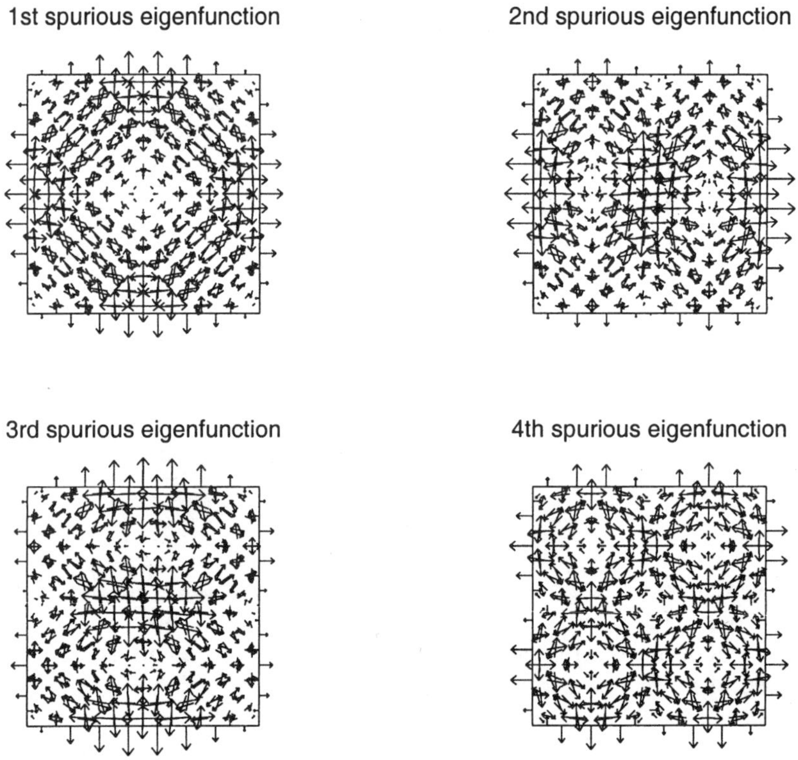

FIG. 1. *Eigenfunctions computed on a criss-cross mesh: spurious modes.*

As we have seen in Subsection 3.1, the first important step consists in the choice of the resolvent operator. The source problem associated with (3.12) is: given $f \in H$, find $(u,p) \in V \times Q$ such that

$$\begin{cases} a(u,v) + b(v,p) = (f,v)_H & \forall v \in V \\ b(u,q) = 0 & \forall q \in Q. \end{cases} \tag{3.13}$$

We suppose that a and b are such that it is possible to define $T : H \to H$, by setting $T\varphi = u$, where u solves (3.13) with $f = \varphi$. With this notation, the eigenproblem (3.12) can be written in the form

$$\lambda T u = u. \tag{3.14}$$

We assume that

$$T : H \to V \text{ is a compact operator.}$$

Given two finite element space sequences $V_h \subset V$ and $Q_h \subset Q$, the discrete counterpart of problem (3.12) is: find $\lambda_h \in \mathbb{R}$ such that there exists

$u_h \in V_h$ with $u_h \neq 0$ satisfying

$$\begin{cases} a(u_h, v) + b(v, p_h) = \lambda_h (u_h, v)_H & \forall v \in V_h \\ b(u_h, q) = 0 & \forall q \in Q_h, \end{cases} \tag{3.15}$$

for some $p_h \in Q_h$. For the definition of the discrete resolvent operator, we proceed as in (3.13) and consider the discrete source problem associated with (3.15); namely, given $f \in H$, find $(u_h, p_h) \in V_h \times Q_h$ such that

$$\begin{cases} a(u_h, v) + b(v, p_h) = (f, v)_H & \forall v \in V_h \\ b(u_h, q) = 0 & \forall q \in Q_h. \end{cases} \tag{3.16}$$

We suppose that it is possible to define $T_h : H \to H$ as $T_h\varphi = u_h$, where $u_h \in V_h$ solves (3.16) with $f = \varphi$. In analogy to (3.14), problem (3.15) can be written in the form $\lambda_h T_h u_h = u_h$.

This section will be devoted to the study of the uniform convergence of T_h to T in a suitable operator norm. We shall prove necessary and sufficient conditions for the convergence with respect to the norm of $\mathcal{L}(H, V)$.

Let $V_0^H \subset V$ and Q_0^H denote the spaces of solutions to (3.13) when f ranges over H. These spaces are endowed with their natural norms, namely

$$\begin{aligned} \|u\|_{V_0^H} &= \inf\{\|\varphi\|_H : u \text{ solves (3.13) with } f = \varphi\} \\ \|p\|_{Q_0^H} &= \inf\{\|\varphi\|_H : p \text{ solves (3.13) with } f = \varphi\}. \end{aligned} \tag{3.17}$$

The continuous and discrete kernels K and K_h are defined as

$$\begin{aligned} K &= \{v \in V : b(v, q) = 0 \ \forall q \in Q\} \\ K_h &= \{v_h \in V_h : b(v_h, q_h) = 0 \ \forall q_h \in Q_h\}. \end{aligned} \tag{3.18}$$

DEFINITION 3.1. *The* weak approximability *of Q_0^H is satisfied if there exists $\omega_1(h)$, tending to zero as h goes to zero, such that for every $p \in Q_0^H$ we have*

$$\sup_{v_h \in K_h} \frac{b(v_h, p)}{\|v_h\|_V} \leq \omega_1(h) \|p\|_{Q_0^H}. \tag{3.19}$$

REMARK 3.2. We notice explicitly that the name of the property introduced in the previous definition arises from the fact that in many cases it is related to suitable approximation of the function p. Typically (3.19) can be proved by subtracting from p a discrete function $p_h \in Q_h$ and by estimating the distance $\|p - p_h\|$.

DEFINITION 3.2. *The* strong approximability *of V_0^H is satisfied if there exists $\omega_2(h)$, tending to zero as h goes to zero, such that for every $u \in V_0^H$ there exists $u^I \in K_h$ with*

$$\|u - u^I\|_V \leq \omega_2(h) \|u\|_{V_0^H}. \tag{3.20}$$

We also the recall the *ellipticity in the discrete kernel* property that holds when there exists $\alpha > 0$ such that

$$a(v_h, v_h) \geq \alpha \|v_h\|_V^2 \quad \forall v_h \in K_h. \tag{3.21}$$

The main result concerning the convergence of eigenmodes for problem written in equilibrium form (see [6]), is stated in the next theorem.

THEOREM 3.1. *If the ellipticity in the discrete kernel* (3.21) *is satisfied, together with the weak* (3.19) *and strong* (3.20) *approximability properties, then the sequence* T_h *converges uniformly to* T *in* $\mathcal{L}(H, V)$*, that is there exists* $\omega_3(h)$*, tending to zero as* h *goes to zero, such that*

$$\|Tf - T_h f\|_V \leq \omega_3(h)\|f\|_H. \tag{3.22}$$

REMARK 3.3. In [6] it has been proved that the three hypotheses of Theorem 3.1 are necessary for the uniform convergence (3.22) under the additional condition that T_h is bounded in $\mathcal{L}(V', V)$.

3.3. Displacement-type eigenvalue problems. Let Σ, U, and H be Hilbert spaces such that the following inclusions hold true

$$U \subset H \simeq H' \subset U'. \tag{3.23}$$

Given a bilinear and continuous symmetric form $a : \Sigma \times \Sigma \to \mathbb{R}$ and a bilinear and continuous form $b : \Sigma \times U \to \mathbb{R}$, we consider the eigenvalue problem: find $\lambda \in \mathbb{R}$ such that there exists $u \in U$ with $u \neq 0$ satisfying

$$\begin{cases} a(\sigma, \tau) + b(\tau, u) = 0 & \forall \tau \in \Sigma \\ b(\sigma, v) = -\lambda(u, v)_H & \forall v \in U. \end{cases} \tag{3.24}$$

for some $\sigma \in \Sigma$.

We suppose that the form a is positive semidefinite, so that we can define the seminorm $|\tau|_a = (a(\tau, \tau))^{1/2}$ with the following estimate

$$a(\sigma, \tau) \leq |\sigma|_a |\tau|_a \quad \forall \sigma, \tau \in \Sigma. \tag{3.25}$$

The source problem associated with (3.24) reads: given $g \in U'$, find $(\sigma, u) \in \Sigma \times U$ such that

$$\begin{cases} a(\sigma, \tau) + b(\tau, u) = 0 & \forall \tau \in \Sigma \\ b(\sigma, v) = -(g, v)_H & \forall v \in U. \end{cases} \tag{3.26}$$

We suppose that problem (3.26) is well posed and the following a priori estimate holds true

$$\|\sigma\|_\Sigma + \|u\|_U \leq C\|g\|_{U'}.$$

We can then define an operator $T : H \to U$ by setting $Tg = u$, where $u \in U$ is the solution to (3.26) with datum g. We suppose that

$$T \text{ is compact in } \mathcal{L}(H, U).$$

Given the discrete space sequences $\Sigma_h \subset \Sigma$ and $U_h \subset U$, we consider the finite element approximation of our problem (3.24): find $\lambda_h \in \mathbb{R}$ such that there exists $u_h \in U$ with $u_h \neq 0$ satisfying

$$\begin{cases} a(\sigma_h, \tau) + b(\tau, u_h) = 0 & \forall \tau \in \Sigma_h \\ b(\sigma_h, v) = -\lambda_h (u_h, v)_H & \forall v \in U_h \end{cases} \tag{3.27}$$

for some $\sigma_h \in \Sigma_h$.

Given $g \in U'$, the discrete source problem reads: find $(\sigma_h, u_h) \in \Sigma_h \times U_h$ such that

$$\begin{cases} a(\sigma_h, \tau) + b(\tau, u_h) = 0 & \forall \tau \in \Sigma_h \\ b(\sigma_h, v) = -(g, v)_H & \forall v \in U_h. \end{cases} \tag{3.28}$$

We suppose that (3.28) is solvable, so that we can define a discrete operator $T_h : H \to U_h$ by $Tg = u_h$ and look for conditions ensuring the uniform convergence of T_h to T. The uniform convergence will indeed guarantee the eigenpair convergence, since problems (3.24) and (3.27) can be written in this framework, respectively, $\lambda T u = -u$ and $\lambda_h T_h u_h = -u_h$.

Let $\Sigma_H^0 \subset \Sigma$ and $U_H^0 \subset U$ denote the spaces of solutions to (3.26) as g ranges in H. These spaces will be endowed with their natural norms.

DEFINITION 3.3. *The* weak approximability *of U_H^0 with respect to a is satisfied if there exists $\omega_1(h)$, tending to zero as h goes to zero, such that for every $\sigma \in K_h$ and every $u \in U_H^0$ it holds*

$$a(\sigma, u) \leq \omega_1(h) |\sigma|_a \|u\|_{U_H^0}$$

DEFINITION 3.4. *The* strong approximability *of U_H^0 is satisfied if there exists $\omega_2(h)$, tending to zero as h goes to zero, such that for every $u \in U_H^0$ there exists $u^I \in U_h$ such that*

$$\|u - u^I\|_U \leq \omega_2(h) \|u\|_{U_H^0}.$$

An important tool for the analysis of mixed problems is the Fortin operator (see [19]) $\Pi_h : \Sigma_H^0 \to \Sigma_h$ which satisfies

$$\begin{aligned} b(\sigma - \Pi_h \sigma, v) &= 0 & \forall v \in U_h \\ \|\Pi_h \sigma\|_\Sigma &\leq C \|\sigma\|_{\Sigma_H^0} \end{aligned}$$

for all $\sigma \in \Sigma_H^0$.

DEFINITION 3.5. *The* Fortid property *is satisfied if there exists* $\omega_3(h)$, *tending to zero as* h *goes to zero, such that for every* $\sigma \in \Sigma_H^0$ *it holds*

$$|\sigma - \Pi_h\sigma|_a \leq \omega_3(h)\|\sigma\|_{\Sigma_H^0}.$$

REMARK 3.4. The name *Fortid* denotes that the **Fort**in operator converges towards the **id**entity.

THEOREM 3.2. *Suppose that the weak approximability of* U_H^0 *with respect to* a *and the strong approximability of* U_H^0 *are satisfied, and that there exists a Fortin operator such that the Fortid property holds. Then the sequence* T_h *converges to* T *in* $\mathcal{L}(H,U)$, *that is there exists* $\omega_4(h)$, *tending to zero as* h *goes to zero, such that*

$$\|Tg - T_hg\|_U \leq \omega_4(h)\|g\|_H. \tag{3.29}$$

REMARK 3.5. In [6] it has been proved that the assumptions of theorem 3.2 are also necessary for eigenvalue convergence. In particular, Example 2 shows that a scheme satisfying the classical inf-sup constants for mixed approximations may provide spurious solutions when applied to eigenvalue problems. In [7] it has beed proved that for such a scheme (3.29) is not satisfied.

4. Applications.

4.1. Time harmonic Maxwell's system. In this section we briefly recall the interior Maxwell's eigenvalue problem and cast it in the framework of our analysis. Indeed, following [10] it can be written as a mixed problem of the displacement type. Moreover, in the spirit of [22] (see also [16]), we recall that it admits a mixed variational formulation of the equilibrium-type as well. Either formulation can be used for the analysis: in [4], using the displacement-type formulation, it has been shown that a Fortin operator can be defined fulfilling the *Fortid* property; in [5] it has been shown that the Fortid property is strictly related to the existence of a link between the de Rham complex of the involved functional spaces and the finite element scheme (see [2]). For a review of the finite element approximation of Maxwell's equations, we refer to [20] and [23] as well.

Given a polyhedral domain Ω with outward normal $\mathbf{n}$, the easiest form of Maxwell's eigenvalue problem reads: find $\lambda \in \mathbb{R}$ such that there exists $\mathbf{u} \neq 0$ which satisfies

$$\begin{aligned} &\mathbf{curl}\,\mathbf{curl}\,\mathbf{u} = \lambda\mathbf{u} && \text{in } \Omega,\\ &\operatorname{div}\mathbf{u} = 0 && \text{in } \Omega,\\ &\mathbf{u}\times\mathbf{n} && \text{on } \partial\Omega \end{aligned} \tag{4.1}$$

In (4.1) we do not consider material coefficients and more complicated (e.g. mixed) boundary conditions. In practical applications, material discontinuities and lack of domain regularity may lead to highly singular solutions. We refer to [15] for the regularity analysis, and to [10, 14] and the references therein for the finite element approximation. To our aim, problem (4.1) contains all the ingredients to show that finite element schemes must be chosen in a careful way in order to avoid spurious modes and to achieve good eigenmodes convergence.

The spaces involved with the variational formulations and the finite element discretization of (4.1) are usually described with the following diagram, known as de Rham complex.

$$
\begin{array}{ccccccccc}
\multicolumn{9}{c}{Q \subset H_0^1(\Omega),\ \Sigma \subset H_0(\mathbf{curl};\Omega),\ T \subset H_0(\mathrm{div};\Omega),\ V \subset L^2(\Omega)} \\
\multicolumn{9}{c}{Q_h \subset H_0^1(\Omega),\ \Sigma_h \subset H_0(\mathbf{curl};\Omega),\ T_h \subset H_0(\mathrm{div};\Omega),\ V_h \subset L^2(\Omega)} \\
0 \to & Q & \xrightarrow{\mathrm{grad}} & \Sigma & \xrightarrow{\mathbf{curl}} & T & \xrightarrow{\mathrm{div}} & V & \to 0 \\
& \downarrow \Pi_h^Q & & \downarrow \Pi_h^\Sigma & & \downarrow \Pi_h^T & & \downarrow \Pi_h^V & \\
0 \to & Q_h & \xrightarrow{\mathrm{grad}} & \Sigma_h & \xrightarrow{\mathbf{curl}} & T_h & \xrightarrow{\mathrm{div}} & V_h & \to 0.
\end{array}
\tag{4.2}
$$

A standard variational formulation of (4.1) is: find $\lambda \in \mathbb{R}$ such that there exists $\mathbf{u} \in H_0(\mathbf{curl}) \cap H(\mathrm{div}^0)$ $(\mathbf{u} \neq 0)$ satisfying

$$(\mathbf{curl}\,\mathbf{u}, \mathbf{curl}\,\mathbf{v}) = \lambda(\mathbf{u}, \mathbf{v}) \quad \forall \mathbf{v} \in H_0(\mathbf{curl}) \cap H(\mathrm{div}^0), \tag{4.3}$$

where $H(\mathrm{div}^0)$ denotes the subspace of $H(\mathrm{div})$ consisting of divergence free functions.

The following *unconstrained* formulation, where the divergence free constraint is substituted by the requirement $\lambda \neq 0$, is known to be equivalent to (4.3): find $\lambda \neq 0$ such that there exists $\mathbf{u} \in H_0(\mathbf{curl})$ $(\mathbf{u} \neq 0)$ satisfying

$$(\mathbf{curl}\,\mathbf{u}, \mathbf{curl}\,\mathbf{v}) = \lambda(\mathbf{u}, \mathbf{v}) \quad \forall \mathbf{v} \in H_0(\mathbf{curl}). \tag{4.4}$$

The mixed equilibrium-type form of (4.1) (see [22]) uses the left part of the diagram (4.2) as follows: find $\lambda \in \mathbb{R}$ such that there exists $\mathbf{u} \in H_0(\mathbf{curl})$ $(\mathbf{u} \neq 0)$ so that for some $p \in H_0^1$ it holds

$$
\begin{aligned}
(\mathbf{curl}\,\mathbf{u}, \mathbf{curl}\,\mathbf{v}) + (\mathrm{grad}\,p, \mathbf{u}) &= \lambda(\mathbf{u}, v) && \forall \mathbf{v} \in H_0(\mathbf{curl}) \\
(\mathrm{grad}\,q, \mathbf{u}) &= 0 && \forall q \in H_0^1.
\end{aligned}
\tag{4.5}
$$

The displacement-type mixed formulation has been presented in [10] and uses the right part of the diagram (4.2): find $\lambda \in \mathbb{R}$ such that there exists $\boldsymbol{\sigma} \in H_0(\mathbf{curl})$ $(\boldsymbol{\sigma} \neq 0)$ so that for some $\mathbf{p} \in H_0(\mathrm{div}^0)$ it holds

$$
\begin{aligned}
(\boldsymbol{\sigma}, \boldsymbol{\tau}) + (\mathbf{curl}\,\boldsymbol{\tau}, \mathbf{p}) &= 0 && \forall \boldsymbol{\tau} \in H_0(\mathbf{curl}) \\
(\mathbf{curl}\,\boldsymbol{\sigma}, \mathbf{q}) &= -\lambda(\mathbf{p}, \mathbf{q}) && \forall \mathbf{q} \in H_0(\mathrm{div}^0).
\end{aligned}
\tag{4.6}
$$

In [10], using again (4.2), it has been shown that all eigenvalues of (4.5) are positive and that the eigenmodes coincides with those of (4.3).

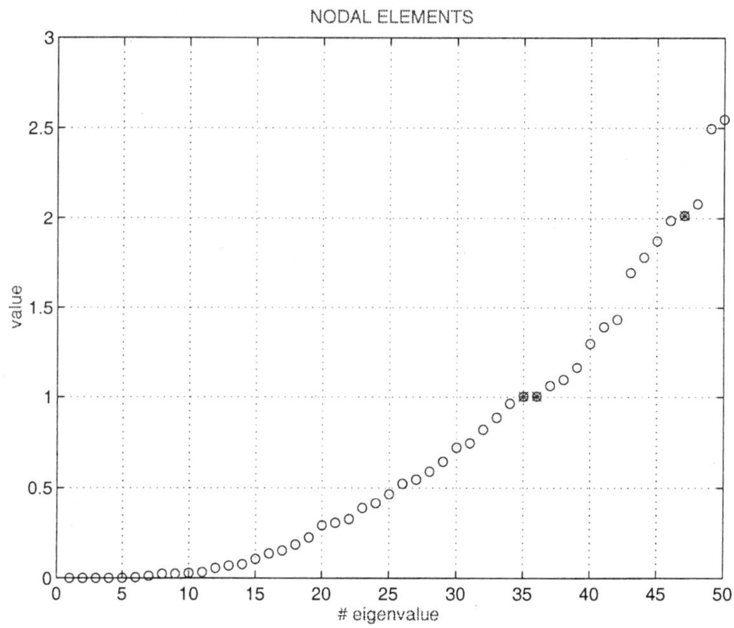

FIG. 2. *Nodal element computation of Maxwell's eigenvalue: spurious modes.*

Let's consider the finite element approximation to (4.4): find $\lambda_h \neq 0$ such that there exists $\mathbf{u}_h \in \Sigma_h$, $(\mathbf{u}_h \neq 0)$ satisfying

$$(\mathbf{curl}\,\mathbf{u}_h, \mathbf{curl}\,\mathbf{v}) = \lambda_h(\mathbf{u}_h, \mathbf{v}) \quad \forall \mathbf{v} \in \Sigma_h. \tag{4.7}$$

Unfortunately, the condition $\lambda_h \neq 0$ is not sufficient to guarantee the divergence free constraint (at a discrete level). Indeed, Figure 2 shows two dimensional numerical results obtained with the use of continuous piecewise linear elements. The correct eigenvalues are approximated by discrete modes which are very difficult to distinguish among other spurious modes. It is evident that many spurious modes are present which make the method unpractical. This bad behavior will be made clearer later on in this section when our analysis will be presented. On the other hand, Figure 3 shows the same computation performed with edge elements, where it is clear the separation between zero frequencies (that have to be discarded) and positive ones which are good approximations to the eigenvalues of (4.3). We point out that the choice of edge finite elements makes the diagram (4.2) commute.

The discretizations to problems (4.5) and (4.6) read, respectively: find $\lambda_h \in \mathbb{R}$ such that there exists $\mathbf{u}_h \in \Sigma_h$ so that for some $p_h \in Q_h$ it holds

$$\begin{aligned} (\mathbf{curl}\,\mathbf{u}_h, \mathbf{curl}\,\mathbf{v}) + (\mathrm{grad}\,p_h, \mathbf{u}) &= \lambda_h(\mathbf{u}_h, v) && \forall \mathbf{v} \in \Sigma_h \\ (\mathrm{grad}\,q, \mathbf{u}_h) &= 0 && \forall q \in Q_h. \end{aligned} \tag{4.8}$$

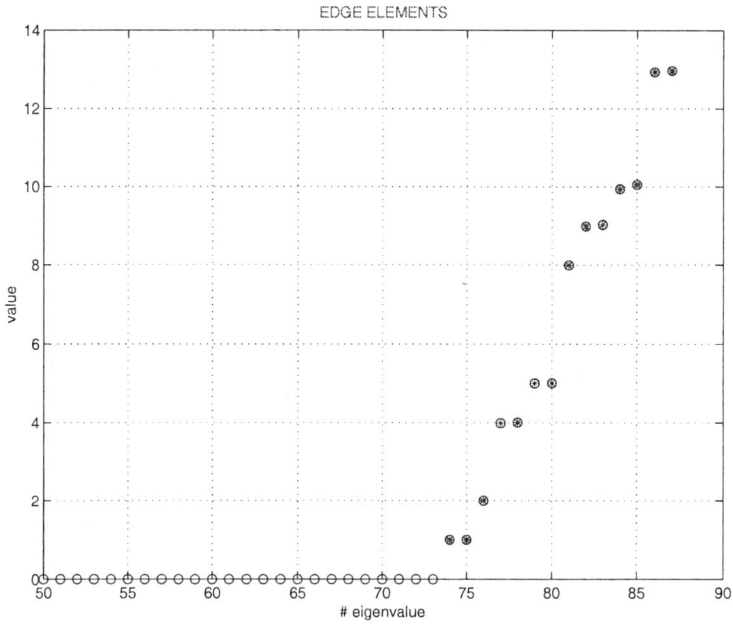

FIG. 3. *Edge element computation of Maxwell's eigenvalue: correct approximation.*

and: find $\lambda_h \in \mathbb{R}$ such that there exists $\boldsymbol{\sigma}_h \in \Sigma_h$ ($\boldsymbol{\sigma}_h \neq 0$) so that for some $\mathbf{p}_h \in T_h$ it holds

$$\begin{aligned} (\boldsymbol{\sigma}_h, \boldsymbol{\tau}) + (\mathbf{curl}\,\boldsymbol{\tau}, \mathbf{p}_h) &= 0 & \forall \boldsymbol{\tau} \in \Sigma_h \\ (\mathbf{curl}\,\boldsymbol{\sigma}_h, \mathbf{q}) &= -\lambda_h(\mathbf{p}_h, \mathbf{q}) & \forall \mathbf{q} \in T_h. \end{aligned} \tag{4.9}$$

The next theorem, which can be proved with standard arguments (see [22, 10] for more details), states the links between problem (4.7) and problems (4.8) and (4.9).

THEOREM 4.1. *Let* $(\lambda_h, \mathbf{u}_h) \in \mathbb{R} \times \Sigma_h$ *be an eigensolution of problem* (4.4). *Then* $(\lambda_h, \mathbf{u}_h)$ *solves problems* (4.8) *(for a suitable* $p_h \in Q_h$*) and* (4.9) *(for a suitable* $\mathbf{p}_h \in T_h$*) as well. Viceversa, if* $(\lambda_h, \mathbf{u}_h, p_h) \in \mathbb{R} \times \Sigma_h \times Q_h$ *solves problem* (4.8), *then* $\lambda_h \neq 0$ *and* $(\lambda_h, \mathbf{u}_h)$ *is also solution to* (4.4). *Analogously, if* $(\lambda_h, \boldsymbol{\sigma}_h, \mathbf{p}_h) \in \mathbb{R}_h \times \Sigma_h \times T_h$ *solves* (4.9), *then* $\lambda_h \neq 0$ *and* $(\lambda_h, \boldsymbol{\sigma}_h)$ *is also solution to* (4.4).

In [10, 4] the displacement-type formulation (4.9) has been used for the analysis of schemes based on edge finite elements; here we detail the use of the equilibrium-type formulation (4.8). According to the theory presented in Section 3.2 (see Theorem 3.1), we need show the validity of three properties: the ellipticity in the kernel (3.21), the weak (3.19), and strong (3.20) approximabilities.

The ellipticity in the kernel property is stated in Proposition 4.6 of [1].

The weak approximability property follows from this lemma.

LEMMA 4.1. *For all* $\mathbf{v}_h \in \Sigma_h$, *with* $(\mathbf{v}_h, \operatorname{grad} q_h) = 0$, $\forall q \in Q_h$, *there exists* $\mathbf{v} \in \Sigma$ *and* $\varepsilon > 0$ *such that*

$$\|\mathbf{v}_h - \mathbf{v}\|_{L^2} \le C h^{1/2+\varepsilon} \|\mathbf{v}_h\|_{H(\mathbf{curl})}$$

Proof. The function $\mathbf{v}_h$ can be presented as the first component of the solution of the following mixed problem

$$\begin{aligned} (\mathbf{v}_h, \mathbf{w}) + (\mathbf{p}_h, \mathbf{curl}\,\mathbf{w}) &= 0 \quad \forall \mathbf{w} \in \Sigma_h \\ (\mathbf{curl}\,\mathbf{v}_h, \mathbf{q}) &= (\mathbf{curl}\,\mathbf{v}_h, \mathbf{q}) \quad \forall \mathbf{q} \in T_h. \end{aligned}$$

Let us define $\mathbf{v} \in \Sigma$ as the first component of the solution to the corresponding continuous problem; that is $\mathbf{v}$ is such that

$$\begin{aligned} (\mathbf{v}, \mathbf{w}) + (\mathbf{p}, \mathbf{curl}\,\mathbf{w}) &= 0 \quad \forall \mathbf{w} \in \Sigma \\ (\mathbf{curl}\,\mathbf{v}, \mathbf{q}) &= (\mathbf{curl}\,\mathbf{v}_h, \mathbf{q}) \quad \forall \mathbf{q} \in T. \end{aligned} \tag{4.10}$$

Since $\mathbf{v} \in \Sigma$ and $\operatorname{div} \mathbf{v} = 0$, it follows that $\mathbf{v} \in H^{1/2+\varepsilon}(\Omega)$, and from the a priori bound of (4.10) we get the estimate $\|\mathbf{v}\|_{H^{1/2+\varepsilon}} \le C \|\mathbf{curl}\,\mathbf{v}_h\|_{L^2}$. We can then use Theorem 1 of [4] and obtain

$$\|\mathbf{v} - \mathbf{v}_h\|_{L^2} \le C h^{1/2+\varepsilon} \|\mathbf{v}\|_{H^{1/2+\varepsilon}} \le C h^{1/2+\varepsilon} \|\mathbf{curl}\,\mathbf{v}_h\|_{L^2}.$$

□

The above lemma is used in order to prove the weak approximability property (3.19) as follows:

$$\sup_{\mathbf{v}_h \in K_h} \frac{(\operatorname{grad} p, \mathbf{v}_h)}{\|\mathbf{v}_h\|_\Sigma} = \sup_{\mathbf{v}_h \in K_h} \frac{(\operatorname{grad} p, \mathbf{v}_h - \mathbf{v})}{\|\mathbf{v}_h\|_\Sigma} \le C h^{1/2+\varepsilon} \|p\|_Q$$

The strong approximability property (3.20) is a consequence of the the interpolation properties of edge finite elements and of the commuting diagram property.

4.2. Photonic band gaps computation. The analysis of the previous section extends to the computation of band gaps for photonic crystals. We do not detail the derivation of our model, which uses standard Bloch–Floquet theory; the interested reader is referred to [17, 9]. In our model the periodicity of the medium is given in terms of the set of relative integer numbers $\mathbb{Z}$. Setting $\Omega = \mathbb{R}^3/\mathbb{Z}^3$ (which can be identified with the unit cube $(0,1)^3$ with periodic boundary conditions), Maxwell's eigenproblem is rewritten as: find λ such that there exists $\mathbf{u} \neq 0$ with periodic boundary conditions satisfying

$$\begin{aligned} \nabla_{\boldsymbol{\alpha}} \times \varepsilon^{-1} \nabla_{\boldsymbol{\alpha}} \times \mathbf{u} &= \lambda \mathbf{u} \quad \text{in } \Omega, \\ \nabla_{\boldsymbol{\alpha}} \cdot \mathbf{u} &= 0 \quad \text{in } \Omega, \end{aligned} \tag{4.11}$$

where the operator $\nabla_{\boldsymbol{\alpha}}$ is defined formally as $\nabla + i\boldsymbol{\alpha}$, i being the imaginary unit and $\boldsymbol{\alpha}$ being a complex vector belonging to the first Brillouin zone K. In practice, one is interested in solving (4.11) for all possible $\boldsymbol{\alpha}$ in K. The real intervals that do not contain any value of λ for all $\boldsymbol{\alpha}$, are called band gaps.

The rigorous analysis of the finite element approximation of (4.11) has been carried on in [9] (see also [17] and [11]). It makes use of an *ad hoc* edge finite element family and of suitable interpolation estimates which improve the classical ones available in the literature. Our modified family of edge elements enjoys a commuting diagram property similar to (4.2) where the symbol ∇ is substituted with the symbol $\nabla_{\boldsymbol{\alpha}}$, so that the operators grad, **curl**, and div become $\nabla_{\boldsymbol{\alpha}}$, $\nabla_{\boldsymbol{\alpha}}\times$, and $\nabla_{\boldsymbol{\alpha}}\cdot$, respectively.

4.3. Evolution problems in mixed form. The results presented in Sections 3.2 and 3.3 have a direct consequence for the analysis of evolution problems in mixed form. In [12] a general theory has been developed, considering evolution problems of the equilibrium and displacement type. For instance, the heat equation can be written as a displacement-type mixed evolution problem and its mixed (continuous and discrete) solutions can be represented as

$$\begin{aligned} u(t) &= \sum_{i=1}^{\infty}\left((u_0, w_i)e^{-\lambda_i t} + \int_0^t (g(s), w_i)e^{-\lambda_i(t-s)}\,ds\right) w_i \\ u_h(t) &= \sum_{i=1}^{N(h)}\left((u_0, w_{i_h})e^{-\lambda_{i,h} t} + \int_0^t (g(s), w_{i,h})e^{-\lambda_{i,h}(t-s)}\,ds\right) w_{i,h}, \end{aligned} \tag{4.12}$$

where λ_i and w_i (resp. $\lambda_{i,h}$ and $w_{i,h}$ for $i = 1, \dots, N(h)$) denote the continuous (resp. discrete) mixed eigensolutions of the Laplace operator.

We refer the interested reader to [12], where a general theory is presented.

In [8] the theory is extended to Maxwell's transient system.

Acknowledgments. I would like to thank the anonymous referee for carefully reading my manuscript and for improving the presentation of the paper and, in particular, of the proof of Proposition 2.1.

REFERENCES

[1] C. Amrouche, C. Bernardi, M. Dauge, and V. Girault, *Vector potentials in three-dimensional non-smooth domains*, Math. Methods Appl. Sci., 21 (1998), pp. 823–864.

[2] D.N. Arnold, *Differential complexes and numerical stability*, in Proceedings of the International Congress of Mathematicians, Vol. I (Beijing, 2002), Beijing, 2002, Higher Ed. Press, pp. 137–157.

[3] I. Babuška and J. Osborn, *Eigenvalue problems*, in Handbook of Numerical Analysis, P. Ciarlet and J. Lions, eds., vol. II, Elsevier Science Publishers B.V., North Holland, 1991, pp. 641–788.

[4] D. Boffi, *Fortin operator and discrete compactness for edge elements*, Numer. Math., 87 (2000), pp. 229–246.

[5] ———, *A note on the de Rham complex and a discrete compactness property*, Appl. Math. Lett., 14 (2001), pp. 33–38.

[6] D. Boffi, F. Brezzi, and L. Gastaldi, *On the convergence of eigenvalues for mixed formulations*, Ann. Scuola Norm. Sup. Pisa Cl. Sci. (4), 25 (1997), pp. 131–154 (1998). Dedicated to Ennio De Giorgi.

[7] ———, *On the problem of spurious eigenvalues in the approximation of linear elliptic problems in mixed form*, Math. Comp., 69 (2000), pp. 121–140.

[8] D. Boffi, A. Buffa, and Gastaldi, *Convergence analysis for hyperbolic evolution problems in mixed form.* submitted.

[9] D. Boffi, M. Conforti, and L. Gastaldi, *Modified edge finite elements for photonic crystals.* submitted.

[10] D. Boffi, P. Fernandes, L. Gastaldi, and I. Perugia, *Computational models of electromagnetic resonators: analysis of edge element approximation*, SIAM J. Numer. Anal., 36 (1999), pp. 1264–1290.

[11] D. Boffi and L. Gastaldi, *Interpolation estimates for edge finite elements and application to band gap computation.* Applied Numerical Mathematics, to appear.

[12] D. Boffi and L. Gastaldi, *Analysis of finite element approximation of evolution problems in mixed form*, SIAM J. Numer. Anal., 42 (2004), pp. 1502–1526 (electronic).

[13] F. Brezzi and M. Fortin, *Mixed and hybrid finite element methods*, vol. 15 of Springer Series in Computational Mathematics, Springer-Verlag, New York, 1991.

[14] S. Caorsi, P. Fernandes, and M. Raffetto, *Spurious-free approximations of electromagnetic eigenproblems by means of Nedelec-type elements*, M2AN Math. Model. Numer. Anal., 35 (2001), pp. 331–354.

[15] M. Costabel and M. Dauge, *Singularities of electromagnetic fields in polyhedral domains*, Arch. Ration. Mech. Anal., 151 (2000), pp. 221–276.

[16] L. Demkowicz and L. Vardapetyan, *Modeling of electromagnetic absorption/scattering problems using hp-adaptive finite elements*, Comput. Methods Appl. Mech. Engrg., 152 (1998), pp. 103–124. Symposium on Advances in Computational Mechanics, Vol. 5 (Austin, TX, 1997).

[17] D.C. Dobson, J. Gopalakrishnan, and J. E. Pasciak, *An efficient method for band structure calculations in 3D photonic crystals*, J. Comput. Phys., 161 (2000), pp. 668–679.

[18] G.J. Fix, M.D. Gunzburger, and R.A. Nicolaides, *On mixed finite element methods for first-order elliptic systems*, Numer. Math., 37 (1981), pp. 29–48.

[19] M. Fortin, *An analysis of the convergence of mixed finite element methods*, R.A.I.R.O., Anal. Numer., 11 (1977), pp. 341–354.

[20] R. Hiptmair, *Finite elements in computational electromagnetism*, Acta Numer., 11 (2002), pp. 237–339.

[21] T. Kato, *Perturbation theory for linear operators*, Classics in Mathematics, Springer-Verlag, Berlin, 1995. Reprint of the 1980 edition.

[22] F. Kikuchi, *Mixed and penalty formulations for finite element analysis of an eigenvalue problem in electromagnetism*, in Proceedings of the first world congress on computational mechanics (Austin, Tex., 1986), vol. 64, 1987, pp. 509–521.

[23] P. Monk, *Finite element methods for Maxwell's equations*, Numerical Mathematics and Scientific Computation, Oxford University Press, New York, 2003.

[24] H.L. Royden, *Real analysis*, Macmillan Publishing Company, New York, second ed., 1988.

CONJUGATED BUBNOV-GALERKIN INFINITE ELEMENT FOR MAXWELL EQUATIONS

IS *THE* OR *AN* EXACT SEQUENCE PROPERTY IMPORTANT?

L. DEMKOWICZ* AND J. KURTZ*

Abstract. We propose a (conjugated) Bubnov-Galerkin Infinite Element (IE) discretization for the time-harmonic Maxwell scattering and radiation problems. The element falls into a family of infinite elements satisfying *an* exact sequence property. The exact sequence results from incorporating the far-field pattern into the anzatz for the solution and the test functions, and it differs from the standard grad-curl-div sequence. We verify the construction with 2D numerical experiments.

Key words. Maxwell equations, infinite element, *hp* finite elements, RCS.

AMS(MOS) subject classifications. 65N30, 35L15.

1. Introduction. The presented work is motivated with the calculation of Radar Cross Sections (RCS) of objects with sharp edges and corners.

Let Ω^{int} be a bounded domain occupied by the scatterer, with Γ denoting its boundary, and let $\Omega = \mathbb{R}^n - \overline{\Omega^{int}}$ denote the exterior domain, with $n = 2, 3$. We truncate the exterior domain with a sphere (circle) $S_a = \{|\boldsymbol{x}| = a\}$ surrounding the scatterer, and split the domain Ω into a *near-field domain* $\Omega^a = \{\boldsymbol{x} \in \Omega : |\boldsymbol{x}| < a\}$, and a *far-field domain* $\Omega_a = \{|\boldsymbol{x}| > a\}$. Assuming for simplicity a *Perfect Conductor* (PEC) scatterer, we formulate the problem as follows.

Find electric field $\boldsymbol{E}$ that satisfies :

- Reduced wave equation in both near-field and far-field domains,

$$\boldsymbol{\nabla} \times (\boldsymbol{\nabla} \times \boldsymbol{E}) - k^2\boldsymbol{E} = \boldsymbol{0}, \quad \boldsymbol{x} \in \Omega^a, \Omega_a \,, \tag{1}$$

- PEC boundary condition on boundary Γ,

$$\boldsymbol{n} \times \boldsymbol{E} = -\boldsymbol{n} \times \boldsymbol{E}^{inc}, \quad \boldsymbol{x} \in \Gamma \,, \tag{2}$$

- Interface boundary conditions on the truncating sphere,

$$\boldsymbol{n} \times [\boldsymbol{E}] = \boldsymbol{0}, \quad \boldsymbol{n} \times [\boldsymbol{\nabla} \times \boldsymbol{E}] = \boldsymbol{0}, \quad |\boldsymbol{x}| = a \,, \tag{3}$$

- Silver-Müller radiation condition at infinity,

$$\boldsymbol{e}_r \times (\boldsymbol{\nabla} \times \boldsymbol{E}) - ik\boldsymbol{E}_t \in L^2(\Omega_a) \,. \tag{4}$$

---

*Institute for Computational Engineering and Sciences, The University of Texas at Austin, Austin, TX 78712. The work has been supported by Air Force under Contract FA9550-04-1-0050.

Here $\boldsymbol{n}$ denotes the normal to the boundary or interface, $k = \omega\sqrt{\epsilon_0\mu_0}$ is the free-space wave number, with ω denoting the angular frequency, and ϵ_0, μ_0 being the free space permittivity and permeability, $[\,]$ denotes the jump across the interface, $\boldsymbol{e}_r$ is radial unit vector corresponding to a spherical (polar) system of coordinates with a center inside of the scatterer, $\boldsymbol{E}^{inc}$ denotes an incident electric field, and $\boldsymbol{E}_t = -\boldsymbol{e}_r \times (\boldsymbol{e}_r \times \boldsymbol{E})$ is the tangential component of $\boldsymbol{E}$.

Infinite Elements. The idea of coupled Finite Element (FE)/Infinite Element (IE) approximations for exterior wave propagation problems dates back to the pioneering contributions of Bettess, and Bettess and Zienkiewicz, see [4] and the literature cited therein.

The works of Astley et al. [2], Cremers *et al.* [8], Givoli [15] and many others recognized the spectral character of the approximation and pointed to the necessity of multipole expansions. Burnett [5] revolutionized the approach from the practical point of view, by introducing a new, *symmetric* unconjugated formulation, and using prolate and oblate spheroidal elements.

Contrary to the concept of Perfectly Matched Layer [3], and other techniques based on Absorbing Boundary Conditions (ABC's), the conjugated element of Astley et al. [2], aims at obtaining the solution in the whole unbounded domain.

A conjugated Petrov-Galerkin infinite element for Maxwell equations was proposed in [11], and studied in [6, 7]. The idea of the construction was based on the assumption that the IE *test functions* come from a space belonging to an exact sequence involving standard grad-curl-div operators. In simple terms, the test functions include gradients of scalar potentials, In this way, any FE/IE solution to the Maxwell equations satisfies automatically the weak form of the continuity equation. The Petrov-Galerkin infinite element was applied to solve difficult three-dimensional scattering problems in [18].

In this contribution, we generalize the concept of a conjugated Bubnov-Galerkin infinite element for Helmholtz equation presented in [13]. The main idea consists of interpreting the integral over the exterior domain in a Cauchy Principal Value (CPV) sense and building the far-field pattern into the ansatz for the approximate solution. The resulting discretization allows for the use of the *same* trial and test functions but the discrete solution no longer satisfies the usual weak form of the continuity equation.

At this point, a full convergence analysis for any infinite element discretizations for either Helmholtz or Maxwell equations is unknown.

Variational formulation. We follow the standard procedure for the near-field domain. Taking a test function such that $\boldsymbol{n} \times \boldsymbol{F} = \boldsymbol{0}$ on Γ, we multiply the reduced wave equation (1) with complex conjugate $\bar{\boldsymbol{F}}$, and integrate by parts to obtain,

$$\int_{\Omega^a} (\nabla\times E)(\nabla\times\bar{F}) - k^2 E\bar{F}\,dx = -\int_{S_a} e_r\times(\nabla\times E)\bar{F}\,dS, \qquad \forall F, n\times F = 0 \ \text{ on } \Gamma\,. \tag{5}$$

Integration over the far-field domain will be interpreted in the Cauchy Principal Value (CPV) sense. We introduce a second truncating sphere $S_R = \{|x| = R\}, R > a$ and consider the truncated far-field domain,

$$\Omega_a^R = \{x \,:\, a < |x| < R\}\,.$$

Performing the same operations for Ω_a^R as for the near-field domain, we obtain,

$$\int_{\Omega_a^R} (\nabla \times E)(\nabla \times \bar{F}) - k^2 E\bar{F}\,dx + \int_{S_R} e_r \times (\nabla \times E)\bar{F}\,dS = \int_{S_a} e_r \times (\nabla \times E)\bar{F}\,dS, \quad \forall F\,. \tag{6}$$

Using the Silver-Müller radiation condition (4), we replace $e_r \times (\nabla \times E)$ in the integral over S_R with ikE_t plus an unknown contribution which, due to the assumption on L^2-integrability, will vanish in the limit when $R \to \infty$,

$$\int_{\Omega_a^R} (\nabla \times E)(\nabla \times \bar{F}) - k^2 E\bar{F}\,dx + ik\int_{S_R} E_t\bar{F}\,dS + (\text{ a term that vanishes at } R\to\infty) = \int_{S_a} e_r\times(\nabla\times E)\bar{F}\,dS, \ \forall F\,. \tag{7}$$

Finally, we sum up contributions (5) and (7). Assuming that both solution E and test function F satisfy the interface conditions (3), we obtain our final variational formulation.

$$\begin{cases} n \times E = -n \times E^{inc}, \\ \displaystyle\int_{\Omega^a} (\nabla \times E)(\nabla \times \bar{F}) - k^2 E\bar{F}\,dx \\ \displaystyle + \lim_{R\to\infty}\left(\int_{\Omega_a^R}(\nabla\times E)(\nabla\times F) - k^2 E\bar{F}\,dx + ik\int_{S_R} E_t\bar{F}\,dS\right) = 0, \\ \qquad\qquad \forall F, n \times F = 0 \quad \text{on } \Gamma\,. \end{cases} \tag{8}$$

Both solution E and test function F are assumed to "live" in $H_{loc}(\Omega, \text{curl})$ with extra assumptions to guarantee the existence of the limit to be discussed next.

2. Infinite element discretization in 3D. In a sense, the three-dimensional setting is more straightforward, and we shall discuss it first, with the two-dimensional one described in the next section.

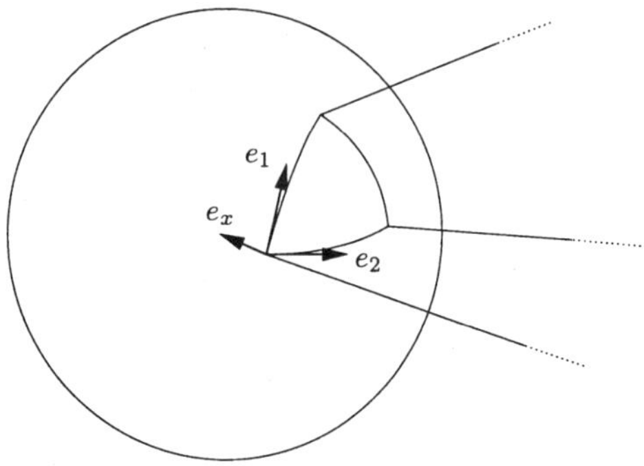

FIG. 1. *Curvilinear system of coordinates on the truncating sphere.*

Infinite element coordinates. The curvilinear system of coordinates used to construct the infinite element discretization combines a FE parametrization of the truncating sphere with the "inverted" radial coordinate,

$$\boldsymbol{x}(\xi_\alpha, x) = x^{-1}\boldsymbol{x}_a(\xi_\alpha), \quad \alpha = 1,2, \quad |\boldsymbol{x}_a(\xi_\alpha)| = a, \quad 0 < x < 1\,. \tag{9}$$

Here $\boldsymbol{x}_a(\xi_\alpha)$ is a parametrization of the sphere of radius a, centered at the origin of a Cartesian system of coordinates x_i.[1], see Fig. 1. The basis vectors are,

$$\boldsymbol{a}_\alpha = x^{-1}\frac{\partial \boldsymbol{x}_a}{\partial \xi_\alpha} = x^{-1}\left|\frac{\partial \boldsymbol{x}_a}{\partial \xi_\alpha}\right| \boldsymbol{e}_\alpha, \qquad \boldsymbol{a}_x = -x^{-2}\boldsymbol{x}_a = x^{-2}a\boldsymbol{e}_x\,. \tag{10}$$

We assume that the parameters α have been denumerated in such a way that $(\boldsymbol{e}_1, \boldsymbol{e}_2, \boldsymbol{e}_x)$ is a right triple. The cobasis vectors are given by,

$$\begin{gathered}\boldsymbol{a}^1 = x\mathrm{jac}_a^{-1}(\frac{\partial \boldsymbol{x}_a}{\partial \xi_2} \times \boldsymbol{e}_x), \quad \boldsymbol{a}^2 = -x\mathrm{jac}_a^{-1}(\frac{\partial \boldsymbol{x}_a}{\partial \xi_1} \times \boldsymbol{e}_x),\\ \boldsymbol{a}^x = x^2a^{-1}\boldsymbol{e}_x\,,\end{gathered} \tag{11}$$

where,

$$\mathrm{jac}_a = (\frac{\partial \boldsymbol{x}_a}{\partial \xi_1} \times \frac{\partial \boldsymbol{x}_a}{\partial \xi_2}) \cdot \boldsymbol{e}_x\,. \tag{12}$$

Denoting by $\hat{\boldsymbol{a}}_\alpha, \hat{\boldsymbol{a}}^\alpha$ basis and cobasis vectors on sphere S_a,

$$\boldsymbol{a}_\alpha = x^{-1}\hat{\boldsymbol{a}}_\alpha, \quad \boldsymbol{a}^\alpha = x\hat{\boldsymbol{a}}^\alpha\,, \tag{13}$$

[1] We shall use Greek letters for indices with range 1, 2, and Roman letters for indices with range 1, 2, 3.

we recall the standard formula for the gradient of a scalar-valued function u,

$$\nabla u = x \frac{\partial u}{\partial \xi_\alpha} \hat{\boldsymbol{a}}^\alpha + x^2 a^{-1} \frac{\partial u}{\partial r} \boldsymbol{e}_x \,. \tag{14}$$

As usual, repeated indices indicate summation.

Consistently with the standard exact sequence property and the construction of parametric finite elements [10], all vector-valued fields will be assumed in the form,

$$\boldsymbol{E} = x E_\alpha \hat{\boldsymbol{a}}^\alpha + x^2 a^{-1} E_x \boldsymbol{e}_x \,. \tag{15}$$

REMARK 1. Notice that, with components $E_\alpha, E_x = O(1)$, the choice of the system of coordinates guarantees the right asymptotic behavior in $r = x^{-1}$, consistent with the formula for the exact solution. Recalling the general transformation rule for the curl vector [10],

$$(\text{curl}\boldsymbol{E})_i = J^{-1} \frac{\partial x_i}{\partial \xi_n} (\hat{\text{curl}} \hat{\boldsymbol{E}})_n \,,$$

where J is the jacobian of transformation $x_i = x_i(\xi_n)$ (in our case $J = x^{-4} a \, \text{jac}_a$), we can compute the curl of vector $\boldsymbol{E}$,

$$\nabla \times \boldsymbol{E} = (x^{-4} a \text{jac}_a)^{-1} \left\{ (\hat{\text{curl}} \hat{\boldsymbol{E}})_\alpha x^{-1} \hat{\boldsymbol{a}}_\alpha + (\hat{\text{curl}} \hat{\boldsymbol{E}})_3 x^{-2} a \boldsymbol{e}_x \right\} \,. \tag{16}$$

Here,

$$\hat{\text{curl}} \hat{\boldsymbol{E}} = \left(\frac{\partial E_x}{\partial \xi_2} - \frac{\partial E_2}{\partial x}, -(\frac{\partial E_x}{\partial \xi_1} - \frac{\partial E_1}{\partial x}), \frac{\partial E_2}{\partial \xi_1} - \frac{\partial E_1}{\partial \xi_2} \right) \,. \tag{17}$$

REMARK 2. The curvilinear coordinates ξ_α discussed here are to be understood in two different ways. On the theoretical side, they provide a basis for the construction of the IE approximation; in this context, one can think about e.g. standard spherical coordinates. On the practical side, they can be directly interpreted as a parametrization corresponding to isoparametric finite elements used to approximate the truncating sphere. In such a case, they will correspond to a *local*, element-wise approximation of the sphere only. Formulas derived here can then be used directly for coding.

Incorporating the far-field pattern. Consistently with the formula for the far-field pattern, we shall postulate the solution $\boldsymbol{E}$ in the form,

$$\boldsymbol{E} := e^{-ika(x^{-1}-1)} \boldsymbol{E} \,, \tag{18}$$

where the symbol $\boldsymbol{E}$ has been "overloaded". Recalling the elementary formula,

$$\nabla \times (\phi \boldsymbol{E}) = \nabla \phi \times \boldsymbol{E} + \phi \, \nabla \times \boldsymbol{E} \,,$$

we obtain,

$$\nabla \times \left(e^{-ika(x^{-1}-1)}\boldsymbol{E}\right) = e^{-ika(x^{-1}-1)}\left(+ik(\boldsymbol{e}_x \times \boldsymbol{E}) + \nabla \times \boldsymbol{E}\right) . \tag{19}$$

Identical substitution for the (conjugated) test function $\bar{\boldsymbol{F}}$, leads to,

$$\nabla \times \left(e^{+ika(x^{-1}-1)}\bar{\boldsymbol{F}}\right) = e^{+ika(x^{-1}-1)}\left(-ik(\boldsymbol{e}_x \times \bar{\boldsymbol{F}}) + \nabla \times \bar{\boldsymbol{F}}\right) . \tag{20}$$

Substituting (19) and (20) into the limit term in (8), we obtain (with overloaded symbols $\boldsymbol{E}$ and $\boldsymbol{F}$)

$$\begin{aligned}\lim_{X\to 0} \Bigg(\int\int_X^1 & \Big\{k^2(\boldsymbol{e}_x \times \boldsymbol{E}) \cdot (\boldsymbol{e}_x \times \bar{\boldsymbol{F}}) + (\nabla \times \boldsymbol{E}) \cdot (\nabla \times \bar{\boldsymbol{F}}) \\ & - ik\left[(\nabla \times \boldsymbol{E}) \cdot (\boldsymbol{e}_x \times \bar{\boldsymbol{F}}) - (\boldsymbol{e}_x \times \boldsymbol{E}) \cdot (\nabla \times \bar{\boldsymbol{F}})\right] \\ & - k^2 \boldsymbol{E} \cdot \bar{\boldsymbol{F}}\Big\}\, x^{-4} a \mathrm{jac}_a \, dx d\boldsymbol{\xi} \\ & + ik \textstyle\int \boldsymbol{E}_t(\cdot, X)\bar{\boldsymbol{F}}_t(\cdot, X) X^{-2} \mathrm{jac}_a \, d\boldsymbol{\xi} \Bigg) .\end{aligned} \tag{21}$$

Here, parameter X in the limit corresponds to R^{-1} compared with the limit in (8). But,

$$k^2(\boldsymbol{e}_x \times \boldsymbol{E}) \cdot (\boldsymbol{e}_x \times \bar{\boldsymbol{F}}) - k^2 \boldsymbol{E} \cdot \bar{\boldsymbol{F}} = k^2 x^2 a^{-2} E_x \bar{F}_x ,$$

so the term reduces to,

$$\begin{aligned}\lim_{X\to 0} \Bigg(\int\int_X^1 & \Big\{(\nabla \times \boldsymbol{E}) \cdot (\nabla \times \bar{\boldsymbol{F}}) - ik\big[(\nabla \times \boldsymbol{E}) \cdot (\boldsymbol{e}_x \times \bar{\boldsymbol{F}}) \\ & -(\boldsymbol{e}_x \times \boldsymbol{E}) \cdot (\nabla \times \bar{\boldsymbol{F}})\big] - k^2 x^2 a^{-2} E_x \bar{F}_x \Big\}\, x^{-4} a \mathrm{jac}_a \, dx d\boldsymbol{\xi} \\ & +ik \int \boldsymbol{E}_t(\cdot, X)\bar{\boldsymbol{F}}_t(\cdot, X) X^{-2} \mathrm{jac}_a \, d\boldsymbol{\xi} \Bigg) .\end{aligned} \tag{22}$$

Substitution of the far-field pattern into the ansatz for the solution and test function yields the limit finite for $E_\alpha, E_x = O(1)$. We obtain,

$$\begin{aligned}\int\int_0^1 & \Big\{(\nabla \times \boldsymbol{E}) \cdot (\nabla \times \bar{\boldsymbol{F}}) - ik\big[(\nabla \times \boldsymbol{E}) \cdot (\boldsymbol{e}_x \times \bar{\boldsymbol{F}}) \\ & -(\boldsymbol{e}_x \times \boldsymbol{E}) \cdot (\nabla \times \bar{\boldsymbol{F}})\big] - k^2 x^2 a^{-2} E_x \bar{F}_x \Big\}\, x^{-4} a \mathrm{jac}_a \, dx d\boldsymbol{\xi} \\ & +ik \int (E_\alpha(\cdot, 0)\boldsymbol{a}^\alpha)(\bar{F}_\alpha(\cdot, 0)\boldsymbol{a}^\alpha) \mathrm{jac}_a \, d\boldsymbol{\xi} .\end{aligned} \tag{23}$$

We record the final formulas necessary for the calculation of the stiffness matrix.

$$\begin{aligned}\nabla \times \boldsymbol{E} &= \mathrm{jac}_a^{-1} \Bigg\{ + \frac{x^3}{a}\left(\frac{\partial E_x}{\partial \xi_2} - \frac{\partial E_2}{\partial x}\right)\hat{\boldsymbol{a}}_1 \\ & \quad - \frac{x^3}{a}\left(\frac{\partial E_x}{\partial \xi_1} - \frac{\partial E_1}{\partial x}\right)\hat{\boldsymbol{a}}_2 + x\left(\frac{\partial E_2}{\partial \xi_1} - \frac{\partial E_1}{\partial \xi_2}\right)\boldsymbol{e}_x \Bigg\} \\ \boldsymbol{e}_x \times \boldsymbol{E} &= x \mathrm{jac}_a^{-1}\left(-E_2 \hat{\boldsymbol{a}}_1 + E_1 \hat{\boldsymbol{a}}_2\right) .\end{aligned} \tag{24}$$

Discretization. Solution components E_α, E_x and test function components F_α, F_x can now be discretized using standard Nédélec hexahedral (or prismatic) elements of the first type [17]. For the hexahedral element of order (p_α, p_x), we have,

$$\begin{aligned} E_1 \in \mathcal{P}^{p_1-1} \otimes \mathcal{P}^{p_2} \otimes \mathcal{P}^{p_x}, \quad E_2 \in \mathcal{P}^{p_1} \otimes \mathcal{P}^{p_2-1} \otimes \mathcal{P}^{p_x}, \\ E_x \in \mathcal{P}^{p_1} \otimes \mathcal{P}^{p_2} \otimes \mathcal{P}^{p_x-1}. \end{aligned} \tag{25}$$

Shape functions depending upon ξ_α have to match shape functions for the standard quadrilateral element, see [1]. The leading term in x, similarly to the Helmholtz case [13], corresponds to term,

$$\int_0^1 x^2 \frac{\partial E_\alpha}{\partial x} \frac{\partial \bar{F}_\beta}{\partial x}\, dx\,, \tag{26}$$

and suggests selecting for the shape functions in x integrals of Jacobi polynomials $P_j^{2,0}$,

$$\psi_j(x) = \begin{cases} 1 & j = 0 \\ \displaystyle\int_x^1 P_{j-1}^{(0,2)}(2t-1)\, dt & j \geq 1\,. \end{cases} \tag{27}$$

3. Infinite element discretization in 2D. The reasoning in two dimensions is very similar.

2D IE coordinates. We use polar-like coordinates,

$$\boldsymbol{x}(x,\xi) = x^{-1}\boldsymbol{x}_a(\xi), \quad |\boldsymbol{x}_a(\xi)| = a, \quad x \in (0,1)\,, \tag{28}$$

where $\boldsymbol{x}_a(\xi)$ is a clockwise parametrization of the truncating circle. Basis and cobasis vectors are defined as,

$$\begin{aligned} \boldsymbol{a}_\xi &= x^{-1}\frac{d\boldsymbol{x}_a}{d\xi} = x^{-1} \underbrace{|\frac{d\boldsymbol{x}_a}{d\xi}|}_{\text{jac}_a} \boldsymbol{e}_\xi, \quad & \boldsymbol{a}_x &= -x^{-2}\boldsymbol{x}_a(\xi) = x^{-2}a\boldsymbol{e}_x, \\ \boldsymbol{a}^\xi &= x\text{jac}_a^{-1}\boldsymbol{e}_\xi, & \boldsymbol{a}^x &= x^2 a^{-1}\boldsymbol{e}_r, \,. \end{aligned} \tag{29}$$

The formula for a gradient of function $u(x,\xi)$ is,

$$\boldsymbol{\nabla} u = \frac{\partial u}{\partial x} x^2 a^{-1} \boldsymbol{e}_x + \frac{\partial u}{\partial \xi} x\text{jac}_a^{-1} \boldsymbol{e}_\xi\,, \tag{30}$$

with a corresponding representation for vector-valued fields,

$$\boldsymbol{E} = E_x x^2 a^{-1} \boldsymbol{e}_x + E_\xi x\,\text{jac}_a^{-1} \boldsymbol{e}_\xi\,. \tag{31}$$

The curl is evaluated using the formula,

$$\text{curl}\boldsymbol{E} = (x^{-3}a\text{jac}_a)^{-1} \underbrace{\left(\frac{\partial E_\xi}{\partial x} - \frac{\partial E_x}{\partial \xi}\right)}_{\hat{\text{curl}}\boldsymbol{E}}\,. \tag{32}$$

Incorporating the far-field pattern. The anzatz incorporating the far field pattern is now slightly different,

$$\boldsymbol{E} := x^{-\frac{1}{2}} e^{-ika(x^{-1}-1)} \boldsymbol{E} \,. \tag{33}$$

Upon the substitution, and cancellation of Lebesgue non-integrable terms, we obtain an expression analogous to (23),

$$\begin{aligned}
&\int \int_0^1 \Big\{ (a\mathrm{jac}_a)^{-1} \Big[x^2 \hat{\mathrm{curl}} \boldsymbol{E} \, \hat{\mathrm{curl}} \bar{\boldsymbol{F}} - \Big(\frac{1}{2}x + ika\Big) \hat{\mathrm{curl}} \boldsymbol{E} \bar{F}_\xi \\
&- \Big(\frac{1}{2}x - ika\Big) E_\xi \hat{\mathrm{curl}} \bar{\boldsymbol{F}} + \frac{1}{4} E_\xi \bar{F}_\xi \Big] - (ka)^2 a^{-3} \mathrm{jac}_a E_x \bar{F}_x \Big\} \, dx d\xi \\
&+ ika \int a^{-1} \mathrm{jac}_a^{-1} E_\xi(\cdot, 0) \bar{F}_\xi(\cdot, 0) \, d\xi \,.
\end{aligned} \tag{34}$$

Leading terms in x are identical as in 3D and suggest the use of the same radial shape functions.

4. Stability. Substituting in (23) $\boldsymbol{F} = \boldsymbol{\nabla} v$, where v is a scalar-valued test function with support in the far-field domain, we learn that any solution to variational formulation (8) satisfies automatically a variational compatibility condition, that implies the corresponding (second order) differential equation and a radiation condition at infinity. These compatibility conditions *do not* coincide with the usual continuity equation obtained directly from the weak form of the Maxwell equations. Equivalently, we could substitute in original variational formulation,

$$\boldsymbol{F} = e^{-ika(x^{-1}-1)} \boldsymbol{\nabla} v = \boldsymbol{\nabla}(e^{-ika(x^{-1}-1)} v) + ike^{-ika(x^{-1}-1)} x^{-2} v \boldsymbol{e}_x \,. \tag{35}$$

Only substitution of a gradient for the test function $\boldsymbol{F}$ results in the weak form of the continuity equation. Notice that the second term in (35) modifies only the radial component of the gradient and, therefore it can also be done in a situation when the infinite elements are coupled with finite elements. In other words, we can extend (35) by a regular gradient into the finite element domain and substitute the resulting test function into the variational formulation extending over the whole domain.

The Nédélec element (25) belongs to the standard family of polynomials satisfying the exact sequence property,

$$W_p \xrightarrow{\mathrm{grad}} \mathbf{Q}_p \xrightarrow{\mathrm{curl}} \mathbf{V}_p \xrightarrow{\mathrm{div}} Y_p \,, \tag{36}$$

where,

$$\begin{aligned}
W_p = &\ \mathcal{P}^{p_1} \otimes \mathcal{P}^{p_2} \otimes \mathcal{P}^{p_x} \\
\mathbf{Q}_p = &\ (\mathcal{P}^{p_1-1} \otimes \mathcal{P}^{p_2} \otimes \mathcal{P}^{p_x}) \times (\mathcal{P}^{p_1} \otimes \mathcal{P}^{p_2-1} \otimes \mathcal{P}^{p_x}) \\
&\times (\mathcal{P}^{p_1} \otimes \mathcal{P}^{p_2} \otimes \mathcal{P}^{p_x-1}) \\
\mathbf{V}_p = &\ (\mathcal{P}^{p_1} \otimes \mathcal{P}^{p_2-1} \otimes \mathcal{P}^{p_x-1}) \times (\mathcal{P}^{p_1-1} \otimes \mathcal{P}^{p_2} \otimes \mathcal{P}^{p_x-1}) \\
&\times (\mathcal{P}^{p_1-1} \otimes \mathcal{P}^{p_2-1} \otimes \mathcal{P}^{p_x}) \\
Y_p = &\ \mathcal{P}^{p_1-1} \otimes \mathcal{P}^{p_2-1} \otimes \mathcal{P}^{p_x-1} .
\end{aligned} \tag{37}$$

Denoting by Au the multiplication by the far-field pattern,

$$Au = e^{-ika(x^{-1}-1)} u \,,$$

we see that the infinite element shape functions incorporating the factor, belong to a modified exact sequence,

$$W_p^{exp} \xrightarrow{\text{Agrad}A^{-1}} \mathbf{Q}_p^{exp} \xrightarrow{\text{Acurl}A^{-1}} \mathbf{V}_p^{exp} \xrightarrow{\text{Adiv}A^{-1}} Y_p^{exp} \,, \tag{38}$$

where spaces $W_p^{exp}, \mathbf{Q}_p^{exp}, \mathbf{V}_p^{exp}, Y_p^{exp}$, have been obtained by multiplying the polynomials from the original spaces with the exponential factor. This situation is similar to the Bloch approximations studied by Dobson and Pasciak [14].

The 2D case is fully analogous.

As usual we can build a stabilized variational formulation by introducing a Lagrange multiplier p,

$$\begin{cases} a(\boldsymbol{E}, \boldsymbol{F}) + \overline{c(\boldsymbol{F}, p)} = l(\boldsymbol{F}), & \forall \boldsymbol{F} \\ c(\boldsymbol{E}, v) = l(\boldsymbol{\nabla} v), & \forall v \,. \end{cases} \tag{39}$$

Here $a(\boldsymbol{E}, \boldsymbol{F})$ and $l(\boldsymbol{F})$ denote the sesquilinear and antilinear forms corresponding to the variational formulation (8), and the sesquilinear form $c(\boldsymbol{E}, v)$ has been obtained by using the substitution (35),

$$c(\boldsymbol{E}, v) := k^{-1} a(\boldsymbol{E}, \boldsymbol{F}) \,. \tag{40}$$

Incorporating one power of k into the Langrange multiplier, improves the stability properties for $k \to 0$ but it does yield a formulation uniformly stable in k as for bounded domains.

REMARK 3. Notice that the space for the Lagrange multiplier *does not* coincide with the corresponding space for the IE discretization for the Helmholtz equation [13]. This is related to the fact that the Lagrange multiplier enters the stabilized formulation only through its gradient. The gradient decays at infinity but the Lagrange multiplier *does not* and, therefore, the corresponding element cannot be used to approximate the Helmholtz problem. For the standard elements, space W is used to solve acoustics in terms of pressure, space $\boldsymbol{Q}$ is for Maxwell equations, and space $\boldsymbol{V}$ of $H(\text{div})$-conforming elements is used when discretizing the acoustics in terms of velocity. This nice analogy for the discussed infinite elements is lost.

5. Numerical experiments. All presented experiments are two-dimensional only.

5.1. Implementation details. The infinite element was implemented within *2Dhp90*, a two-dimensional code supporting automatic *hp*-adaptivity for both H^1- and $H(\text{curl})$- conforming hybrid meshes consisting of isoparametric quads and triangles [9]. Integration of infinite element stiffness matrix was done using the standard Gauss-Legendre quadrature, with the number of integration points equal to $p+1$ in the tangential, and $N+1$ in the radial directions, where p and N denote the order of the infinite element in the tangential and the radial directions, resp.

Automatic hp-adaptivity. We refer to [12] for details on the algorithm executing the automatic *hp*-adaptivity. Starting with an initial *coarse mesh*, we refine the mesh globally in both h and p, and solve the problem on the *fine mesh*. The next optimal coarse mesh is obtained then by minimizing the *projection-based interpolation error* of the fine grid solution with respect to the coarse grid. More precisely, the optimal coarse mesh is obtained by maximizing the rate with which the interpolation error decreases, as the current coarse grid undergoes selective, local *hp*-refinements. One of the primary goals of the presented research is to determine whether the strategy can be used for the scattering problems. The mesh optimization is restricted to the near-field (truncated) domain only, i.e. the infinite elements are treated as an implicit implementation of ABC's of arbitrary order.

Choice of radial order N. The infinite elements in the initial mesh are assumed to be *isotropic*, i.e. order N in the radial direction is set to the corresponding order p of the edge on the truncating circle. We always begin with elements of second order.

During the *hp*-refinements, edges on the truncating circle get p- or h-refined. Every time, the edge is p-refined, its IE radial order is updated, i.e. N is increased to $N+1$. We also increase the IE order when the edge is h-refined. Therefore, in presence of h-refinements, we encounter infinite elements with radial order N *greater* than the FE order p. This reflects the philosophy that any improvement in the approximation properties on the truncating circle, should be accompanied with the corresponding improvement in the radial direction as well.

In the presented experiments (due to software related limitations), the IE radial order has been restricted to $N < 9$.

5.2. Evaluation of the error. In all reported experiments, the error is computed in the $H(\text{curl})$–norm, integrated *only* over the near-field domain. This is in line with treating the infinite element as an implicit implementation of an Absorbing Boundary Condition (ABC) only. Evaluation of the error over the whole exterior domain should be done in a weighted Sobolev norm. Since, at present, we cannot prove any convergence result for the IE discretization, we shall restrict ourselves to the near-field domain only and will not claim any convergence over the whole exterior domain. The error is reported in percent of the semi-norm of the solution, defined over the same domain. The exact solution for the cylinder problem is computed using the standard formula involving Hankel functions [16]. For the wedge problem, the unknown solution is replaced with the solution on the hp-refined mesh, see [12].

Scattering of a plane wave on a PEC cylinder. Verification of the code. We begin with the standard example of scattering a plane wave,

$$\boldsymbol{E}^{inc} = -\frac{1}{ik}\boldsymbol{\nabla} \times e^{-ik\boldsymbol{e}\cdot\boldsymbol{c}}, \tag{41}$$

on a unit PEC cylinder. Here $\boldsymbol{e}$ specifies the direction of the incident wave, in our case $\boldsymbol{e} = (-1, 0)$.

We set the wave number to $k = \pi$, and truncate the infinite domain with a circle of radius $a = 3$. Fig. 2 displays convergence history for p-uniform and hp-adaptive refinements, starting with a mesh of 16 quadratic elements. The horizontal axis corresponds to the number of d.o.f. n displayed on the algebraic scale $n^{1/3}$, with the vertical axis presenting the error on the logarithmic scale. A straight line indicates the exponential convergence,

$$\text{error} \approx Ce^{\beta n^{1/3}},$$

predicted by the theory. Notice that the actual numbers displayed on the axes correspond to the quantities being displayed, i.e. the relative error in percent of the norm of the solution on the vertical axis, and the number of d.o.f. on the horizontal axis.

As expected, the uniform p refinements deliver exponential convergence, with the adaptive refinements delivering slightly worse results but the same rates. Fig. 3 shows the optimal hp mesh, corresponding to an error of 1 percent. Different colors indicate different (locally changing) order of approximation p ranging from $p = 1$ to $p = 8$ (color scale on the right). The distribution of orders clearly reflects the pattern of the solution.

Fig. 4 presents contour lines of the real part of the error function, for a uniform mesh of quartic elements. The values, indicated by the color scale on the right, range from -0.02 to 0.02. Along with the FE mesh, the graph displays a portion of the infinite elements corresponding to $0.5 < x < 1$. The solution in the IE domain seems to be actually better than in the finite

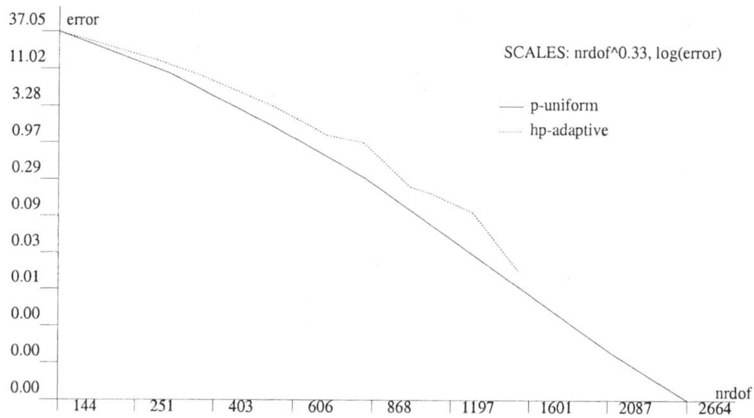

FIG. 2. *Scattering of a plane wave on a PEC cylinder. Convergence history for p-uniform and hp-adaptive refinements.*

element domain which indicates that lower order infinite elements would have been sufficient.

5.3. Scattering of a plane wave on a PEC wedge. The second example illustrates the resolution of singularities using the coupled *hp*-adaptive FE/IE discretizations. We have divided the cylinder from the previous example into four equal quadrants and kept just one of them. We set the wave number $k = \pi$, i.e. the distance between the truncating boundary and the object is equal to one wavelength. We start with an example of a typical, automatically obtained *hp* mesh corresponding to the incident wave coming from the NE direction (angle $\theta = 45^o$), and rather academic error level of 0.1 percent. Figures 5 and 6 present the optimal mesh, with three consecutive zooms showing details of the mesh around the lower corner. Fig. 7 presents convergence history for the problem. We start with an initial mesh of just seven elements of second order that clearly do not resolve the wave pattern (one second order element per wave length) to illustrate the principle that only the fine grid must be in the asymptotic convergence region. Consequently, the convergence curve consists roughly of two straight lines, the first one corresponding to the preasymptotic region, and the second one reflecting the actual exponential convergence.

5.4. Evaluation of RCS. We come to the final experiment reflecting the impact of adaptivity on evaluation of Radar Cross Section (RCS). For two-dimensional problems, the monostatic RCS reflects simply the far-field

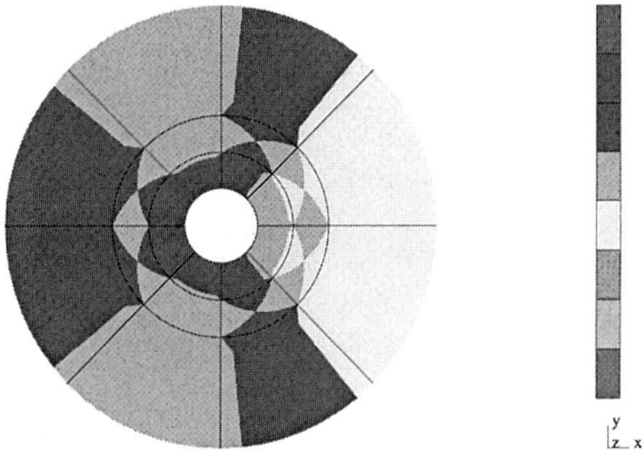

FIG. 3. *Scattering of a plane wave on a cylinder. Optimal hp mesh corresponding to 1 percent error. Different colors indicate different (locally changing) order of approximation p ranging from p = 1 to p = 8 (color scale on the right.*

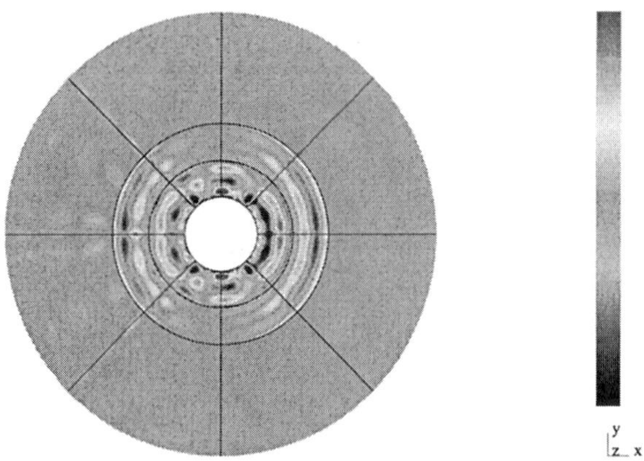

FIG. 4. *Scattering of a plane wave on a cylinder. Real part of the second component of the error function for a uniform mesh of quartic elements.*

pattern of the solution, and it is defined as follows

$$\lim_{r\to\infty} |\boldsymbol{E}(r\hat{\boldsymbol{x}})| r^{\frac{1}{2}} \,. \tag{42}$$

Here $\hat{\boldsymbol{x}}$ is a point on the unit circle corresponding to the direction of the incoming plane wave $\boldsymbol{e} = -\hat{\boldsymbol{x}}$, and $r^{\frac{1}{2}}$ compensates the decay rate of the solution.

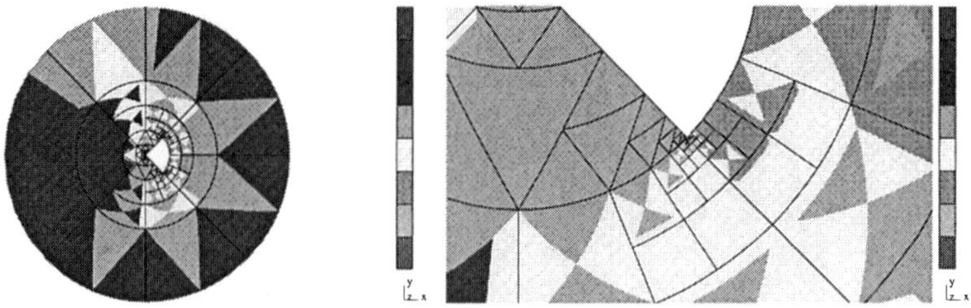

FIG. 5. *Scattering of a plane wave on a PEC wedge, $\theta = 45^o$. Optimal hp mesh for 0.1 percent error, with a 10 times zoom on the lower corner.*

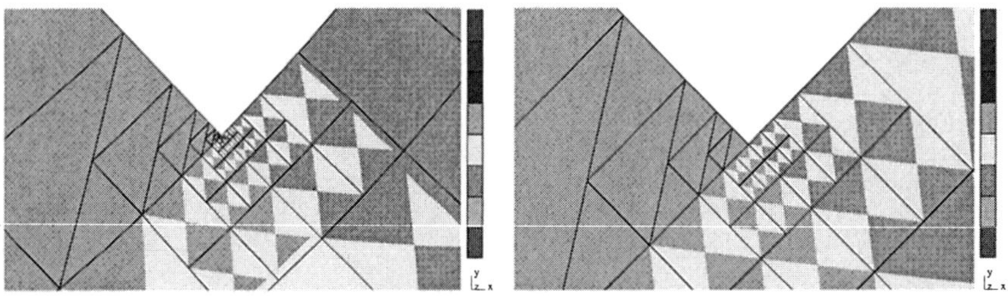

FIG. 6. *Scattering of a plane wave on a PEC wedge, $\theta = 45^o$. Optimal hp mesh for 0.1 percent error. Zooms on the lower corner with 100 and 1000 magnifications.*

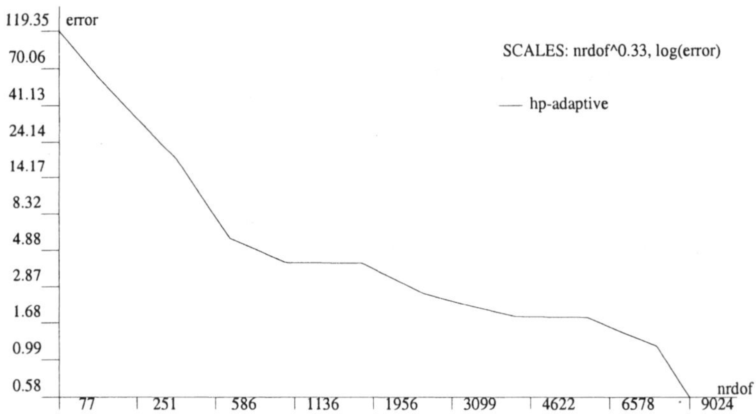

FIG. 7. *Scattering of a plane wave on a PEC wedge, $\theta = 45^o$. Convergence history for adaptive hp refinements.*

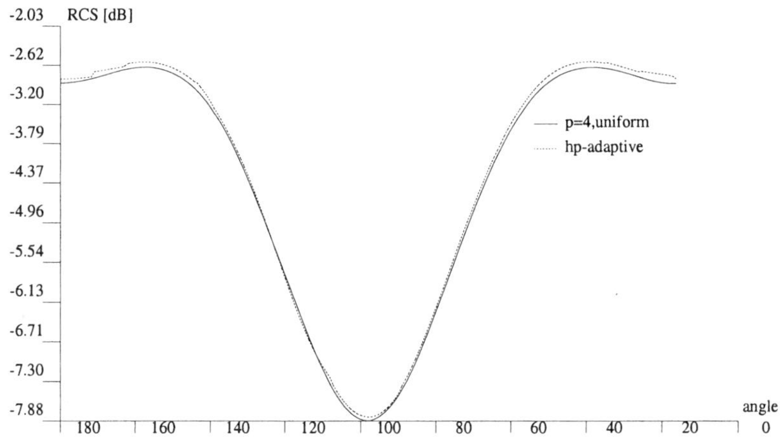

FIG. 8. *Scattering of a plane wave on a wedge. RCS in dB vs. the direction of the incident wave in degrees, for the uniform mesh of quartic elements (3-4 percent error range level) and hp-adaptive mesh (2 percent error).*

The infinite element discretization offers an inexpensive way of computing the RCS without the usual costly postprocessing involving integration of surface electric and magnetic currents on any closed surface surrounding the scatterer. As the far-field pattern is built into the ansatz for the approximate solution, one simply evaluates the approximate solution at $r = \infty (x = 0)$.

Fig. 8 presents RCS for the wedge problem evaluated using a uniform mesh of quartic elements and and hp-adaptive mesh. The choice of the uniform mesh reflects the usual practice of selecting a mesh that reproduces the wave form of the solution (two quartic elements per wavelength) and delivers an error in the range of 3-4 percent. The second curve corresponds to RCS evaluated using hp-adaptivity.

The hp meshes were obtained by requesting a two percent error level, at which several levels of hp-refinements resolve the structure of the singularities in the solution. For each direction of the incoming wave ($\theta = 180, 179, \ldots, 0$, left to right), the hp-adaptive algorithm was run starting with the optimal mesh for the previous direction, with the optimization procedure restarted from the initial mesh every 10 degrees. Except for a slight shift in the RCS level, the results are practically identical. Resolution of the singularities seems to have no impact on quality of the RCS computations.

6. Conclusions. We have presented a novel construction of an infinite element for the time-harmonic Maxwell scattering/radiation problems. The main idea consists in interpreting the integral over the exterior domain in the CPV sense, and building the known far-field pattern of the solution into the ansatz for the approximate solution. The CPV interpretation allows then for canceling the Lebesgue non-integrable terms.

The element space of shape functions including the far-field pattern, belongs to a family of spaces forming an exact sequence with modified operators reflecting the solution ansatz.

The presented numerical experiments indicate stability. The infinite element has successfully been coupled with *hp* finite elements in context of the energy driven, automatic *hp*-adaptivity. As expected, the method converges exponentially.

The Bubnov-Galerkin discretization offers a simplicity of the formulation and, perhaps, may be explored in a theoretical convergence analysis for the exterior wave propagation problems. For the two simple scattering problems studied in the paper, results obtained using the new formulation and the earlier Petrov-Galerkin discretization [6, 7] are identical and it is not clear that either formulation offers an advantage over the other one from the computational point of view. Both formulations result in a stiffness matrix that is neither hermitian nor complex-symmetric, forcing the use of a general solver for complex, non-symmetric matrices. The structure of the stiffness matrix does reflect though the physics of the problem - the domain contribution in (8) is hermitian and the term corresponding to the surface integral at infinity reflecting the radiation damping is in the form of the product of a hermitian matrix and imaginary impedance term ik. The same structure of the stiffness matrix is encountered when studying problems in bounded domains with conductive materials or impedance boundary conditions.

We hope to report soon results of analogous numerical experiments in three dimensions.

The question asked in the title of this paper: *Is the or an exact sequence property important?*, remains unanswered. Both formulations of the infinite element: the one based on the standard exact sequence [6], and the one discussed in this paper, seem to work. Both deliver exponential convergence for the cylinder test case, and both behave well when coupled with adaptive *hp* finite elements. Unfortunately, as mentioned in the Introduction, in neither case, so far, we can prove the convergence.

REFERENCES

[1] M. Ainsworth and J. Coyle. Hierarchic *hp*-edge element families for maxwell's equations on hybrid quadrilateral/triangular meshes. *Comput. Methods Appl. Mech. Engrg.*, 190: 6709–6733, 2001.

[2] R.J. Astley, G.J. Macaulay, and J.P. Coyette. Mapped wave envelope elements for

acoustical radiation and scattering. *Journal of Sound and Vibration*, 170(1): 97–118, 1994.

[3] J.-P. Bérenger. A perfectly matched layer for the absorption of electromagnetic waves. *Journal of Computational Physics*, 114: 185–200, 1994.

[4] P. Bettess. *Infinite Elements.* Penshaw Press, 1992.

[5] D.S. Burnett. A three-dimensional acoustic infinite element based on a prolate spheroidal multipole expansion. *Journal of the Acoustical Society of America*, 96: 2798–2816, 1994.

[6] W. Cecot, L. Demkowicz, and R. Rachowicz. A two-dimensional infinite element for Maxwell's equations. *Comput. Methods Appl. Mech. Engrg.*, 188: 625–643, 2000.

[7] W. Cecot, L. Demkowicz, and W. Rachowicz. An *hp*-adaptive finite element method for electromagnetics. part 3: A three-dimensional infinite element for Maxwell's equations. *Int. J. Num. Meth. Eng.*, 57: 899–921, 2003.

[8] L. Cremers, K.R. Fyfe, and J.P. Coyette. A variable order infinite acoustic wave envelope element. *Journal of Sound and Vibrations*, 17(4): 483–508, 1994.

[9] L. Demkowicz. 2D *hp*-adaptive finite element package (2Dhp90). version 2.0. Technical Report 06, TICAM, 2002.

[10] L. Demkowicz. Finite element methods for Maxwell equations. In E. Stein R. de Borst, T.J.R. Hughes, editor, *Encyklopedia of Computational Mechanics*, chapter 26, pages 723–737. John Wiley & Sons, Ltd, Chichester, 2004.

[11] L. Demkowicz and M. Pal. An infinite element for Maxwell's equations. *Comput. Methods Appl. Mech. Engrg.*, 164: 77–94, 1998.

[12] L. Demkowicz, W. Rachowicz, and Ph. Devloo. A fully automatic *hp*-adaptivity. *Journal of Scientific Computing*, 17(1-3): 127–155, 2002.

[13] L. Demkowicz and J. Shen. A few new (?) facts about infinite elements. Technical Report 60, ICES, 2004.

[14] D.C. Dobson and J.E. Pasciak. Analysis of an algorithm for computing electromagnetic Bloch modes using Nedelec spaces. *Comp. Meth. Appl. Math.*, 1(2): 138–153, 2001.

[15] D. Givoli. *Numerical Methods for Problems in Infinite Domains.* Elsevier, Amsterdam, 1992.

[16] D.S. Jones. *Acoustic and Electromagntic Waves.* Oxford Science Publications, 1986.

[17] J.C. Nedelec. Mixed finite elements in $\mathbb{R}^3$. *Numer. Math.*, 35: 315–341, 1980.

[18] W. Rachowicz and A. Zdunek. An *hp*-adaptive finite element method for scattering problems in computational electromagnetics. *International Journal for Numerical Methods in Engineering*, 2005.

COVOLUME DISCRETIZATION OF DIFFERENTIAL FORMS

R.A. NICOLAIDES* AND K.A. TRAPP†

1. Introduction. The language of differential forms provides the most compact expression of many partial differential equations occurring in physical applications. We have in mind Maxwell's equations and the Laplace-Beltrami equation as instances. On the other hand, discretization of equations in differential forms is not so well studied. In this paper we construct a theory of discrete differential forms and apply it to solving some basic equations.

The theory presented below is an abstraction of a class of computational techniques collectively designated as *covolume algorithms*. Covolume algorithms are a class of compatible discretizations for computing vector fields from partial differential equations. By this we mean that basic vector identities such as curlgrad = 0 and divcurl = 0 are preserved and that scalar and vector potentials exist, all within the framework of the discrete calculus. In the covolume setting such identities and relations appear in a natural, almost obvious way. A basic reference to covolume techniques is [6] where they are applied to systems of the form

$$\left\{ \begin{array}{ccc} \operatorname{div}\vec{u} = \rho & \text{in} & \Omega \\ \operatorname{curl}\vec{u} = \vec{\omega} & \text{in} & \Omega \\ \vec{u}\cdot\vec{n} = 0 & \text{on} & \partial\Omega \end{array} \right\}$$

with appropriate solvability conditions and where Ω may be multiply connected. The three dimensional version of this system is treated in [8].

In differential form notation the system above is

$$\left\{ \begin{array}{c} d * u = \rho \\ du = \omega \end{array} \right\}$$

where d denotes the usual differential operator d and u, ρ, and ω are to be interpreted respectively as 1, 3, and 2-forms. The Hodge star operator $*$ is a linear operator that takes k-forms to $(n-k)$-forms. It depends on the inner product on the space of forms. Because the divergence and curl operators naturally act on 2 and 1-forms respectively we require the Hodge star operator to transform u into a 2-form. We explain later how the covolume approach relates the d and $d*$ operators through the use of dual mesh systems. There is a somewhat surprising and beautiful relationship between these apparently distinct concepts.

*Carnegie Mellon University, Pittsburgh, PA 15213 (nic@cmu.edu).
†University of Richmond, Richmond, VA 23173 (ktrapp@richmond.edu).

Another very useful application of covolume techniques is to Maxwell's equations. In standard form these equations are

$$\left\{\begin{array}{ll} -\operatorname{curl}\vec{E} = \dfrac{d\vec{B}}{dt} & \text{in} \quad \Omega \\ \operatorname{curl}\vec{H} = \dfrac{d\vec{D}}{dt} & \text{in} \quad \Omega \\ \operatorname{div}\vec{D} = \rho & \text{in} \quad \Omega \\ \operatorname{div}\vec{B} = 0 & \text{in} \quad \Omega \end{array}\right\}.$$

From this basic system, many different wave equations can be derived by differentiation and elimination of variables. For instance we obtain the standard wave equations for the $\vec{E}$ and $\vec{H}$ fields in that way. Further, we can obtain the usual wave equations for the scalar and vector potentials in various gauges. The covolume framework is sufficiently imitative (mimetic) that similar manipulations can be performed on the covolume Maxwell system to produce standard discretizations of the different wave equations. In particular there are discrete scalar and vector potentials which satisfy appropriate discrete wave equations. In [7] traditional error estimates are provided for this covolume algorithm.

There is more than one way to write Maxwell's equations in differential form notation, depending on whether time is included as a differential form variable and depending on the description of the unknowns. For instance they may be lumped together in a single variable. In our context, where spatial discretizations only are under consideration, the differential forms expression of Maxwell's equations is just

$$\left\{\begin{array}{c} -dE = (*H)^{\cdot} \\ dH = (*E)^{\cdot} \\ d*E = \rho \\ d*H = 0 \end{array}\right\}.$$

A good description of the other possible formats can be found, for instance in [4,5] or numerous other sources.

Turning now to the material presented below, one result of our work is that for any equation in differential forms a recipe is given (and justified) for translating it over to a discrete framework for computer implementation. This recipe is almost as simple as it could be: just replace d and $d*$ and their relatives with certain discrete operators that we provide. The d and $d*$ operators are fully compatible in that we have analogs of the main theorems of differential forms, such as Stokes' theorem, Poincaré's lemma, Hodge star operations and so on. Moreover, these relations are quite transparent in our framework.

Applying the discretizations to standard equations expressed in differential forms notation delivers the known covolume approximations. The error estimates for these discretizations can then be taken over directly to justify the differential forms approximations. This fact encourages belief in the correctness of the discretization technique in situations where error estimates have not yet been obtained.

Section 3 of the paper contains, among other things, a novel application of our results to discretizing the Laplace-Beltrami operator on a compact manifold. In this application the new theory is used in its fullest form, in that many of the techniques we develop come into play, including exact sequences, the discrete Hodge operation and their related circle of ideas. The section which precedes sets the groundwork for the Laplace-Beltrami discretization. It develops a coherent theory of discrete differential forms which resembles the classical continuous treatment, at least in those aspects which do not involve statements about coordinates.

2. Discrete differential forms. The essential ingredient in discretizing differential forms is the Delaunay-Voronoi mesh system. Most obviously, it defines the discrete edges, surfaces, and volumes over which the discrete forms can be integrated. It is from the geometric duality of simplex and complex that the dual forms, dual differential operator d, and the Hodge map are defined. From these, in turn, follows a discrete Laplace-Beltrami operator and other discretizations.

2.1. Simplices, orientation, volume, boundary operator. Let Ω be an n-dimensional differentiable manifold equipped with a Delaunay-Voronoi mesh system [3]. The primal mesh is made up of simplices which in one, two and three dimensions are edges, triangles and tetrahedra respectively. For each dimension $k \leq n$ the set of k-dimensional simplices, or k-simplices, is written $\{\sigma_i^k\}_{i=1}^{N_k}$ and the circumcenter of the simplex σ_i^k is denoted p_i^k.

Each simplex, σ_i^k, has an associated dual complex, $\tilde{\sigma}_i^k$, which is an $(n-k)$-dimensional polyhedron. It is important to note that while σ_i^k is k-dimensional, $\tilde{\sigma}_i^k$ is $(n-k)$-dimensional. From the Delaunay-Voronoi property, it follows that each simplex and its dual complex are orthogonal and intersect at the circumcenter of the simplex, i.e. $\sigma_i^k \bigcap \tilde{\sigma}_i^k = p_i^k$. See Figure 1.

An orientation is placed on the highest-order simplices and then each successive lower dimensional simplex inherits a relational orientation to each simplex of one degree higher that it bounds. The boundary operator ∂ then acts on a k-simplex to yield a chain of $(k-1)$-simplices in the usual way. We write this $\partial\sigma_i^k = \sum\limits_j \sigma_{i_j}^{k-1}$ where the double subscript i_j denotes a signed simplex corresponding to the orientation and connectivity of the i and i_j simplices. It follows from the connectivity of the mesh that the boundary operator applied twice to a simplex yields an empty chain (i.e. $\partial\partial\sigma_i^k = 0$).

An orientation is also placed on the dual complexes. From the connectivity of these complexes the dual boundary operator $\tilde{\partial}$ takes a (dual) k-complex to a chain of $(k+1)$-complexes in a similar fashion, i.e.

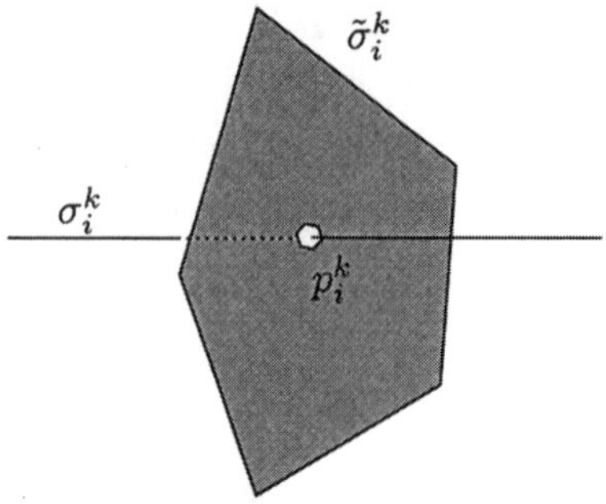

FIG. 1. *The intersection of* σ_i^k *and* $\tilde{\sigma}_i^k$ *at* p_i^k.

$\tilde{\partial}\tilde{\sigma}_i^k = \sum_j \tilde{\sigma}_{i_j}^{k+1}$.

We denote the area of a simplex σ_i^k by $A(\sigma_i^k)$ and for 0-simplices we define $A(\sigma_i^0) = 1$. Similarly, the area of a dual complex is written $A(\tilde{\sigma}_i^k)$. Depending on orientation, the area of $\sigma_{i_j}^k$ may be positive or negative.

For each circumcenter p_i^k we define the volume $V_i^k := A(\sigma_i^k)A(\tilde{\sigma}_i^k) > 0$. Thus for each $k \leq n$ we have a discrete volume form on the manifold, $V^k = \sum_i V_i^k$.

2.2. Discrete forms, d operator, integration and Stokes theorem. A *discrete* differential k-form, ω^k can act only on k-simplices. The collection of discrete differential k-forms on the discrete differentiable manifold (the Delaunay-Voronoi mesh) will be denoted $\mathcal{A}^k$. Each element $\omega^k \in \mathcal{A}^k$ is indexed by the k-simplices on which it acts. We define ω^k as the formal sum $\sum_i^{N_k} u_i^k \sigma_i^k$ where u^k is a vector in $\mathbb{R}^{N_k}$. A 0-form is just a function defined on the nodes of the mesh.

The inner product on the space of k-forms $\mathcal{A}^k$ is defined using the volume form V^k. Thus for two elements $\omega^k = \sum_i u_i^k \sigma_i^k$ and $\eta^k = \sum_i v_i^k \sigma_i^k$ in $\mathcal{A}^k$, their inner product is defined to be

$$(\omega^k, \eta^k)_{V^k} := \sum_i u_i^k v_i^k V_i^k.$$

We borrow the integral sign to describe the action of a discrete k-form on a k-simplex. Thus, the "integral" of a discrete k-form, ω^k, over a k-simplex, σ_i^k, is defined to be

$$\int_{\sigma_i^k} \omega^k := u_i^k A(\sigma_i^k).$$

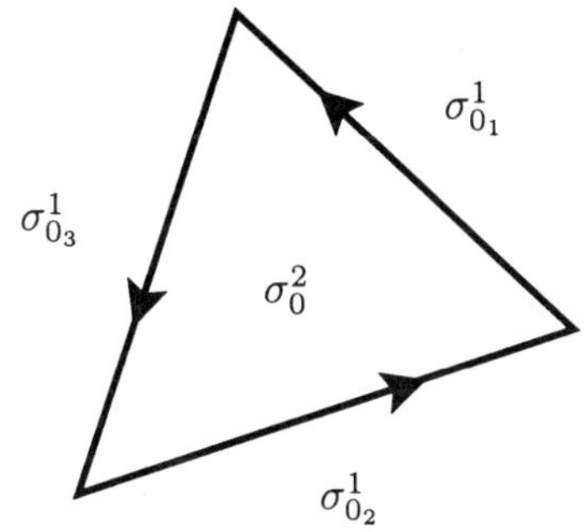

FIG. 2. *The circulation* $\int_{\sigma_0^2} d\omega^1 = u_0^2 A(\sigma_0^2)$.

Integrating over the boundary of a $(k+1)$-simplex yields

$$\int_{\partial\sigma_i^{k+1}} \omega^k = \sum_j \int_{\sigma_{i_j}^k} \omega^k = \sum_j u_{i_j}^k A(\sigma_{i_j}^k),$$

where, it is important to recall that the double subscript i_j denotes a signed simplex and thus $A(\sigma_{i_j}^k)$ is a signed area.

We can now introduce the, also misappropriated, discrete differential operator $d : \mathcal{A}^k \longrightarrow \mathcal{A}^{k+1}$. For $\omega^k = \sum_i u_i^k \sigma_i^k$ we define

$$d\omega^k = \sum_i u_i^{k+1}\sigma_i^{k+1} \quad \text{where} \quad u_i^{k+1} := \sum_j \frac{u_{i_j}^k A(\sigma_{i_j}^k)}{A(\sigma_i^{k+1})}.$$

To see how this easily translates back into the covolume framework, consider the triangle σ_0^2. The d operator acting on one-forms is just the differential operator curl and the integral $\int_{\sigma_0^2} d\omega^1 = u_0^2 A(\sigma_0^2)$ is just the circulation $u_{0_1}^1 A(\sigma_{0_1}^1) + u_{0_2}^1 A(\sigma_{0_2}^1) + u_{0_3}^1 A(\sigma_{0_3}^1)$ about the triangle, or surface, σ_0^2. See Figure 2.

Now, with this definition of the d operator, the integration of forms on simplices, and the boundary operator ∂ acting on simplices we have the following calculation for all forms and simplices:

$$\int_{\sigma_i^{k+1}} d\omega^k = u_i^{k+1} A(\sigma_i^{k+1}) = \sum_j \frac{u_{i_j}^k A(\sigma_{i_j}^k)}{A(\sigma_i^{k+1})} A(\sigma_i^{k+1})$$

$$= \sum_j u_{i_j}^k A(\sigma_{i_j}^k) = \int_{\partial\sigma_i^{k+1}} \omega^k.$$

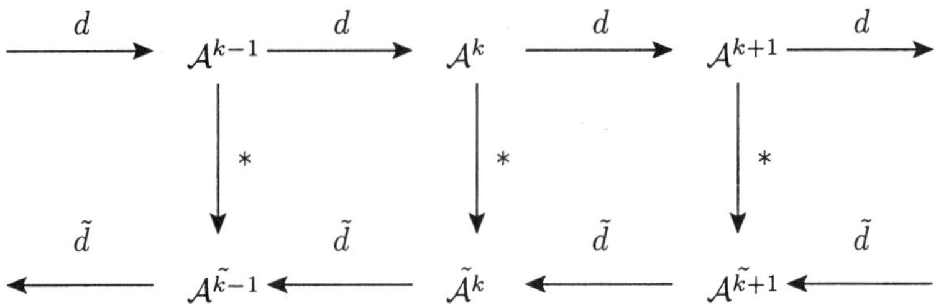

FIG. 3. *The operators d, $*$, and $\tilde{d}$ acting on the spaces of discrete differential forms.*

Thus, we have the discrete analog of Stokes' Theorem:

$$\int_{\sigma_i^{k+1}} d\omega^k = \int_{\partial\sigma_i^{k+1}} \omega^k.$$

Together with the boundary operator property $\partial\partial\sigma_i^k = 0$, the discrete Stokes' Theorem yields the appropriate discrete analog of Poincaré's Lemma: for every form ω^k, $dd\omega^k = 0$.

2.3. Dual forms and the Hodge star and decomposition. The set of dual discrete differential k-forms will be denoted $\tilde{\mathcal{A}}^k$ and we write an element $\tilde{\omega}^k$ as a sum over the dual k-complexes. Thus $\tilde{\omega}^k = \sum_i \tilde{u}_i^k \tilde{\sigma}_i^k$. Recall that a dual k-complex $\tilde{\sigma}_i^k$ is an $(n-k)$-dimensional polyhedron therefore a dual k-form should be considered an $(n-k)$-form on the dual mesh.

The dual differential operator $\tilde{d} : \tilde{\mathcal{A}}^k \longrightarrow \tilde{\mathcal{A}}^{k-1}$, is defined to act on a dual k-form in the following way:

$$\tilde{d}\tilde{\omega}^k = \sum_i \tilde{u}_i^{k-1} \tilde{\sigma}_i^{k-1} \quad \text{where} \quad \tilde{u}_i^{k-1} := \sum_j \frac{\tilde{u}_{i_j}^k A(\tilde{\sigma}_{i_j}^k)}{A(\tilde{\sigma}_i^{k-1})}.$$

It is important to note that Poincaré's Lemma also applies to the dual differential operator. Thus, $\tilde{d}\tilde{d}\tilde{\omega}^k = 0$ for all dual k-forms.

Just as the Delaunay-Voronoi mesh system provides duality between the simplices and complexes, the Hodge star operator creates a dual association between forms. The Hodge star mapping, $* : \mathcal{A}^k \longrightarrow \tilde{\mathcal{A}}^k$, takes the form $\omega^k = \sum_i u_i^k \sigma_i^k$ to the dual k-form

$$*\omega^k = \tilde{\omega}^k = \sum_i u_i^k \tilde{\sigma}_i^k.$$

The diagram in Figure 3 summarizes the behavior of the three maps d, $*$, and $\tilde{d}$.

Because the Hodge star operator is invertible we can define a second differential operator δ which takes $\mathcal{A}^k$ to $\mathcal{A}^{k-1}$. Define $\delta\omega^k := *^{-1}\tilde{d} * \omega^k$. This operator is the formal adjoint of the differential operator d with respect to the volume form inner product $(\cdot,\cdot)_{V^k}$, i.e.

$$(\omega^k, d\eta^{k-1})_{V^k} = (\delta\omega^k, \eta^{k-1})_{V^{k-1}}. \tag{2.3.1}$$

We can see this by considering $\omega^k \in \mathcal{A}^k$ and $\eta^{k-1} \in \mathcal{A}^{k-1}$. For $\omega^k = \sum_i u_i^k \sigma_i^k$ we have

$$\delta\omega^k = *^{-1}\tilde{d} * \omega^k = *^{-1}\tilde{d}\sum_i u_i^k \tilde{\sigma}_i^k = *^{-1}\sum_i u_i^{k-1}\tilde{\sigma}_i^{k-1} = \sum_i u_i^{k-1}\sigma_i^{k-1}$$

where $u_i^{k-1} = \sum_j \frac{u_{i_j}^k A(\tilde{\sigma}_{i_j}^k)}{A(\tilde{\sigma}_i^{k-1})}$. If we let $\eta^{k-1} = \sum_i v_i^{k-1}\sigma_i^{k-1}$ then $d\eta^{k-1} = \sum_i v_i^k \sigma_i^k$ where $v_i^k = \sum_j \frac{v_{i_j}^{k-1} A(\sigma_{i_j}^{k-1})}{A(\sigma_i^k)}$.

Putting this together with the definition of the volume forms we have

$$\begin{aligned}
(\omega^k, d\eta^{k-1})_{V^k} &= (\sum_i u_i^k \sigma_i^k, \sum_i v_i^k \sigma_i^k)_{V^k} \\
&= \sum_i u_i^k v_i^k A(\sigma_i^k) A(\tilde{\sigma}_i^k) \\
&= \sum_i u_i^k [\sum_j v_{i_j}^{k-1} A(\sigma_{i_j}^{k-1})] A(\tilde{\sigma}_i^k) \\
&= \sum_i u_i^k A(\tilde{\sigma}_i^k)[\sum_j v_{i_j}^{k-1} A(\sigma_{i_j}^{k-1})] \\
&= \sum_n v_n^{k-1} A(\sigma_n^{k-1})[\sum_j u_{n_j}^k A(\tilde{\sigma}_{n_j}^k)] \\
&= \sum_n v_n^{k-1} A(\sigma_n^{k-1}) u_n^{k-1} A(\tilde{\sigma}_n^{k-1}) \\
&= (\sum_n u_n^{k-1}\sigma_n^{k-1}, \sum_n v_n^{k-1}\sigma_n^{k-1})_{V^{k-1}} \\
&= (\delta\omega^k, \eta^{k-1})_{V^{k-1}}.
\end{aligned}$$

Composing the d and δ operators yield discrete differential operators from the space of discrete k-forms to itself. The Laplace-Beltrami operator L is the sum of these two compositions: $L := d\delta + \delta d$. The volume form inner product ensures that a discrete k-form ω^k satisfies $L\omega^k = 0$ if and only if $\delta\omega^k = 0$ and $d\omega^k = 0$. The kernel of L (contained in $\mathcal{A}^k$) is the space of discrete harmonic k-forms and we denote it by H^k.

We now have an appropriate discrete analog of the Hodge decomposition theorem which is the differential forms analog to the Helmholtz decomposition theorem for vector fields. The space of differential forms $\mathcal{A}^k$ with inner product V^k can be written as the direct sum of three different subspaces:

$$\mathcal{A}^k = d\mathcal{A}^{k-1} \oplus \delta\mathcal{A}^{k+1} \oplus H^k.$$

This decomposition follows directly from (2.3.1) and Poincaré's Lemma.

3. Applications to differential equations. As an application of this structure of Delaunay-Voronoi meshes and discrete differential forms we consider the discretization of Poisson's equation, $\Delta u = f$.

In this first example we consider the dual Laplace-Beltrami operator $\tilde{L}$ which acts on discrete dual forms. Just as above, the differential operator $\tilde{L}$ takes the space of dual k-forms to itself. It is the sum of the dual operator compositions $\tilde{d}\tilde{\delta}$ and $\tilde{\delta}\tilde{d}$ where $\tilde{\delta} := *d*^{-1}$. See Figure 3.

The discrete differential equation is written

$$\tilde{L}\tilde{\psi}^0 = \tilde{f}^0$$

where $\tilde{\psi}^0 = \sum_i u_i^0 \tilde{\sigma}_i^0$ and $\tilde{f}^0 = \sum_i f_i^0 \tilde{\sigma}_i^0$. Since $\tilde{\psi}^0$ is a dual 0-form and $\tilde{d} : \tilde{\mathcal{A}}^0 \longrightarrow \tilde{\mathcal{A}}^{-1}(:= 0)$ the operator $\tilde{L}$ is simply $\tilde{L} = \tilde{d}\tilde{\delta} + \tilde{\delta}\tilde{d} = \tilde{d}\tilde{\delta}$. The discretization can be written

$$\begin{aligned}
\tilde{d}\tilde{\delta}\tilde{\psi}^0 &= \tilde{d} * d *^{-1} \tilde{\psi}^0 \\
&= \tilde{d} * d\psi^0 \\
&= \tilde{d} * (\sum_i u_i^1 \sigma_i^1) \qquad \text{where } u_i^1 := \sum_j \frac{u_{i_j}^0 A(\sigma_{i_j}^0)}{A(\sigma_i^1)} = \frac{u_{i_1}^0 + u_{i_2}^0}{A(\sigma_i^1)} \\
&= \tilde{d}(\sum_i u_i^1 \tilde{\sigma}_i^1) \\
&= \sum_i U_i^0 \tilde{\sigma}_i^0 \qquad \text{where } U_i^0 = \sum_j \frac{u_{i_j}^1 A(\tilde{\sigma}_{i_j}^1)}{A(\tilde{\sigma}_i^0)}.
\end{aligned}$$

Therefore, integrating over a dual 0-complex $\tilde{\sigma}_i^0$, the equation $\int_{\tilde{\sigma}_i^0} \tilde{L}\tilde{\psi}^0 = \int_{\tilde{\sigma}_i^0} \tilde{f}^0$ is just $U_i^0 A(\tilde{\sigma}_i^0) = f_i^0 A(\tilde{\sigma}_i^0)$. In terms of the unknowns $\{u_i^0\}$ this i^{th} equation can be written:

$$\begin{aligned}
&\left(\sum_j \frac{u_{i_j}^1 A(\tilde{\sigma}_{i_j}^1)}{A(\tilde{\sigma}_i^0)} \right) A(\tilde{\sigma}_i^0) = f_i^0 A(\tilde{\sigma_i^0}) \\
\Rightarrow &\sum_j u_{i_j}^1 A(\tilde{\sigma}_{i_j}^1) = f_i^0 A(\tilde{\sigma_i^0}) \\
\Rightarrow &\sum_j \frac{u_{i_{j1}}^0 + u_{i_{j2}}^0}{A(\sigma_{i_j}^1)} A(\tilde{\sigma}_{i_j}^1) = f_i^0 A(\tilde{\sigma_i^0}) \ .
\end{aligned}$$

In $\mathbb{R}^2$ and $\mathbb{R}^3$ this would reduce to the standard covolume Laplacian stencil, where $A(\sigma^1_{i_j})$ is just the length of the edge connecting the two nodes $\sigma^0_{i_{j_1}}$ and $\sigma^0_{i_{j_2}}$, $A(\tilde{\sigma}^1_{i_j})$ is the area of the coface intersecting that edge, and finally $A(\tilde{\sigma}^0_i)$ is the volume of the covolume with the coface as part boundary. The values $u^0_{i_j}$ are signed so that the sum $u^0_{i_{j_1}} + u^0_{i_{j_2}}$ may be a difference.

As another example of the applications of these discrete differential forms, we have the means to discretize the vector Laplace-Beltrami operator, which in $\mathbb{R}^3$ reduces to $\operatorname{grad}\operatorname{div}\vec{\phi} - \operatorname{curl}\operatorname{curl}\vec{\phi} = \vec{f}$.

For a 1-form $\phi^1 = \sum_i u^1_i \sigma^1_i$ and 1-form $f^1 = \sum_i f^1_i \sigma^1_i$ we can discretize the equation $L\phi^1 = f^1$ as follows:

$$
\begin{aligned}
L\phi^1 &= d\delta\phi^1 + \delta d\phi^1 \\
&= *^{-1}\tilde{d} * \phi^1 + *^{-1}\tilde{d} * d\phi^1 \\
&= d *^{-1} \tilde{d} \sum_i u^1_i \tilde{\sigma}^1_i + *^{-1}\tilde{d} * \sum_i u^2_i \sigma^2_i \\
&\qquad\qquad \text{where } u^2_i := \textstyle\sum_{j=1}^3 \frac{u^1_{i_j} A(\sigma^1_{i_j})}{A(\sigma^2_i)} \\
&= d *^{-1} \sum_i u^0_i \tilde{\sigma}^0_i + *^{-1}\tilde{d} \sum_i u^2_i \tilde{\sigma}^2_i \\
&\qquad\qquad \text{where } u^0_i := \textstyle\sum_j \frac{u^1_{i_j} A(\tilde{\sigma}^1_{i_j})}{A(\tilde{\sigma}^0_i)} \\
&= d \sum_i u^0_i \sigma^0_i + *^{-1} \sum_i W^1_i \tilde{\sigma}^1_i \\
&\qquad\qquad \text{where } W^1_i := \textstyle\sum_j \frac{u^2_{i_j} A(\tilde{\sigma}^2_{i_j})}{A(\tilde{\sigma}^1_i)} \\
&= \sum_i U^1_i \sigma^1_i + \sum_i W^1_i \sigma^1_i \\
&\qquad\qquad \text{where } U^1_i := \textstyle\sum_{j=1}^2 \frac{u^0_{i_j} A(\sigma^0_{i_j})}{A(\sigma^1_i)} = \frac{u^0_{i_1} + u^0_{i_2}}{A(\sigma^1_i)}.
\end{aligned}
$$

If we integrate these forms, $L\phi^1$ and f^1 over each edge σ_i then the ith equation is just $(U^1_i + W^1_i)A(\sigma^1_i) = f^1_i A(\sigma^1_i)$ where

$$
U^1_i = \frac{1}{A(\sigma^1_i)} \left[\sum_j \frac{u^1_{i_{1_j}} A(\tilde{\sigma}^1_{i_{1_j}})}{A(\tilde{\sigma}^0_{i_1})} + \sum_j \frac{u^1_{i_{2_j}} A(\tilde{\sigma}^1_{i_{2_j}})}{A(\tilde{\sigma}^0_{i_2})} \right]
$$

and

$$
W^1_i = \frac{1}{A(\tilde{\sigma}^1_i)} \left[\sum_j \left(\frac{u^1_{i_{j_1}} A(\sigma^1_{i_{j_1}}) + u^1_{i_{j_2}} A(\sigma^1_{i_{j_2}}) + u^1_{i_{j_3}} A(\sigma^1_{i_{j_3}})}{A(\sigma^2_{i_j})} \right) A(\tilde{\sigma}^2_{i_j}) \right].
$$

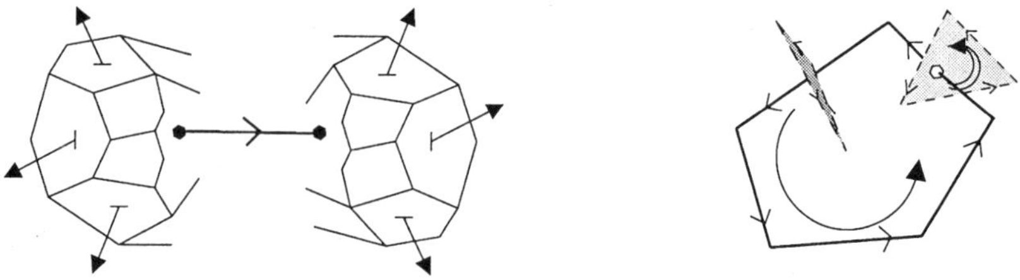

FIG. 4. *In* $\mathbb{R}^3$ *the discrete covolume operators that approximate* grad div *and* curl curl.

In $\mathbb{R}^3$ the U_i^1 terms corresponds to the grad div $\vec{\phi}$ term of the Laplace-Beltrami operator and we see that this expression, in terms of the covolume method, can be understood as the difference of two nodal values (at the endpoints of the edge σ_i^1) divided by the length of the edge thus yielding a gradient. The two nodes have associated covolumes and the value at each node is a sum over the faces of the covolume yielding a divergence.

The W_i^1 term in this context corresponds to the curl curl $\vec{\phi}$ term of the Laplace-Beltrami operator. Here a circulation around the coface $\tilde{\sigma}_i^1$ is summed and each coedge $\tilde{\sigma}_{i_j}^2$ contributes a circulation of its own around the triangle which is dual to it. See Figure 4.

4. Conclusions. We have presented a discrete calculus of differential forms and applied it to several partial differential equations of current interest. It is of interest that our techniques apply on smooth manifolds in any finite number of dimensions. Interesting possibilities remain for future work, including applications to manifolds with indefinite inner products – related to time discretization – and deriving new error estimates in the differential forms setting.

REFERENCES

[1] P. BOCHEV. A discourse on variational and geometric aspects of stability of discretizations. In: *33rd Computational Fluid Dynamics Lecture Series*, VKI LS 2003-05, edited by H. Deconinck, ISSN0377-8312. Von Karman Institute for Fluid Dynamics, Chaussee de Waterloo, 72, B-1640 Rhode Saint Genese, Belgium. 90 pages.

[2] W. BOOTHBY. *An Introduction to Differentiable Manifolds and Riemannian Geometry*, Second Edition. Volume 120 in Pure and Applied Mathematics. Editors: S. Eilenberg and H. Bass. Academic Press, Inc., New York, 1986.

[3] Q. DU, V. FABER, AND M. GUNZBURGER. Centroidal Voronoi Tessellations: Applications and Algorithms. *SIAM Review* 41(4):637–676, 1999.

[4] H. FLANDERS. *Differential Forms with Applications to the Physical Sciences*. Academic Press, Inc., New York, 1963.

[5] T. FRANKEL. *The Geometry of Physics: An Introduction*, Second Edition. Cambridge University Press, Cambridge, UK, 1997.

[6] R.A. NICOLAIDES. Direct discretization of planar div-curl problems. *SIAM J. Numer. Anal.* 29(1):32–56, 1992.

[7] R.A. Nicolaides and D.Q. Wang. Convergence analysis of a covolume scheme for Maxwell's equations in three dimensions. *Mathematics of Computation* 67(223):947–963, 1998.

[8] R.A. Nicolaides and X. Wu. Covolume Solutions of three-dimensional div-curl equations. *SIAM J. Numer. Anal.* 34(6): 2195–2203, 1997.

MIMETIC RECONSTRUCTION OF VECTORS

J. BLAIR PEROT*, DRAGAN VIDOVIC†, AND PIETER WESSELING†‡

Abstract. Compatible or mimetic numerical methods typically use vector components as the primary unknowns in the discretization. It is frequently necessary or useful to be able to recover vectors from these spatially dispersed vector components. In this paper we discuss the relationship between a number of low order vector reconstruction methods and some preliminary results on higher order vector reconstruction. We then proceed to demonstrate how explicit reconstruction can be used to define discrete Hodge star interpolation operators, and how some reconstruction approaches can lead to local conservation statements for vector derived quantities such as momentum and kinetic energy.

Key words. Vector, reconstruction, interpolation, conservation.

AMS(MOS) subject classifications. 65D05, 65N30, 65M60, 76M12, 76M10.

1. Background. Many numerical methods for the solution of Partial Differential Equations use point values or cell averages as the primary discrete unknowns. For scalar equations, such as the Poisson equation, the heat equation, or the scalar wave equation, this is a very appropriate starting point. However, for vector equations, such as Maxwell's equations or the Navier-Stokes equations, there is considerable evidence now suggesting that advantageous numerical properties can be obtained, by using integral averages of vector *components* as the primary discrete unknowns.

In Finite Elements these are often referred to as edge or face elements. They were originally discussed in 2D by Raviart and Thomas [1] and in 3D by Nedelec [2]. In the Finite Volume or Finite Difference context, this type of approach is often referred to as a staggered mesh method. The staggered mesh approach was first proposed for Cartesian meshes by Harlow and Welch [3] in 1965, and has since been generalized to unstructured and curvilinear meshes [4–7]. Face and edge elements are becoming increasingly popular in electromagnetic wave propagation. These methods appear to be the only way to capture difficult physical effects such as resonant frequencies (eigenmodes) [8]. Staggered mesh methods are attractive in incompressible fluid dynamics because they allow the exact satisfaction of the continuity constraint [9], and the satisfaction of a number of local conservation properties (conservation of kinetic energy being perhaps the most important) [10].

Having stated that vector components, not vectors themselves, should be the primary variables of interest when solving vector partial differential

*Department of Mechanical & Industrial Engineering, University of Massachusetts, Amherst, MA 01003 (perot@ecs.umass.edu).

†Applied Mathematics, TU Delft, Netherlands (D.Vidovic@ewi.tudelft.nl).

‡(P.Wesseling@ewi.tudelft.nl).

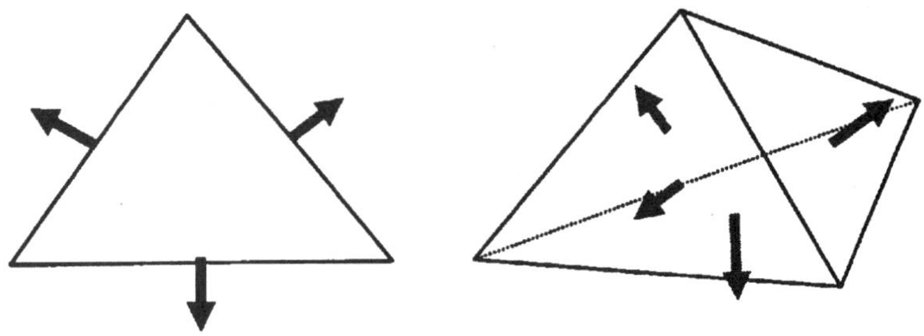

FIG. 1. *Representation of unknowns for low order face-based mimetic methods.*

equations, this paper will now proceed to discuss how vector quantities can be obtained in these schemes. In the case of the Navier-Stokes equations, the need for the velocity vector is obvious since the convective term requires a velocity vector. On the surface, it is far less clear why this might be a useful procedure for Maxwell's (or Stokes') equations. These equations can, and probably should, be discretized entirely in terms of vector components that are edge or face averages. Nevertheless, even in these discretization schemes there is the necessity to interpolate edge averages to face averages and vice-versa. Vector reconstruction can (though certainly does not have to) be used to construct numerically attractive interpolation schemes. Vector interpolation is also useful for graphical output.

We note that there is a more precise terminology emanating from Algebraic Topology for describing many of the concepts described in this paper. However, in order to keep the potential audience broad, and in order to discuss vectors (which fit less well in the formalism of differential forms), we will continue to use the more primitive vector calculus.

2. Lowest order face-based reconstruction methods. The lowest (first) order faced-based mimetic methods all use $u_f = \frac{1}{A}\int \mathbf{v}\cdot\mathbf{n}dA$, the face-normal average vector component on element faces as the primary unknown (see Fig. 1). Throughout this paper the formulas and text refer to the three-dimensional case. This means that in two-dimensions 'cells' refers to 2D polygonal regions (often triangles in the figures), 'faces' are the boundaries of the cells and are actually 1D objects (frequently referred to in other texts as edges), and 'edges' coincide with faces in 2D.

The face-based FE method for simplices assumes a piecewise polynomial for the vector field of the form $\mathbf{v}(\mathbf{x}) = \mathbf{a} + b\mathbf{x}$ where $\mathbf{a}$ is a constant vector and b a constant scalar in each element (or cell). For Cartesian grids, the polynomial is assumed to be $\mathbf{v}(\mathbf{x}) = \mathbf{a} + B\mathbf{x}$ where B is a diagonal matrix. Note that in both cases the normal component of the vector field is constant along each face of the element (or cell) and therefore also continuous across the face. This means that at lowest order the integral average of the normal vector component over the face, u_f, can also be associated

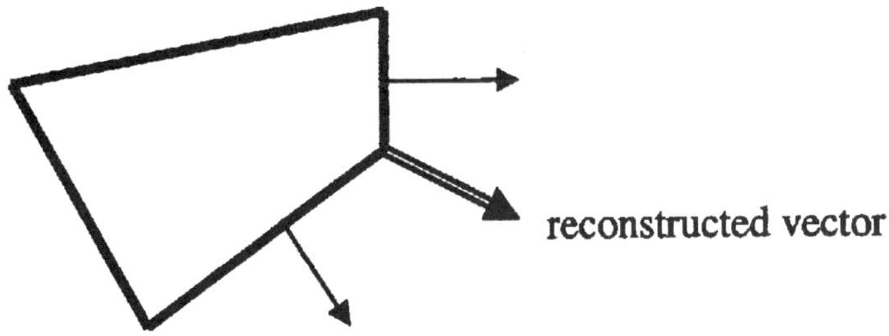

FIG. 2. *Reconstruction of corner vector from face normal components assuming no variation in the component values along each cell/element face.*

with a pointwise value on that face (often the midpoint value of the normal component is cited as the primary unknown). Also note that when the vector field is divergence free (which is frequently true in both the fluid dynamic and electromagnetic contexts), then the lowest order reconstruction on simplices assumes that the vector field is piecewise constant.

Least squares reconstruction of the vector field was proposed by Nicolaides [4]. In that method one finds the constant vector field, $\mathbf{v}_{\text{cell}}$ that best satisfies all the face equations $\mathbf{v}_{\text{cell}} \cdot \mathbf{n}_f = u_f$ for all the faces of a cell/element. For a divergence free field on a simplex, the result is the same as the FE reconstruction and $\mathbf{v}_{\text{cell}} = \mathbf{a}$.

Hyman and Shashkov [5] and Shashkov *et al.* [11] proposed reconstruction at the corners of each element using the immediately neighboring face unknowns (see Fig. 2). Because the low order FE approximation assumes the normal velocity is constant on faces, the corner velocities recovered by this method are identical to the low order FE reconstruction. To obtain the vector value at the cell center a simple average of the node velocities is suggested, $\mathbf{v}^{\text{CG}} = \frac{1}{\text{NCN}} \sum_{\text{nodes}} \mathbf{v}^n$ where *NCN* is the number of nodes in the cell or element.

Since $\mathbf{v}^n = \mathbf{v}^C + b(\mathbf{x}^n - \mathbf{x}^C)$ the simple average gives $\frac{1}{NCN} \sum_{\text{nodes}} \mathbf{v}^n = \mathbf{v}^C + b\frac{1}{NCN} \sum_{\text{nodes}} (\mathbf{x}^n - \mathbf{x}^C)$. If the cell center is defined to be the average of the cell corners the last term is zero, and we see that the simple average is the value at the cell center. For simplices and Cartesian meshes, the average of the cell corners equals the center of gravity (or centroid). Unlike the FE reconstruction, this approach is explicit and does not require a matrix inversion (which is as large as 6x6 for 3D Cartesian meshes). In addition, in contrast to the FE reconstruction this method can easily be generalized to arbitrary polygons, since no explicit piecewise polynomial form for the vector field is assumed.

Finally, Perot and Nallapati [12] suggest a reconstruction formula derived from Gauss' Divergence Theorem and the position vector, $\mathbf{x}$. In particular it is noted that the exact relation

$$\int \mathbf{v}dV + \int \mathbf{x}(\nabla \cdot \mathbf{v})dV = \sum_{\text{faces}} \int \mathbf{x}\mathbf{v} \cdot \mathbf{n}dA \tag{2.1}$$

applies in each element or cell. Making the same assumptions as the low order FE reconstruction (constant normal velocity along each face, constant dilatation, and a linear velocity field), gives the discrete interpolation formula

$$\mathbf{v}_c^{\mathrm{CG}} = \frac{1}{V_c} \sum_{\text{cell faces}} \pm u_f A_f (\mathbf{x}_f^{\mathrm{CG}} - \mathbf{x}_c^{\mathrm{CG}}) \tag{2.2}$$

where CG stands for the (cell or face) center of gravity (or centroid) and the $\pm$ is to account for the fact that u_f should point out of the cell in question. The cell volume is v_c and the face areas are A_f. This formula is directly equivalent to the low order FE reconstruction (since the assumptions are the same). However, like the method of Hyman and Shashkov it easily generalizes to arbitrary polygons.

We can see that the method of Hyman and Shashkov is fully equivalent to the FE interpolation but returns the vector value at the average of the element corner positions (which is not equal to the centroid position on arbitrary polygons). The method of Perot is also always equivalent to the FE method but returns the centroid value for the vector no mater what the element shape. The method of Perot is also a simple average of the primary unknowns, u_f, whereas the method of Hyman and Shashkov requires the intermediate step of corner velocity reconstruction. However, the corner reconstruction approach may be easier to generalize to higher order.

3. Higher order face-based reconstruction methods. For n^{th} order faced-based methods on simplices, the FE interpolation is generalized to $\mathbf{v}(\mathbf{x}) = \mathbf{a}(\mathbf{x}) + b(\mathbf{x})\mathbf{x}$, where $\mathbf{a}$ and b are $n-1$ order polynomials. The normal velocity component on each face is also an $n-1$ order polynomial (and remains continuous across the face). For Cartesian meshes, the polynomial is assumed to be $\mathbf{v}(\mathbf{x}) = \mathbf{a}(\mathbf{x}) + B(\mathbf{x})\mathbf{x}$ where B is a diagonal matrix. As with all FE methods, the underlying interpolation changes for every possible element shape. The generalization to quads, hexahedra, prisms, and pyramids is non-trivial but possible [13, 14], and the FE generalization to arbitrary polygons appears to be extremely difficult.

At the next higher order, $\int \mathbf{v}dV$ and $\int \mathbf{x}\mathbf{v} \cdot \mathbf{n}dA$ are primary unknowns (along with u_f) of face-based mimetic methods. This means there is now a total of ND unknowns per face and ND unknowns per cell/element, where ND is the number of dimensions. The terminology face element is now less appropriate (since there are also cell unknowns), but it is still used. Staggered mesh and finite volume methods typically obtain higher order by enlarging the interpolation stencil rather than increasing the number of unknowns within a cell. Higher order staggered mesh methods for Cartesian meshes using a larger stencil have been proposed [15, 16]. However, larger than nearest neighbor stencils on arbitrary 3D polygonal meshes are difficult to formulate and program, very difficult to implement efficiently on parallel computers, and create complex issues at domain boundaries. The common FE practice of more unknowns per cell is not commonly practiced in FV methods but is perfectly possible and is the approach discussed herein.

The exact integral relation (Eq. (2.1)) now provides the first order dilatation moments $\int \mathbf{x}\nabla \cdot \mathbf{v} dV$ immediately from the primary data. The zeroth order dilatation moment is also directly known $\int \nabla\cdot\mathbf{v} dV = \sum u_f A_f$. On a simplex the first order dilatation moments are enough to rapidly recover the centroid velocity vector. To see how, note that for a simplex the polynomial form is known and $\nabla \cdot \mathbf{v} = \nabla \cdot \mathbf{a} + \mathbf{x} \cdot \nabla b + bND$ where ND is the number of dimensions. Switching to index notion for clarity, this implies that $\int x_k \nu_{i,i} dV = (ND+1) b_{,i} \int x_k x_i dV$. In addition we can use the polynomial form to write $\int \nu_k dV = V \nu_k^{\mathrm{CG}} + b_{,i} \int x_k x_i dV$. So in the case of simplices the point value of the vector at the center of gravity is given by the expression $\mathbf{v}_c^{\mathrm{CG}} = \frac{1}{V_c} \int \mathbf{v} dV - \frac{1}{(ND+1)V_c} \int \mathbf{x}\nabla\cdot\mathbf{v} dV$ or in terms of primary variables

$$\mathbf{v}_c^{\mathrm{CG}} = \frac{ND}{(ND+1)V_c} \int \mathbf{v} dV - \frac{1}{(ND+1)V_c} \sum_{\text{faces}} \int \mathbf{x}\mathbf{v} \cdot \mathbf{n} dA. \tag{3.1}$$

This is entirely equivalent to the FE reconstruction, though explicit and simpler than inverting an 8×8 matrix (in 2D) or a 15×15 matrix (in 3D). Note however, that this reconstruction expression for the centroid velocity vector does not appear to be general. It does not equal the FE reconstruction on Cartesian meshes.

Elements of the method of Hyman and Shashkov (in particular the corner velocity reconstruction) can be extended to higher order in 2D and 3D. Assuming linear variation of the normal vector component on a face, the primary variables u_f and $\int \mathbf{x}\mathbf{v} \cdot \mathbf{n} dA$ contain enough information to specify the face normal velocity at face corners and therefore the velocity vector at element corners.

Note that the reconstruction of the element corner velocities is straightforward only if the element corners only have ND faces meeting at every corner. The top corner of a pyramid is an instance that violates this condition. This type of corner is also degenerate for FE polynomial reconstructions. It is anticipated that a unique corner solution still exists even though the problem appears to be over specified.

Because the vector field is now piecewise quadratic, a simple average of the corner velocities is no longer sufficient to recover the centroid vector value. However, the cell value can be recovered from the corner velocities and the cell average value. For example, it can be shown that

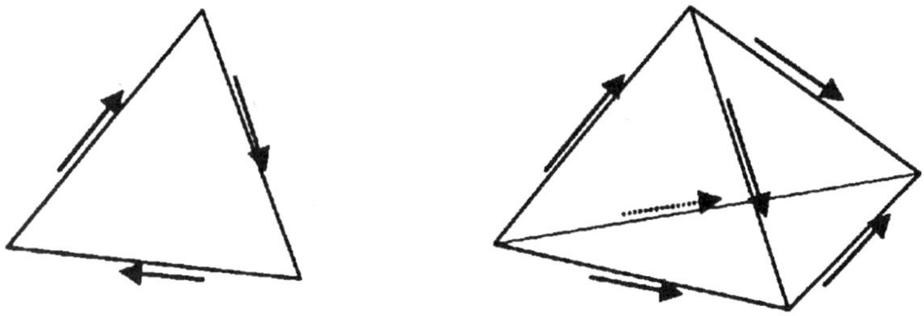

FIG. 3. *Edge-based primary unknowns in 2D and 3D.*

$\mathbf{v}_c^{CG} = \frac{4}{3}\frac{1}{V_c}\int \mathbf{v}dV - \frac{1}{3}\frac{1}{3}\sum_{\text{nodes}} \mathbf{v}^n$ is satisfied on triangles, and on rectangles, $\mathbf{v}_c^{CG} = \frac{3}{2}\frac{1}{V_c}\int \mathbf{v}dV - \frac{1}{2}\frac{1}{4}\sum_{\text{nodes}} \mathbf{v}^n$ holds true. A general formula is not available at this time.

4. Lowest order edge-based reconstruction methods. The lowest (first) order edge-based mimetic methods use $u_e = \frac{1}{L_e}\int \mathbf{v}\cdot d\mathbf{l}$, the edge-tangential average vector component on element edges as the primary unknown (see Fig. 3). The edge-based FE method for simplices then assumes a piecewise polynomial for the vector field of the form $\mathbf{v}(\mathbf{x}) = \mathbf{a} + \mathbf{b}\times\mathbf{x}$ where $\mathbf{a}$ and $\mathbf{b}$ are constant vectors in each element. Whereas the lowest order face-elements are vorticity free (except between elements), the lowest order edge-based elements are divergence free (except between elements). The tangential velocity is constant along each edge and is therefore continuous. The velocity tangential to an element face is given by $\mathbf{v}\times\mathbf{n} = \mathbf{a}\times\mathbf{n} - (\mathbf{x}\cdot\mathbf{n})\mathbf{b} + (\mathbf{b}\cdot\mathbf{n})\mathbf{x}$ and varies linearly on the face in a fashion akin to a rotated face-based vector. The vorticity in the simplictical FE reconstruction is given by $\nabla\times\mathbf{v} = \mathbf{b}(ND-1)$ where ND is the number of dimensions.

Edge elements have more degrees of freedom than face elements. In FE the choice of which element is appropriate depends on the physical nature of the vector in question and its inherent continuity requirements and natural boundary conditions. Because FE are restricted to certain element shapes, the primary mesh must define the elements/cells. However, in methods that handle arbitrary polygons, there is an additional choice because it is also possible for cells/elements to be associated with the dual mesh. This means edges could also be associated with the lines connecting the tetrahedra cell centers.

In the context of finite volume or finite difference methods there is far less published work on vector reconstruction of edge-based vectors. While it is not discussed in their papers the basic idea of Hyman & Shashkov of corner reconstruction is still valid. Again, some degeneracy may occur on cells that have more than three edges meeting at a corner (such as the top of a pyramid). And as before, the sum of the corner velocities equals the velocity at the cell center (defined to be the average of the corner positions).

On arbitrary polygonal meshes this is not equal to the cell center of gravity but is an equally well defined center.

Zhang *et al.* [17] presents an analog of Eqs. (2.1) and (2.2) for edge based vectors. This is based on the application of Stokes' Curl Theorem. Note that for each face the Curl Theorem states that

$$\begin{aligned} n_i \int (\varepsilon_{ijk} \nu_{k,j} x_m + \varepsilon_{imk} \nu_k) dA &= n_i \int \varepsilon_{ijk} (\nu_k x_m)_{,j} dA \\ &= \sum_{\text{edges}} \int x_n \nu_k dl_k. \end{aligned} \tag{4.1}$$

Index notation is used for clarity and ε_{ijk} is the standard permutation symbol. If we assume, consistent with the FE polynomials, that the vorticity is constant in each cell and the velocity component along each edge is constant and the tangential velocity varies linearly then this gives the formula,

$$\mathbf{v}_f^{\text{CG}} A_f = \int \mathbf{v} \times \mathbf{n} dA = \sum_{\text{edges}} \pm (\mathbf{x}_e^{\text{CG}} - \mathbf{x}_f^{\text{CG}}) u_e L_e \tag{4.2}$$

where L_e is the length of each edge and $\pm$ indicates counterclockwise (right hand rule) integration around the edges of the face with respect to the face normal, $\mathbf{n}$. In this way the tangential velocity at the center of gravity of each face can be recovered. In 2D the reconstruction is complete since a face corresponds to a the cell/element. In 3D we note that sometimes the tangential velocity on faces is sufficient and the cell velocity vector is not actually required. This is the case for the rotational form of the convective term $(\nabla \times \mathbf{b}) \times \mathbf{v} + \nabla(\frac{1}{2}\mathbf{v} \cdot \mathbf{v})$ [11].

The face tangential velocity can be used to quickly recover the vorticity in the cell. Using the divergence theorem we note that,

$$\int \varepsilon_{ijk} \nu_{k,j} dV = \sum_{\text{faces}} \int \varepsilon_{ijk} \nu_k n_j dA. \tag{4.3}$$

Assuming the vorticity is constant in each cell we see that the sum of the face tangential velocities equals the cell vorticity.

$$\nabla \times \mathbf{v} = -\frac{1}{V_c} \sum_{\text{faces}} \int \mathbf{v} \times \mathbf{n} dA. \tag{4.4}$$

Remember that for the lowest order face-based reconstructions the cell vorticity is always zero and vorticity is confined to thin sheets between the elements/cells.

In 3D the cell velocity can be obtained from the relation,

$$\begin{aligned} \int (\varepsilon_{ijk} \nu_{k,j} x_m + \varepsilon_{imk} \nu_k) dV &= \int \varepsilon_{ijk} (\nu_k x_m)_{,j} dV \\ &= \sum_{\text{faces}} \int x_m \varepsilon_{ijk} \nu_k n_j dA. \end{aligned} \tag{4.5}$$

Assuming constant vorticity, and linear velocity we obtain

$$\varepsilon_{imk}\mathbf{v}_k^{\mathrm{CG}} = -\frac{1}{V_c}\sum_{\text{faces}}\int x_m[\mathbf{v}_f^{\mathrm{CG}} \times \mathbf{n} + (\mathbf{b}\cdot\mathbf{n})(\mathbf{x} - \mathbf{x}_f^{\mathrm{CG}})]_i dA. \tag{4.6}$$

Note that $0 = \int(\delta_{ik}x_j + \delta_{jk}x_i)dV = \int(x_i x_j)_{,k}dV = \sum_{\text{faces}}\int x_i x_j n_k dA$ if we assume the position origin is at the cell center of gravity. Then the second term of (4.6) is seen to be zero and

$$\varepsilon_{imk}\nu_k^{\mathrm{CG}} = -\frac{1}{V_c}\sum_{\text{faces}} A_f r_m^f(\mathbf{v}_f^{\mathrm{CG}} \times \mathbf{n})_i - b_k\sum_{\text{faces}} A_f r_m^f r_m^f n_k \tag{4.7}$$

where $\mathbf{r}^f = (\mathbf{x}_f^{\mathrm{CG}} - \mathbf{x}_c^{\mathrm{CG}})$ is the distance between the face and cell center of gravities and $\mathbf{b} = \frac{\nabla\times\mathbf{v}}{ND-1}$. These formulas were developed for simplicies but appear to generalize naturally to arbitrary polygons.

5. Higher order edge-based reconstruction methods. For n^{th} order edged-based methods on simplices, the FE interpolation is generalized to $\mathbf{v}(\mathbf{x}) = \mathbf{a}(\mathbf{x}) + \mathbf{b}(\mathbf{x})\times\mathbf{x}$, where $\mathbf{a}$ and $\mathbf{b}$ are $n-1$ order polynomial vectors.

The primary unknowns for the next order edge-based discretizations are $\frac{1}{A_f}\int \mathbf{v}\times\mathbf{n}dA$ the average tangential velocity on faces, $\frac{1}{L_e}\int \mathbf{x}\mathbf{v}\cdot\mathbf{n}dl$ the moment of the tangential velocity component, as well as lowest order unknown $\frac{1}{L_e}\int \mathbf{v}\cdot\mathbf{n}dl$. Eqn. (4.1) now becomes an exact relation for the gradients or the face-normal vorticity (which are assumed constant) on each face,

$$\int \mathbf{x}(\mathbf{n}\cdot\nabla\times\mathbf{v})dA = \sum_{\text{edges}}\int \mathbf{x}\mathbf{v}\cdot d\mathbf{l} - \int \mathbf{v}\times\mathbf{n}dA \tag{5.1}$$

and Eqn. (4.3) now becomes an exact expression for the average vorticity in the cell

$$\int \nabla\times\mathbf{v}dV = -\sum_{\text{faces}}\int \mathbf{v}\times\mathbf{n}dA \tag{5.2}$$

or its value at the center of gravity (since it is now assumed to vary linearly).

The corner reconstruction method still works for edge elements. The average tangential component and its first moment provide enough information to reconstruct the corner velocities exactly. However, as with the face-based elements, a simple average of the corner velocities is no longer sufficient to recover the vector at any cell center pointwise location, and a general averaging formula for arbitrary polygons is not known at this time. Corner velocities are discontinuous at the cell nodes and do not provide a unique output for the velocity at nodes (often desired for graphical output).

6. Mass matrices and the discrete Hodge star operator. In mimetic methods, it is frequently necessary to convert a face-based set of unknowns to edge-based or vice versa. This occurs because when primary unknowns are face-based the evolution equations are posed on a dual mesh that is edge-based. While mimetic FE methods avoid the explicit definition of a dual mesh, it is still present and implicitly defined by the functional form of the weighting functions in the weak statement of the equations. See Mattiussi [18] for a detailed explanation. This process of converting one type of vector field to another is sometimes referred to as a discrete Hodge star operator. This operator is symmetric and positive definite for Galerkin FE and for many low order mimetic methods, but it is not clear that symmetry is absolutely necessary. One possible method for explicitly converting a face-based vector structure to an edge based one is to reconstruct the piecewise polynomials in each cell/element based on existing face-based data and then use high enough order numerical quadrature on the piecewise polynomials to compute the necessary edge-based integrals. This is what implicitly happens in the Galerkin FE methods.

Consider the transpose of the low order Perot interpolation method for determining the cell centroid vector value (Eq. (2.2)). The transpose operation applied to those centroid values is $\sum_{\text{face cells}} \pm \mathbf{v}^{\mathrm{CG}} \cdot (\mathbf{x}_f^{\mathrm{CG}} - \mathbf{x}_c^{\mathrm{CG}})$. This is a first order accurate (like the reconstruction itself) integration along the median dual edge connecting two cell centroids. Note that the median dual edge consists of two line segments each joining the face centroid to the neighboring cell centroids. We can therefore write to first order

$$\int \mathbf{v} \cdot d\mathbf{l} \approx \mathbf{R}^T \frac{1}{V_c} \mathbf{R} \int \mathbf{v} \cdot \mathbf{n} dA \tag{6.1}$$

where the line integral is along the median dual edge and the area integral over the corresponding face. The reconstruction operator is defined as $\mathbf{R}\nu_f = \sum_{\text{cell faces}} (\mathbf{x}_f^{\mathrm{CG}} - \mathbf{x}_c^{\mathrm{CG}})\nu_f$. On uniform (or nearly uniform) meshes, errors cancel out during the integration and despite the first order nature of the reconstruction and integration, this approximation is found to be second order accurate. The discrete Hodge star operator that converts from face-based to edge based vectors is $\mathbf{R}^T \frac{1}{V_c} \mathbf{R}$.

Exact (rather than approximate) integration over a simplex median dual mesh and the low order piecewise approximation $\mathbf{v} = \mathbf{a} + \mathbf{b}\mathbf{x}$ gives a slightly modified formula

$$\int \mathbf{v} \cdot d\mathbf{l} = \sum_{\text{face cells}} \pm \mathbf{v}^{\mathrm{CG}} \cdot (\mathbf{x}_f^{\mathrm{CG}} - \mathbf{x}_c^{\mathrm{CG}}) \pm \frac{(\nabla \cdot \mathbf{v})}{2ND} (\mathbf{x}_f^{\mathrm{CG}} - \mathbf{x}_c^{\mathrm{CG}})^2 \tag{6.2}$$

which is only symmetric on a uniform mesh, but which is probably always positive definite.

The distributed two-step nature of the low order corner reconstruction approaches makes it difficult to evaluate the properties of their effective discrete Hodge star operators. However, due to the demonstrated equivalence of these methods with the reconstruction method of Perot, it can be demonstrated that for simplices and Cartesian grids, these methods also produce symmetric positive definite discrete Hodge star operators.

7. Conservation properties of the Navier-Stokes equations. The attractive conservation properties of Cartesian staggered mesh methods have been known for some time [19]. On Cartesian meshes the orthogonal structure of the mesh allows non-overlapping staggered control volumes to be defined in which conservation is relatively easy to demonstrate. However, on general unstructured meshes demonstrating that local conservation properties (such as those obtained in standard Finite Volume methods) exist is extremely difficult. The problem lies in the fact that only velocity components and transport equations for velocity components exist so it is difficult to make conservation statements about vector quantities.

Two conservation statements of particular interest are conservation of momentum and conservation of kinetic energy (in the incompressible limit). Conservation of vorticity or circulation (the curl of the momentum) is also possible and is discussed in Perot et al. in Refs. [7, 10, 17]. Finite Element methods frequently have a global conservation statement that can be associated with them, but one attraction of mimetic methods is their ability to correctly represent physics at the local (cell) level as well.

In the following sections we focus on the conservation properties of low order face-based discretization schemes of the Navier-Stokes equations. Integrating along the two line segments connecting the cell centers and the face center (a dual mesh edge) gives a discrete equation for each dual edge.

$$\frac{\partial}{\partial t}[\mathbf{R}^T\mathbf{m}_c] + \mathbf{R}^T\mathbf{a}_c = -\mathbf{G}p_c \tag{7.1}$$

where $\mathbf{m}_c = \frac{\rho_c}{V}\sum_{\text{faces}}(\mathbf{x}_f^{CG} - \mathbf{x}_c^{CG})u_f A_f$ is the cell momentum, and $\mathbf{a}_c = \frac{1}{V_c}\sum_{\text{faces}}\{\mathbf{u}u_f - \mu(\nabla\mathbf{u}+\mathbf{u}\nabla)\cdot\mathbf{n} - \lambda(\nabla\cdot\mathbf{u})\mathbf{n}\}_f A_f$ is a standard finite volume flux representation of the advection-diffusion term in each cell. The term with the second coefficient of viscosity, λ can also be directly absorbed into the pressure term instead of into $\mathbf{a}$. The exact operator $\mathbf{G}$ is the difference between the pressure at the two end points of the line segment.

On Dirichlet boundaries the normal velocity is fixed and this equation does not exist. On variable-boundaries (such as an outflow), the pressure on the boundary is fixed and only one segment of the dual mesh edge has non-zero length.

8. Conservation of momentum. In order to show conservation of momentum, we must be able to show that linear combinations of the existing discrete edge based equations can be constructed such that those combinations look like a local discrete vector conservation statement.

Consider a single cell. We have update equations for the normal component on each face of that cell. Let us associate each line segment of the dual edge equation with the cell in which it resides. Ultimately we will multiply both line segments by the same scaling factor, so this splitting is really for accounting purposes only. If the cell face is also a domain boundary the associated dual edge only has a single segment (associated with the interior cell).

For a single cell, multiplying each segment equation by the face normal vector and face area and summing over the cell faces gives (assuming outward normals for convenience).

$$\sum_{\text{faces}} \mathbf{n}_f A_f \left\{ \frac{\partial}{\partial t}\mathbf{m}_c + \mathbf{a}_c \right\} \cdot (\mathbf{x}_f^{\text{CG}} - \mathbf{x}_c^{\text{CG}}) = - \sum_{\text{faces}} \mathbf{n}_f A_f (p_f - p_c). \quad (8.1)$$

A number of geometric identities allow this equation to be simplified. In particular,

$$\sum_{\text{faces}} \mathbf{n}_f A_f = 0 \qquad \text{and} \qquad \mathbf{I} = \frac{1}{V} \sum_{\text{faces}} (\mathbf{x}_f^{\text{CG}} - \mathbf{x}_c^{\text{CG}}) \mathbf{n}_f A_f. \quad (8.2)$$

These expressions (like many of the paper's formulas) are a result of Gauss' Divergence Theorem. They both start from the exact expression, $\int a_{i,j} dV = \sum_{\text{faces}} \int a_i n_j dA$. If a_i is constant and the faces are planar then $0 = \sum_{\text{faces}} \int n_j dA = \sum_{\text{faces}} \mathbf{n} A_f$. If $a_i = x_i$ then $\int_V \delta_{ij} dV = \sum_{\text{faces}} \int x_i n_i dA$ and if the faces are planar the second relation is derived. These expression simplify the previous momentum vector equation on each cell to,

$$V_c \left(\frac{\partial}{\partial t}\mathbf{m}_c + \mathbf{a}_c \right) = - \sum_{\text{faces}} \mathbf{n}_f A_f p_f. \quad (8.3)$$

Since the advection diffusion term is also represented as a sum of fluxes we see that this is a statement of local momentum conservation for the discrete momentum, $\mathbf{m}_c$. One key distinction with standard finite volume methods is that the conserved quantity is a derived, not a primary variable. Conservation of momentum places restrictions on the form of the advection-diffusion term but does not restrict how the discrete momentum $\mathbf{m}_c$ must be defined.

The derivation of momentum conservation is possible because the integration operator (the square root of the discrete Hodge star operator) $\mathbf{R}^T$, has an explicit geometric inverse. Global conservation is a result of the traditional telescoping property where internal fluxes cancel out.

9. Conservation of kinetic energy. Taking the dot product of the incompressible momentum equation with the velocity (and assuming constant viscosity for simplicity) gives the kinetic energy equation,

$$\frac{\partial(\frac{1}{2}u^2)}{\partial t} + \nabla \cdot \left(\mathbf{u} \frac{1}{2} u^2 \right) = -\nabla \cdot (\mathbf{u}p) + \nabla \cdot \nu \nabla \left(\frac{1}{2} u^2 \right) - \nu u_{i,j} u_{i,j}. \quad (9.1)$$

This equation shows that in the incompressible limit kinetic energy is convected and diffused. It is also transported by pressure and removed by velocity gradients, but it is never created. In the inviscid, incompressible limit, kinetic energy is a conserved variable. In the viscous limit, we would like the total kinetic energy to decrease at the correct rate (and never increase).

Numerical methods with numerical diffusion decrease kinetic energy more quickly than the physics would suggest ($\nu u_{i,j} u_{i,j}$). Numerical diffusion excessively smears solutions and can be detrimental in some situations, such as DNS and LES simulations of turbulence where energy dissipation is a critical physical process controlling the turbulence. Kinetic energy conservation is a statement that numerical diffusion is not present in the method. It is also a statement of stability.

To demonstrate kinetic energy conservation, each segment of the dual-edge equation within a cell is multiplied by the area weighted normal velocity component and summed over the cell faces to obtain

$$\sum_{\text{faces}} u_f A_f \mathbf{R}^T \left\{ \frac{1}{V_c} \mathbf{R} A_f \frac{\partial u_f}{\partial t} + \mathbf{a}_c \right\} = - \sum_{\text{faces}} u_f A_f (p_f - p_c). \tag{9.2}$$

Focusing first on the time derivative term we see that this is an approximation for the cell average kinetic energy because

$$u_f A_f \mathbf{R}^T \frac{1}{V_c} \mathbf{R} A_f \frac{\partial u_f}{\partial t} = V_c \mathbf{v}_c \cdot \frac{\partial \mathbf{v}_c}{\partial t} = V_c \frac{\partial \frac{1}{2}(\mathbf{v}_c)^2}{\partial t}. \tag{9.3}$$

If the system is fully discrete, this result still holds as long as we multiply each equation by the half-time velocity $u_f^{n+1/2} = \frac{1}{2}(u_f^n + u_f^{n+1})$. Then,

$$\begin{aligned} \frac{u_f^{n+1} + u_f^n}{2} A_f \mathbf{R}^T \frac{1}{V_c} \mathbf{R} A_f \frac{u_f^{n+1} - u_f^n}{\Delta t} &= V_c \frac{\mathbf{v}_c^{n+1} + \mathbf{v}_c^n}{2} \cdot \frac{\mathbf{v}_c^{n+1} - \mathbf{v}_c^n}{\Delta t} \\ &= V_c \frac{\frac{1}{2}(\mathbf{v}_c^{n+1})^2 - \frac{1}{2}(\mathbf{v}_c^n)^2}{\Delta t}. \end{aligned} \tag{9.4}$$

Due to incompressibility $\sum_{\text{faces}} u_f A_f = 0$ the second part of the pressure term is zero and the pressure term becomes a faced based conservative flux term, $-\sum_{\text{faces}} u_f^{n+1/2} A_f p_f$.

The advection-diffusion term becomes,

$$\sum_{\text{faces}} u_f^{n+1/2} A_f \mathbf{R}^T \mathbf{a}_c = \mathbf{a}_c \sum_{\text{faces}} u_f^{n+1/2} A_f \mathbf{R}^T = \mathbf{a}_c \cdot \mathbf{v}_c^{n+1/2} V_c. \tag{9.5}$$

Expanding the advection-diffusion term gives,

$$V_c \mathbf{v}_c^{n+1/2} \cdot \mathbf{a}_c = \mathbf{v}_c^{n+1/2} \cdot \sum_{\text{faces}} \{\mathbf{u} u_f - \nu(\nabla \mathbf{u}) \cdot \mathbf{n}\}_f A_f. \tag{9.6}$$

Considering the convective term first,

$$\mathbf{v}_c^{n+1/2} \cdot \sum_{\text{faces}} \mathbf{u} u_f A_f = \mathbf{v}_c^{n+1/2} \cdot \sum_{\text{faces}} \frac{1}{2}(\mathbf{v}_c + \mathbf{v}_{c\_n}) u_f A_f. \tag{9.7}$$

Here we have assumed that the velocity vector in the advective flux calculation is the simple average of the neighboring two cell velocities. One cell is the cell in question and the other is the nearest neighbor. The velocity in the first term can come out of the summation leaving the incompressibility condition (which is zero), so finally the advective term becomes $\sum_{\text{faces}}(\frac{1}{2}\mathbf{v}_c^{n+1/2} \cdot \mathbf{v}_{c\_n}) u_f A_f$. This is also a flux term. Note that the kinetic energy fluxing through the cell faces is quite specific. It is one half of the dot-product of the two neighboring cell velocities. To obtain correct symmetry conservation also requires that the advection velocity be the half-time velocity. This implies that true conservation (in unsteady flows) occurs only if the advection term is semi implicit. The normal flux can be time lagged. This is an example of the implicit midpoint rule which is known to be a symplectic integrator. Other symplectic time integration schemes may also be possible. There appears to be a close connection between mimetic discretization schemes and symplectic time integration which should be explored more fully.

The diffusion term becomes

$$\mathbf{v}_c^{n+1/2} \cdot \sum_{\text{faces}} \nu(\nabla \mathbf{u}) \cdot \mathbf{n} A_f = \mathbf{v}_c^{n+1/2} \cdot \sum_{\text{faces}} \nu \frac{\partial \mathbf{u}}{\partial \mathbf{n}} A_f. \tag{9.8}$$

Using a very simple approximation for the normal derivative gives, $\sum_{\text{faces}} \nu \mathbf{v}_c^{n+1/2} \cdot (\mathbf{v}_{c\_n} - \mathbf{v}_c) \frac{A_f}{L_f}$ which can be expanded in two parts as

$$\begin{aligned} &= \sum_{\text{faces}} \nu \frac{1}{2}(\mathbf{v}_{c\_n}^{n+1/2} + \mathbf{v}_c^{n+1/2}) \cdot (\mathbf{v}_{c\_n} - \mathbf{v}_c) \frac{A_f}{L_f} \\ &\quad - \sum_{\text{faces}} \nu \frac{1}{2}(\mathbf{v}_{c\_n}^{n+1/2} - \mathbf{v}_c^{n+1/2}) \cdot (\mathbf{v}_{c\_n} - \mathbf{v}_c) \frac{A_f}{L_f} \end{aligned} \tag{9.9}$$

this simplifies to a viscous diffusion of kinetic energy term and a negative definite dissipation term.

$$\begin{aligned} &= \sum_{\text{faces}} \nu \left(\frac{1}{2}\mathbf{v}_{c\_n}^{n+1/2} \cdot \mathbf{v}_{c\_n} - \frac{1}{2}\mathbf{v}_c^{n+1/2} \cdot \mathbf{v}_c \right) \frac{A_f}{L_f} \\ &\quad - \sum_{\text{faces}} \nu \frac{1}{2}(\mathbf{v}_{c\_n}^{n+1/2} - \mathbf{v}_c^{n+1/2}) \cdot (\mathbf{v}_{c\_n} - \mathbf{v}_c) \frac{A_f}{L_f}. \end{aligned} \tag{9.10}$$

To see that this latter term is an approximation of the dissipation term consider the divergence theorem applied to $\int (x_n \nu_{i,n} \nu_{i,m})_{,m} dV =$

$\sum_{\text{faces}} \int x_n \nu_{i,n} \nu_{i,m} n_m dA$. Then assuming the velocity gradients are constant in the volume gives

$$\nu_{i,m}\nu_{i,m}V_c = \sum_{\text{faces}} (x_f - x_c)_n \nu_{i,n}\nu_{i,m} n_m A_f. \tag{9.11}$$

With the approximation $(\mathbf{v}_f - \mathbf{v}_c) \approx \frac{1}{2}(\mathbf{v}_{c\_n} - \mathbf{v}_c)$ this becomes the dissipation in a cell,

$$= \sum_{\text{faces}} \frac{1}{2}(\mathbf{v}_{c\_n} - \mathbf{v}_c) \cdot \frac{(\mathbf{v}_{c\_n} - \mathbf{v}_c)}{L_f} A_f. \tag{9.12}$$

The final statement of local energy conservation is,

$$\begin{aligned} V_c \frac{\frac{1}{2}(\mathbf{v}_c^{n+1})^2 - \frac{1}{2}(\mathbf{v}_c^n)^2}{\Delta t} &+ \sum_{\text{faces}} \left(\frac{1}{2}\mathbf{v}_c^{n+1/2} \cdot \mathbf{v}_{c\_n} \right) u_f A_f \\ = - \sum_{\text{faces}} u_f^{n+1/2} A_f p_f &+ \sum_{\text{faces}} \nu \frac{\partial}{\partial n}\left(\frac{1}{2}\mathbf{v}_{c\_n}^{n+1/2} \cdot \mathbf{v}_{c\_n} \right) A_f \\ - \sum_{\text{faces}} \nu \frac{1}{2} (\mathbf{v}_{c\_n}^{n+1/2} &- \mathbf{v}_c^{n+1/2}) \cdot (\mathbf{v}_{c\_n} - \mathbf{v}_c) \frac{A_f}{L_f}. \end{aligned} \tag{9.13}$$

For strict negative definite dissipation, the viscous diffusion term should use the half-time velocity (implicit midpoint rule) as well. Note that this equation is not solved in the numerical code. It is a rearrangement of the numerical equations that demonstrates that a discrete analog of kinetic energy conservation holds under certain fairly strict assumptions about the form of the advection and diffusion terms.

Global kinetic energy conservation follows from the internal cancellation of fluxes. The symmetry of the discrete Hodge star operator is useful for deriving kinetic energy conservation. However, a positive definite discrete Hodge star would be sufficient to formulate a strictly positive kinetic energy.

In order to test the kinetic energy conservation property a problem was chosen that has zero mass flux at the boundaries, but is inherently unsteady. The initial flow field of this problem involves a Rankine vortex located in the bottom left quadrant of a box. Although the problem is tested in a 3D domain (1.0m × 1.0m × 0.1m) and using an unstructured tetrahedral mesh, it is a two-dimensional flow since the motion only occurs in X-Y plane and only the Z component of the vorticity vector is nonzero. The domain is meshed with 7578 tetrahedra. The viscosity of the fluid is 0.01m^2/s and the maximum initial velocity magnitude is 0.16m/s. The initial tangential velocity reaches its maximum at radius $R = 0.01$m for an initial circulation Reynolds number of 1.

In numerical tests of the vortex motion in the absence of viscosity, the total discrete kinetic energy remained constant to within six significant

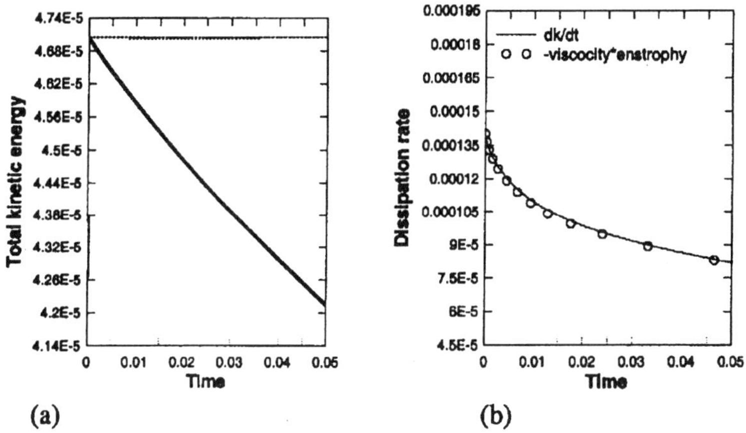

FIG. 4. *Kinetic energy conservation test. (a) Total kinetic energy vs. time. (b) Rate of change of kinetic energy versus time (solid line) and total physical dissipation versus time (circles).*

digits after 5000 time steps (0.05 Seconds). This is about as constant as can be expected given the tolerance prescribed for the iterative solver and is shown as the dotted line in Fig. 4(a). When viscosity is present (0.01 m^2/s), the total discrete kinetic energy as a function of time is also shown in Fig. 4(a). The rate of change of the kinetic energy obtained by differentiating this curve is compared with the calculated physical dissipation. A perfect match is shown in Fig. 4(b). This test indicates that the theoretical analysis of this section is well founded and that there is no artificial dissipation in the method.

Acknowledgements. This work was supported in part by the Office of Naval Research (Grant # N00014-04-1-0267), the Air Force Office of Scientific Research (Grant # FA9550-04-1-0023), and the National Science Foundation (SGER Grant).

REFERENCES

[1] P.A. RAVIART AND J.M. THOMAS, *A mixed finite element method for second order elliptic problems*, Springer Lecture Notes in Mathematics Vol. **606**, Springer-Verlag, 292–315, 1977.
[2] J.-C. NEDELEC, *Mixed finite elements in R^3*, Numer. Math., **50**, 315–341, 1980.
[3] F.H. HARLOW AND J.E WELCH, *Numerical calculations of time dependent viscous incompressible flow of fluid with a free surface*, Phys. Fluids, **8** (12), 2182–2189, 1965.
[4] R.A. NICOLAIDES, *The covolume approach to computing incompressible flow,* Incompressible Computational Fluid Dynamics, M.D. Gunzburger & R.A. Nicolaides, eds., Cambridge University Press, 295–234, 1993.
[5] J.M. HYMAN AND M. SHASHKOV, *The orthogonal decomposition theorems for*

mimetic finite difference methods, SIAM J. on Num. Anal., **36** (3), 788–818, 1999.

[6] P. Wesseling, A. Segal, C.G.M. Kassels, and H. Bijl, *Computing flows on general two-dimensional nonsmooth staggered grids*, J. of Engin. Math., **34**, 21–44, 1998.

[7] J.B. Perot, *Conservation properties of unstructured staggered mesh schemes,* J. Comput. Phys., **159**, 58–89, 2000.

[8] D. White, *Orthogonal vector basis functions for time domain finite element solution of the vector wave equation*, 8th Biennial IEEE Conference on Electromagnetic Field Computation, Tucson, AZ. UCRL-JC-129188, 1998.

[9] W. Chang, F. Giraldo and J.B. Perot, *Analysis of an Exact Fractional Step Method*, J. Comput. Phys., **179**, 1–17, 2002.

[10] J.B. Perot and X. Zhang, *Reformulation of the unstructured staggered mesh method as a classic finite volume method*, Finite Volumes for Complex Applications II, Hermes Science Publications, pp. 263–270, 1999.

[11] M. Shashkov, B. Swartz, and B. Wendroff, *Local reconstruction of a vector field from its normal components on the faces of grid cells.* J. Comput. Phys., **139**, 406–409, 1998.

[12] J.B. Perot and R. Nallapati, *A Moving Unstructured Staggered Mesh Method for the Simulation of Incompressible Free-Surface Flows*, J. Comput. Phys., **184**, 192–214, 2003.

[13] P. Castillo, J. Koning, R.Rieben, M. Stowell, and D. White, *Discrete Differential Forms: A Novel Methodology for Robust Computational Electromagnetics*, LLNL report UCRL-ID-151522, January 2003.

[14] R. Rieben, *A Novel High Order Time Domain Vector Finite Element Method for the Simulation of Electromagnetic Device*, Ph.D. dissertation, University of California at Davis, Livermore, CA, 2004 UCRL-TH-205466.

[15] Y. Morinishi, T.S. Lund, O.V. Vasilyev, and P. Moin, *Fully Conservative Higher Order Finite Difference Schemes for Incompressible Flow*, J. Comput. Phys, **143**, 90–124, 1998.

[16] O.V. Vasilyev, Hi*gh Order Finite Difference Schemes on Non-uniform Meshes with Good Conservation Properties*, J. Comput. Phys, **157**, 746–761, 2000.

[17] X. Zhang, D. Schmidt, and J. B. Perot, *Accuracy and Conservation Properties of a Three-Dimensional Unstructured Staggered Mesh Scheme for Fluid Dynamics*, J. Comput. Phys., **175**, 764–791, 2002.

[18] C. Mattiussi, *An analysis of finite volume, finite element, and finite difference methods using some concepts from algebraic topology*, J. Comput. Phys, **133**, 289–309, 1997.

[19] D.K. Lilly, *On the computational stability of numerical solutions of time-dependent non-linear geophysical fluid dynamics problems*, Mon. Weather Rev, **93 (1)**, 11–26, 1965.

A CELL-CENTERED FINITE DIFFERENCE METHOD ON QUADRILATERALS

MARY F. WHEELER* AND IVAN YOTOV†

Abstract. We develop a cell-centered finite difference method for elliptic problems on curvilinear quadrilateral grids. The method is based on the lowest order Brezzi-Douglas-Marini (BDM) mixed finite element method. A quadrature rule gives a block-diagonal mass matrix and allows for local flux elimination. The method is motivated and closely related to the multipoint flux approximation (MPFA) method. An advantage of our method is that it has a variational formulation. As a result finite element techniques can be employed to analyze the algebraic system and the convergence properties. The method exhibits second order convergence of the scalar variable at the cell-centers and of the flux at the midpoints of the edges. It performs well on problems with rough grids and coefficients, which is illustrated by numerical experiments.

Key words. Mixed finite element, multipoint flux approximation, cell centered finite difference, tensor coefficient.

AMS(MOS) subject classifications. 65N06, 65N12, 65N15, 65N30, 76S05.

1. Introduction. Cell-centered finite difference (CCFD) methods have been widely used in flow in porous media modeling, especially in the petroleum industry [5]. They combine local mass conservation and accuracy for discontinuous coefficients with relatively easy, compared to finite element methods, implementation and computational efficiency. CCFD methods, however, have certain accuracy limitations on irregular grids.

A relationship between CCFD methods and mixed finite element (MFE) methods was established by Russell and Wheeler [17] for rectangular grids and diagonal tensor coefficients. They noted that a special quadrature rule diagonalizes the velocity mass matrix and the MFE method reduces to CCFD for the pressure. This relation was exploited by Weiser and Wheeler [21] to obtain optimal convergence and superconvergence for both pressure and velocity in CCFD methods on rectangular grids. These results were extended to full tensor coefficients and triangular and logically rectangular grids by Arbogast et al. in [4, 3] by introducing the expanded mixed finite element (EMFE) method (see also related results by Vassilevski et al. [19], Baranger et al. [6], and Micheletti et al. [15] for triangular grids and diagonal tensor coefficients).

*Institute for Computational Engineering and Sciences (ICES), Department of Aerospace Engineering & Engineering Mechanics, and Department of Petroleum and Geosystems Engineering, The University of Texas at Austin, Austin, TX 78712 (mfw@ices.utexas.edu). Partially supported by NSF grant DMS 0411413 and the DOE grant DE-FGO2-04ER25617.

†Department of Mathematics, University of Pittsburgh, Pittsburgh, PA 15260 (yotov@math.pitt.edu). Supported in part by the DOE grant DE-FG02-04ER25618, and by the NSF grants DMS 0107389 and DMS 0411694.

The EMFE method is superconvergent for smooth grids and coefficients, but loses accuracy near discontinuities. Pressure Lagrange multipliers can be introduced along discontinuous interfaces to recover higher order convergence [3], which, however, leads to a hybrid cell-centered – face-centered formulation. Two other closely related methods that handle accurately rough grids and coefficients are the control volume mixed finite element (CVMFE) method, see Cai et al. [9], and the mimetic finite difference (MFD) methods, see Hyman et al. [12]. Each of these, however, as in the case of MFE methods, leads to an algebraic saddle-point problem. The multipoint flux approximation (MPFA) method, see Aavatsmark et al. [2, 1] has been developed as a finite volume method and combines the advantages of the above mentioned methods, i.e., it is accurate for rough grids and coefficients and reduces to a cell-centered stencil for the pressures. However, due to its non-variational formulation, the theoretical understanding of its convergence properties is limited. Relationships between the above methods have been studied by Russell and Klausen in [13].

Our goal in this paper is to develop and analyze an accurate cell-centered finite difference method for elliptic problems with full discontinuous tensor coefficients on curved quadrilateral grids. We base our approach on a mixed finite element method that reduces to a cell-centered stencil for the pressures via a special quadrature rule and local velocity elimination. Motivated by the MPFA method [1] where sub-edge fluxes are introduced, we consider the lowest order Brezzi-Douglas-Marini BDM_1 mixed finite element method [7, 8]. The BDM_1 velocity space on quadrilaterals has two degrees of freedom per edge. A special quadrature rule is employed that for each corner couples only the four associated degrees of freedom. The CCFD method is obtained by inverting the block-diagonal velocity mass matrix.

We develop the method for a second order elliptic problem that models single phase flow in porous media. The problem can be written as a system of two first order equations

$$\mathbf{u} = -K\nabla p \quad \text{in } \Omega, \tag{1.1}$$

$$\nabla \cdot \mathbf{u} = f \quad \text{in } \Omega, \tag{1.2}$$

$$p = g \quad \text{on } \Gamma_D, \tag{1.3}$$

$$\mathbf{u}\cdot\mathbf{n} = 0 \quad \text{on } \Gamma_N, \tag{1.4}$$

where the domain $\Omega \subset \mathbf{R}^2$ has a boundary $\partial\Omega = \overline{\Gamma}_D \cup \overline{\Gamma}_N$, $\Gamma_D \cap \Gamma_N = \emptyset$, measure$(\Gamma_D) > 0$, $\mathbf{n}$ is the outward unit normal on $\partial\Omega$, and K is a symmetric, uniformly positive definite tensor satisfying, for some $0 < k_0 \leq k_1 < \infty$,

$$k_0\xi^T\xi \leq \xi^T K(x)\xi \leq k_1\xi^T\xi \quad \forall x \in \Omega,\ \forall \xi \in \mathbf{R}^2. \tag{1.5}$$

In the above equations p is the pressure, $\mathbf{u}$ is the Darcy velocity, and K represents the permeability divided by the viscosity

REMARK 1.1. The choice of homogeneous Neumann boundary conditions and the assumption measure(Γ_D) > 0 are made for the sake of simplicity of the presentation. Non-homogeneous and full Neumann boundary conditions can also be handled.

We will use the following standard notation. For a subdomain $G \subset \mathbf{R}^2$, the $L^2(G)$ inner product (or duality pairing) and norm are denoted $(\cdot,\cdot)_G$ and $\|\cdot\|_G$, respectively, for scalar and vector valued functions. The norms of the Sobolev spaces $W^k_\infty(G)$, $k \in \mathbf{R}$ are denoted $\|\cdot\|_{k,\infty,G}$. Let $\|\cdot\|_{k,G}$ be the norm of the Hilbert space $H^k(G)$. We omit G in the subscript if $G = \Omega$. For a section of the domain or an element boundary $S \subset \mathbf{R}^1$ we write $\langle\cdot,\cdot\rangle_S$ and $\|\cdot\|_S$ for the $L^2(S)$ inner product (or duality pairing) and norm, respectively. We will also make use of the space

$$H(\text{div};\Omega) = \{\mathbf{v} \in (L^2(\Omega))^2 : \nabla\cdot\mathbf{v} \in L^2(\Omega)\}$$

equipped with the norm

$$\|\mathbf{v}\|_{\text{div}} = (\|\mathbf{v}\|^2 + \|\nabla\cdot\mathbf{v}\|^2)^{1/2}.$$

The weak formulation of (1.1)–(1.4) is: find $\mathbf{u} \in \mathbf{V}$ and $p \in W$ such that

$$(K^{-1}\mathbf{u},\mathbf{v}) = (p,\nabla\cdot\mathbf{v}) - \langle g,\mathbf{v}\cdot\mathbf{n}\rangle_{\Gamma_D}, \quad \mathbf{v} \in \mathbf{V}, \tag{1.6}$$

$$(\nabla\cdot\mathbf{u},w) = (f,w), \quad w \in W, \tag{1.7}$$

where

$$\mathbf{V} = \{\mathbf{v} \in H(\text{div};\Omega) : \mathbf{v}\cdot\mathbf{n} = 0 \text{ on } \Gamma_N\}, \quad W = L^2(\Omega).$$

It is well known [8, 16] that (1.6)–(1.7) has a unique solution.

The rest of the paper is organized as follows. The numerical method and its analysis are developed in Section 2 and Section 3, respectively. As part of the analysis we establish approximation properties for the BDM_1 velocity spaces on curved quadrilaterals. We prove that the method converges with rate $O(h)$ in the L^2-norm for the pressure and the velocity and with rate $O(h^2)$ for the pressure at the cell centers. Numerical experiments confirming the theoretical results and comparisons with the EMFE method are presented in Section 4.

2. The numerical method.

2.1. Definition of the finite element partition. Let $\mathcal{T}_h$ be a shape regular and quasiuniform [10] finite element partition of Ω, consisting of small curvilinear perturbations (to be made precise later) of convex quadrilaterals. If an element has curved edges, we refer to it as curved quadrilateral. We assume that for each element $E \in \mathcal{T}_h$ there exists a bijection mapping $F_E : \hat{E} \to E$ where $\hat{E}$ is the reference unit square with vertices

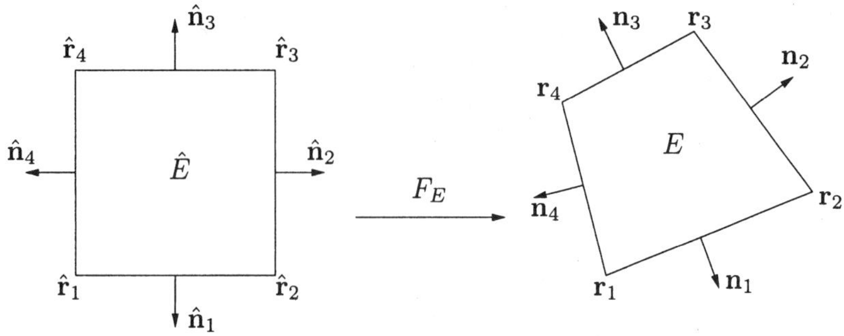

FIG. 1. *Bilinear mapping and orientation of normal vectors.*

$\hat{\mathbf{r}}_1 = (0,\,0)^T$, $\hat{\mathbf{r}}_2 = (1,\,0)^T$, $\hat{\mathbf{r}}_3 = (1,\,1)^T$ and $\hat{\mathbf{r}}_4 = (0,\,1)^T$. Denote by $\mathbf{r}_i = (x_i,\,y_i)^T$, $i = 1,\ldots,4$, the four corresponding vertices of element E as shown in Figure 1. The outward unit normal vectors to the edges of E and $\hat{E}$ are denoted by $\mathbf{n}_i$ and $\hat{\mathbf{n}}_i$, $i = 1,\ldots,4$, respectively. Let DF_E be the Jacobi matrix and let J_E be its Jacobian. We denote the inverse mapping by F_E^{-1}, its Jacobi matrix by DF_E^{-1}, and its Jacobian by $J_{F_E^{-1}}$. We have that

$$DF_E^{-1}(x) = (DF_E)^{-1}(\hat{x}), \qquad J_{F_E^{-1}}(x) = \frac{1}{J_E(\hat{x})}.$$

It is easy to check that

$$\mathbf{n}_i = \frac{1}{J_{n_i}} J_E (DF_E^{-1})^T \hat{\mathbf{n}}_i, \quad \text{where} \quad J_{n_i} = J_E |(DF_E^{-1})^T \hat{\mathbf{n}}_i|_{\mathbf{R}^2} \tag{2.1}$$

and $|\cdot|_{\mathbf{R}^2}$ is the Euclidean vector norm in $\mathbf{R}^2$.

If E is a quadrilateral, then F_E is the bilinear mapping given by

$$\begin{aligned} F_E(\hat{\mathbf{r}}) &= \mathbf{r}_1\,(1-\hat{x})(1-\hat{y}) + \mathbf{r}_2\,\hat{x}(1-\hat{y}) + \mathbf{r}_3\,\hat{x}\hat{y} + \mathbf{r}_4\,(1-\hat{x})\hat{y} \\ &= \mathbf{r}_1 + \mathbf{r}_{21}\hat{x} + \mathbf{r}_{41}\hat{y} + (\mathbf{r}_{21} - \mathbf{r}_{34})\hat{x}\hat{y}, \end{aligned} \tag{2.2}$$

where $\mathbf{r}_{ij} = \mathbf{r}_i - \mathbf{r}_j$. In this case DF_E and J_E are linear functions of $\hat{x}$ and $\hat{y}$:

$$\begin{aligned} DF_E &= [(1-\hat{y})\,\mathbf{r}_{21} + \hat{y}\,\mathbf{r}_{34},\ (1-\hat{x})\,\mathbf{r}_{41} + \hat{x}\,\mathbf{r}_{32}] \\ &= [\mathbf{r}_{21}, \mathbf{r}_{41}] + [(\mathbf{r}_{21} - \mathbf{r}_{34})\hat{y}, (\mathbf{r}_{21} - \mathbf{r}_{34})\hat{x}], \end{aligned} \tag{2.3}$$

$$J_E = 2|T_1| + 2(|T_2| - |T_1|)\hat{x} + 2(|T_4| - |T_1|)\hat{y}, \tag{2.4}$$

where $|T_i|$ is the area of the triangle formed by the two edges sharing $\mathbf{r}_i$. Note that $J_E > 0$ for convex quadrilaterals. It is also easy to see that $J_{n_i} = |e_i|$ on any edge e_i.

If E is a curved quadrilateral, we assume that it is an $O(h^2)$-perturbation of a quadrilateral, i.e.,

$$F_E = \tilde{F}_E + R(\hat{x}, \hat{y}), \quad \|R\|_{j,\infty,\hat{E}} \le Ch^2, \; j = 0, 1, 2, \tag{2.5}$$

where $\tilde{F}_E$ is a bilinear map. We call such elements h^2-quadrilaterals.

Let $a \sim b$ mean that there exist positive constants c_0 and c_1 independent of h such that $c_0 a \le b \le c_1 a$. For shape-regular and quasi-uniform quadrilateral grids, (2.3) and (2.4) imply that for all elements E

$$\|DF_E\|_{\infty,\hat{E}} \sim h, \quad \|J_E\|_{\infty,\hat{E}} \sim h^2, \quad \text{and} \quad \|J_{F_E^{-1}}\|_{\infty,\hat{E}} \sim h^{-2}. \tag{2.6}$$

Moreover, (2.6) also holds for any curved quadrilateral satisfying (2.5).

For the remainder of the paper we will restrict our attention to curved quadrilateral elements that are $O(h^2)$-perturbations of parallelograms. We assume that

$$\|\mathbf{r}_{21} - \mathbf{r}_{34}\| \le Ch^2. \tag{2.7}$$

Following the terminology adopted in [11], we call such elements h^2-parallelograms.

REMARK 2.1. Note that the notion of h^2-parallelograms from [11] is extended to elements with curved edges, i.e., elements that satisfy (2.5), where $\tilde{F}_E$ satisfies (2.7).

Using (2.3), (2.5), and (2.7), a simple direct calculation shows that for h^2-parallelograms

$$J_E = a + b(\hat{x}, \hat{y}) + d(\hat{x}, \hat{y}), \tag{2.8}$$

where $|a| \le Ch^2$ is a constant, $|b(\hat{x}, \hat{y})| \le Ch^3$ is a bilinear function, and $|d(\hat{x}, \hat{y})| \le Ch^4$.

2.2. The BDM$_1$ spaces on curved quadrilaterals. Let $\mathbf{V}_h \times W_h$ be the lowest order BDM$_1$ mixed finite element spaces [7, 8]. On the reference unit square these spaces are defined as

$$\begin{aligned}\hat{\mathbf{V}}(\hat{E}) &= P_1(\hat{E})^2 + r\,\mathrm{curl}(\hat{x}^2\hat{y}) + s\,\mathrm{curl}(\hat{x}\hat{y}^2) \\ &= \begin{pmatrix} \alpha_1\hat{x} + \beta_1\hat{y} + \gamma_1 + r\hat{x}^2 + 2s\hat{x}\hat{y} \\ \alpha_2\hat{x} + \beta_2\hat{y} + \gamma_2 - 2r\hat{x}\hat{y} - s\hat{y}^2 \end{pmatrix}, \\ \hat{W}(\hat{E}) &= P_0(\hat{E}) = \alpha,\end{aligned} \tag{2.9}$$

where $\alpha, \alpha_1, \alpha_2, \beta_1, \beta_2, \gamma_1, \gamma_2, s, r$ are real constants and P_k denotes the space of polynomials of degree $\le k$. Note that $\hat{\nabla} \cdot \hat{\mathbf{V}}(\hat{E}) = \hat{W}(\hat{E})$ and that for all $\hat{\mathbf{v}} \in \hat{\mathbf{V}}(\hat{E})$ and for any edge $\hat{e}$ of $\hat{E}$

$$\hat{\mathbf{v}} \cdot \hat{\mathbf{n}}_{\hat{e}} \in P_1(\hat{e}).$$

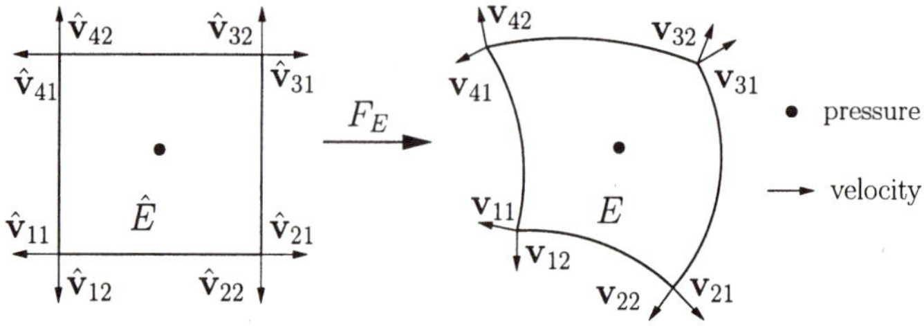

FIG. 2. *Degrees of freedom and basis functions for the BDM_1 spaces.*

It is well known [7, 8] that the degrees of freedom for $\hat{\mathbf{V}}(\hat{E})$ can be chosen to be the values of $\hat{\mathbf{v}} \cdot \hat{\mathbf{n}}_{\hat{e}}$ at any two points on each edge $\hat{e}$. *We choose these points to be the vertices of* $\hat{e}$, see Figure 2. This choice is motivated by the requirement of accuracy and certain orthogonalities for the quadrature rule introduced in the next section.

The velocity space on any element E is defined via the Piola transformation

$$\mathbf{v} \leftrightarrow \hat{\mathbf{v}} : \quad \mathbf{v} = \frac{1}{J_E} DF_E \hat{\mathbf{v}} \circ F_E^{-1}$$

and the pressure space is defined via the standard change of variables

$$w \leftrightarrow \hat{w} : \quad w = \hat{w} \circ F_E^{-1}.$$

The BDM_1 spaces on $\mathcal{T}_h$ are given by

$$\begin{aligned} \mathbf{V}_h &= \{\mathbf{v} \in \mathbf{V}: \quad \mathbf{v}|_E \leftrightarrow \hat{\mathbf{v}},\ \hat{\mathbf{v}} \in \hat{\mathbf{V}}(\hat{E}) \quad \forall E \in \mathcal{T}_h\}, \\ W_h &= \{w \in W: \quad w|_E \leftrightarrow \hat{w},\ \hat{w} \in \hat{W}(\hat{E}) \quad \forall E \in \mathcal{T}_h\}. \end{aligned} \tag{2.10}$$

The Piola transformation preserves the normal components of the velocity vectors on the edges and satisfies [8]

$$(\nabla \cdot \mathbf{v}, w)_E = (\hat{\nabla} \cdot \hat{\mathbf{v}}, \hat{w})_{\hat{E}} \qquad \text{and} \qquad \langle \mathbf{v} \cdot \mathbf{n}_i, w \rangle_{e_i} = \langle \hat{\mathbf{v}} \cdot \hat{\mathbf{n}}_i, \hat{w} \rangle_{\hat{e}_i}. \tag{2.11}$$

Moreover, (2.1) implies

$$\mathbf{v} \cdot \mathbf{n}_i = \frac{1}{J_E} DF_E \hat{\mathbf{v}} \cdot \frac{1}{J_{n_i}} J_E (DF_E^{-1})^T \hat{\mathbf{n}}_i = \frac{1}{J_{n_i}} \hat{\mathbf{v}} \cdot \hat{\mathbf{n}}_i. \tag{2.12}$$

Let $\hat{\Pi} : (H^1(\hat{E}))^2 \to \hat{\mathbf{V}}(\hat{E})$ be the reference element projection operator satisfying

$$\forall\, \hat{e}_i \subset \partial \hat{E}, \qquad \langle (\hat{\Pi} \hat{\mathbf{q}} - \hat{\mathbf{q}}) \cdot \hat{\mathbf{n}}_i, \hat{p}_1 \rangle_{\hat{e}_i} = 0 \qquad \forall\, \hat{p}_1 \in P_1(\hat{e}_i). \tag{2.13}$$

The divergence theorem and (2.13) imply that

$$(\hat{\nabla} \cdot (\hat{\Pi} \hat{\mathbf{q}} - \hat{\mathbf{q}}), \hat{w})_{\hat{E}} = 0 \qquad \forall\, \hat{w} \in \hat{W}(\hat{E}). \tag{2.14}$$

Following [18, 20, 3], the operator Π is defined locally on each element E by

$$\Pi \mathbf{q} \leftrightarrow \widehat{\Pi \mathbf{q}}, \quad \widehat{\Pi \mathbf{q}} := \hat{\Pi} \hat{\mathbf{q}} \quad \forall \mathbf{q} \in (H^1(E))^2. \tag{2.15}$$

It is shown in [20] that in the case of quadrilaterals Π is a well defined operator from $\mathbf{V} \cap (H^1(\Omega))^d$ onto $\mathbf{V}_h$ satisfying

$$(\nabla \cdot (\Pi \mathbf{q} - \mathbf{q}), w) = 0 \qquad \forall w \in W_h \tag{2.16}$$

and

$$\|\Pi \mathbf{q}\|_{\text{div}} \leq C \|\mathbf{q}\|_1. \tag{2.17}$$

Due to (2.11), property (2.16) extends trivially to the case of curved quadrilaterals. The continuity bound (2.17) follows from the argument for proving the approximation properties of Π, which is given in Lemma A.1 in the Appendix.

Using an argument due to Fortin (see [8]) and properties (2.16)–(2.17), it can be shown that the BDM_1 spaces on curved quadrilaterals satisfy the inf-sup condition

$$\inf_{\substack{w \in W_h \\ w \neq 0}} \sup_{\substack{\mathbf{v} \in \mathbf{V}_h \\ \mathbf{v} \neq 0}} \frac{(\nabla \cdot \mathbf{v}, w)}{\|\mathbf{v}\|_{\text{div}} \|w\|} \geq \beta, \tag{2.18}$$

where β is a positive constant independent of h.

The following auxiliary estimate will be used in the analysis of the method.

LEMMA 2.1. *If $E \in \mathcal{T}_h$ and $\mathbf{q} \in (L^2(E))^2$, then*

$$\|\mathbf{q}\|_E \sim \|\hat{\mathbf{q}}\|_{\hat{E}}. \tag{2.19}$$

Proof. The statement of the lemma follows immediately from the relations

$$\int_E \mathbf{q} \cdot \mathbf{q} \, d\mathbf{x} = \int_{\hat{E}} \frac{1}{J_E} DF_E \hat{\mathbf{q}} \cdot \frac{1}{J_E} DF_E \hat{\mathbf{q}} \, J_E \, d\hat{\mathbf{x}},$$

$$\int_{\hat{E}} \hat{\mathbf{q}} \cdot \hat{\mathbf{q}} \, d\hat{\mathbf{x}} = \int_E \frac{1}{J_{F_E^{-1}}} DF_E^{-1} \mathbf{q} \cdot \frac{1}{J_{F_E^{-1}}} DF_E^{-1} \mathbf{q} \, J_{F_E^{-1}} \, d\mathbf{x},$$

and bounds(2.6). □

The BDM_1 mixed finite element method is based on approximating the variational formulation (1.6)–(1.7) in the discrete spaces $\mathbf{V}_h \times W_h$: find $\mathbf{u}_h^{bdm} \in \mathbf{V}_h$ and $p_h^{bdm} \in W_h$ such that

$$(K^{-1} \mathbf{u}_h^{bdm}, \mathbf{v}) = (p_h^{bdm}, \nabla \cdot \mathbf{v}) - \langle g, \mathbf{v} \cdot \mathbf{n} \rangle_{\Gamma_D}, \quad \mathbf{v} \in \mathbf{V}_h, \tag{2.20}$$

$$(\nabla \cdot \mathbf{u}_h^{bdm}, w) = (f, w), \quad w \in W_h. \tag{2.21}$$

It has been shown in [20] that on quadrilaterals the above method has a unique solution and that it is second order accurate for the velocity and first order accurate for the pressure in the L^2-norm. These results can be extended to h^2-quadrilaterals in light of the approximation results of Lemma A.1. The method handles well discontinuous coefficients due to the presence of K^{-1} in the mass matrix. However, the resulting algebraic system is a coupled velocity-pressure system and it can be quite large. Moreover, it is of a saddle-point problem type. Our goal is to design a quadrature rule that allows for local elimination of the velocities and results in a positive definite cell-centered pressure matrix.

2.3. A quadrature rule. For $\mathbf{q}$, $\mathbf{v} \in \mathbf{V}_h$, define the global quadrature rule

$$(K^{-1}\mathbf{q}, \mathbf{v})_Q \equiv \sum_{E \in \mathcal{T}_h} (K^{-1}\mathbf{q}, \mathbf{v})_{Q,E}.$$

The integration on any element E is performed by mapping to the reference element $\hat{E}$. The quadrature rule is defined on $\hat{E}$. Using the definition (2.10) of the finite element spaces we have

$$\begin{aligned}\int_E K^{-1}\mathbf{q} \cdot \mathbf{v}\, d\mathbf{x} &= \int_{\hat{E}} \hat{K}^{-1} \frac{1}{J_E} DF_E \hat{\mathbf{q}} \cdot \frac{1}{J_E} DF_E \hat{\mathbf{v}}\, J_E\, d\hat{\mathbf{x}} \\ &= \int_{\hat{E}} \frac{1}{J_E} DF_E^T \hat{K}^{-1} DF_E \hat{\mathbf{q}} \cdot \hat{\mathbf{v}}\, d\hat{\mathbf{x}} \equiv \int_{\hat{E}} \mathcal{K}^{-1}\hat{\mathbf{q}} \cdot \hat{\mathbf{v}}\, d\hat{\mathbf{x}},\end{aligned}$$

where

$$\mathcal{K} = J_E DF_E^{-1} \hat{K} (DF_E^{-1})^T. \tag{2.22}$$

It is easy to see that bounds (2.6) imply

$$\|\mathcal{K}\|_{\infty,\hat{E}} \sim \|K\|_{\infty,E} \quad \text{and} \quad \|\mathcal{K}^{-1}\|_{\infty,\hat{E}} \sim \|K^{-1}\|_{\infty,E}. \tag{2.23}$$

The quadrature rule on an element E is defined as

$$(K^{-1}\mathbf{q}, \mathbf{v})_{Q,E} \equiv (\mathcal{K}^{-1}\hat{\mathbf{q}}, \hat{\mathbf{v}})_{\hat{Q},\hat{E}} \equiv \frac{|\hat{E}|}{4} \sum_{i=1}^{4} \mathcal{K}^{-1}(\hat{\mathbf{r}}_i)\hat{\mathbf{q}}(\hat{\mathbf{r}}_i) \cdot \hat{\mathbf{v}}(\hat{\mathbf{r}}_i). \tag{2.24}$$

Note that on $\hat{E}$ this is the trapezoidal quadrature rule.

The corner vector $\hat{\mathbf{q}}(\hat{\mathbf{r}}_i)$ is uniquely determined by its normal components to the two edges that share that vertex. Recall that we chose the velocity degrees of freedom on any edge $\hat{e}$ to be the the normal components at the vertices of $\hat{e}$. Therefore, there are two degrees of freedom associated with each corner $\hat{\mathbf{r}}_i$ and they uniquely determine the corner vector $\hat{\mathbf{q}}(\hat{\mathbf{r}}_i)$. More precisely,

$$\hat{\mathbf{q}}(\hat{\mathbf{r}}_i) = \sum_{j-1}^{2} (\hat{\mathbf{q}} \cdot \hat{\mathbf{n}}_{ij})(\hat{\mathbf{r}}_i)\hat{\mathbf{n}}_{ij},$$

where $\hat{\mathbf{n}}_{ij}$, $j = 1, 2$, are the outward unit normal vectors to the two edges intersecting at $\hat{\mathbf{r}}_i$, and $(\hat{\mathbf{q}} \cdot \hat{\mathbf{n}}_{ij})(\hat{\mathbf{r}}_i)$ are the velocity degrees of freedom associated with this corner. Let us denote the basis functions associated with $\hat{\mathbf{r}}_i$ by $\hat{\mathbf{v}}_{ij}$, $j = 1, 2$ (see Figure 2), i.e.,

$$(\hat{\mathbf{v}}_{ij} \cdot \hat{\mathbf{n}}_{ij})(\hat{\mathbf{r}}_i) = 1, \quad (\hat{\mathbf{v}}_{ij} \cdot \hat{\mathbf{n}}_{ik})(\hat{\mathbf{r}}_i) = 0, \quad k \neq j, \text{ and}$$
$$(\hat{\mathbf{v}}_{ij} \cdot \hat{\mathbf{n}}_{lk})(\hat{\mathbf{r}}_l) = 0, \quad l \neq i, k = 1, 2.$$

Clearly the quadrature rule (2.24) only couples the two basis functions associated with a corner. For example,

$$(\mathcal{K}^{-1}\hat{\mathbf{v}}_{11}, \hat{\mathbf{v}}_{11})_{\hat{Q},\hat{E}} = \frac{\mathcal{K}_{22}^{-1}(\hat{\mathbf{r}}_1)}{4}, \quad (\mathcal{K}^{-1}\hat{\mathbf{v}}_{11}, \hat{\mathbf{v}}_{12})_{\hat{Q},\hat{E}} = \frac{\mathcal{K}_{12}^{-1}(\hat{\mathbf{r}}_1)}{4}, \tag{2.25}$$

and

$$(\mathcal{K}^{-1}\hat{\mathbf{v}}_{11}, \hat{\mathbf{v}}_{ij})_{\hat{Q},\hat{E}} = 0 \quad \forall ij \neq 11, 12. \tag{2.26}$$

REMARK 2.2. On quadrilaterals the quadrature rule can be defined directly on an element E. It is easy to see from (2.4) that

$$(K^{-1}\mathbf{q}, \mathbf{v})_{Q,E} = \frac{1}{2}\sum_{i=1}^{4} |T_i| K^{-1}(\mathbf{r}_i)\mathbf{q}(\mathbf{r}_i) \cdot \mathbf{v}(\mathbf{r}_i). \tag{2.27}$$

The above quadrature rule is closely related to an inner product used in the mimetic finite difference methods [12]. We note that it is simpler to evaluate the quadrature rule on the reference element $\hat{E}$.

Denote the element quadrature error by

$$\sigma_E(K^{-1}\mathbf{q}, \mathbf{v}) \equiv (K^{-1}\mathbf{q}, \mathbf{v})_E - (K^{-1}\mathbf{q}, \mathbf{v})_{Q,E} \tag{2.28}$$

and define the global quadrature error by $\sigma(K^{-1}\mathbf{q}, \mathbf{v})|_E = \sigma_E(K^{-1}\mathbf{q}, \mathbf{v})$.

2.4. The multipoint flux mixed finite element method. We are now ready to define our method. We seek $\mathbf{u}_h \in \mathbf{V}_h$ and $p_h \in W_h$ such that

$$(K^{-1}\mathbf{u}_h, \mathbf{v})_Q = (p_h, \nabla \cdot \mathbf{v}) - \langle g, \mathbf{v} \cdot \mathbf{n} \rangle_{\Gamma_D}, \quad \mathbf{v} \in \mathbf{V}_h, \tag{2.29}$$
$$(\nabla \cdot \mathbf{u}_h, w) = (f, w), \quad w \in W_h. \tag{2.30}$$

REMARK 2.3. We call the method (2.29)–(2.30) a multipoint flux mixed finite element (MFMFE) method due to its relation to the MPFA method.

To establish solvability of (2.29)–(2.30) we need the following coercivity result.

LEMMA 2.2. *There exists a constant C independent of h such that*

$$(K^{-1}\mathbf{q}, \mathbf{q})_Q \geq C\|\mathbf{q}\|^2 \quad \forall \mathbf{q} \in \mathbf{V}_h. \tag{2.31}$$

Proof. Let $\mathbf{q} \leftrightarrow \hat{\mathbf{q}}$ and $\hat{\mathbf{q}} = \sum_{i=1}^4 \sum_{j=1}^2 \hat{q}_{ij}\hat{\mathbf{v}}_{ij}$. We have

$$\begin{aligned}(K^{-1}\mathbf{q},\mathbf{q})_{Q,E} &= \frac{|\hat{E}|}{4}\sum_{i=1}^4 \mathcal{K}^{-1}(\hat{\mathbf{r}}_i)\hat{\mathbf{q}}(\hat{\mathbf{r}}_i)\cdot\hat{\mathbf{q}}(\hat{\mathbf{r}}_i) \\ &\geq \frac{C}{k_1}\sum_{i=1}^4 \hat{\mathbf{q}}(\hat{\mathbf{r}}_i)\cdot\hat{\mathbf{q}}(\hat{\mathbf{r}}_i) = \frac{C}{k_1}\sum_{i=1}^4\sum_{j=1}^2 \hat{q}_{ij}^2,\end{aligned}$$

where we used (2.23) and (1.5) in the inequality, and the location of the degrees of freedom at the vertices in the last equality. On the other hand, using (2.19),

$$\|\mathbf{q}\|_E^2 \leq C(\hat{\mathbf{q}},\hat{\mathbf{q}})_{\hat{E}} = C\left(\sum_{i=1}^4\sum_{j=1}^2 \hat{q}_{ij}\hat{\mathbf{v}}_{ij}, \sum_{k=1}^4\sum_{l=1}^2 \hat{q}_{kl}\hat{\mathbf{v}}_{kl}\right) \leq C\sum_{i=1}^4\sum_{j=1}^2 \hat{q}_{ij}^2.$$

The assertion of the lemma follows from the above two estimates. □

REMARK 2.4. Lemma 2.2 implies that $(K^{-1}\cdot,\cdot)_Q^{1/2}$ is a norm in $\mathbf{V}_h$. Let us denote this norm by $\|\cdot\|_Q$. It is easy to see that $\|\cdot\|_Q$ is equivalent to $\|\cdot\|$. Indeed, using (2.23), (1.5), the equivalence of norms on reference element $\hat{E}$, and (2.19), we have that for all $\mathbf{q}\in\mathbf{V}_h$

$$(K^{-1}\mathbf{q},\mathbf{q})_{Q,E} = (\mathcal{K}^{-1}\hat{\mathbf{q}},\hat{\mathbf{q}})_{\hat{Q},\hat{E}} \leq \frac{C}{k_0}\|\hat{\mathbf{q}}\|_{\hat{E}}^2 \leq C\|\mathbf{q}\|_E^2,$$

which, combined with (2.31), implies that

$$c_0\|\mathbf{q}\| \leq \|\mathbf{q}\|_Q \leq c_1\|\mathbf{q}\| \tag{2.32}$$

for some positive constants c_0 and c_1.

LEMMA 2.3. *The multipoint flux mixed finite element method* (2.29)–(2.30) *has a unique solution.*

Proof. Since (2.29)–(2.30) is a square system, it is enough to show uniqueness. Letting $f = 0$ and $g = 0$ and taking $\mathbf{v} = \mathbf{u}_h$ and $w = p_h$, we conclude that $(K^{-1}\mathbf{u}_h,\mathbf{u}_h)_Q = 0$, and therefore $\mathbf{u}_h = 0$, due to (2.31). Let ϕ be the solution to

$$\begin{aligned} -\nabla\cdot K\nabla\phi &= -p_h && \text{in } \Omega, \\ \phi &= 0 && \text{on } \Gamma_D, \\ -K\nabla\phi\cdot\mathbf{n} &= 0 && \text{on } \Gamma_N. \end{aligned}$$

Taking $\mathbf{v} = \Pi K\nabla\phi \in \mathbf{V}_h$ in (2.29) and using (2.16), we obtain

$$0 = (p_h, \nabla\cdot\Pi K\nabla\phi) = (p_h,\nabla\cdot K\nabla\phi) = \|p_h\|^2,$$

implying $p_h - 0$. □

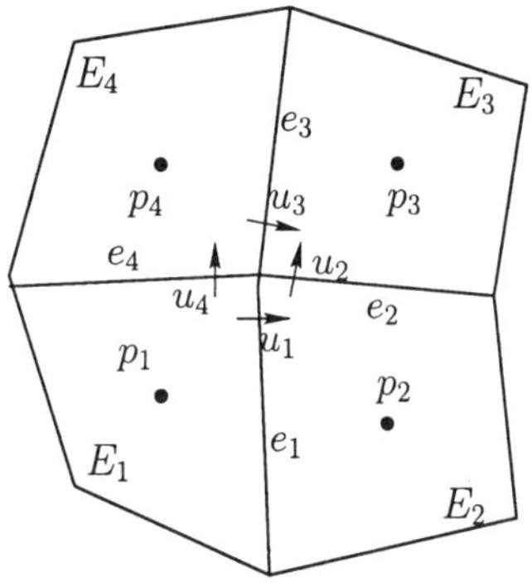

FIG. 3. *Four elements sharing a vertex.*

2.5. Reduction to a cell-centered finite difference method. The multipoint flux mixed finite element method presented above reduces to a cell-centered system for the pressures. Let us consider any interior vertex $\mathbf{r}$ and suppose that it is shared by elements $E_1, \ldots, E_4$, see Figure 3. We denote the edges that share the vertex by $e_1, \ldots, e_4$, the velocity basis functions on these edges that are associated with the vertex by $\mathbf{v}_1, \ldots, \mathbf{v}_4$, and the corresponding values of the normal components of $\mathbf{u}_h$ by $u_1, \ldots, u_4$. Note that for clarity the normal velocities on Figure 3 are drawn at a distance from the vertex.

Due to the locality of the basis functions interaction in the quadrature rule $(K^{-1}\cdot, \cdot)_Q$ in (2.25)–(2.26), taking, for example, $\mathbf{v} = \mathbf{v}_1$ in (2.29) will only lead to coupling u_1 with u_4 and u_2. Therefore, the four equations obtained from taking $\mathbf{v} = \mathbf{v}_1, \ldots, \mathbf{v}_4$ form a linear system for $u_1, \ldots, u_4$. Note that the coefficients of this linear system are

$$a_{ij} = (K^{-1}\mathbf{v}_i, \mathbf{v}_j)_Q, \qquad i, j = 1, \ldots, 4.$$

The local linear system is symmetric and, due to (2.31), positive definite, and it is therefore invertible. Solving the 4×4 linear system allows to express the velocities u_i in terms of the cell-centered pressures p_i, $i = 1, \ldots, 4$. Substituting these expressions into the mass conservation equation (2.30) leads to a cell-centered stencil. The pressure in each element E is coupled with the pressures in the elements that share a vertex with E. On logically rectangular grids this is a 9-point stencil.

We give the equation obtained by taking $\mathbf{v} = \mathbf{v}_1$ in (2.29). On the left hand side we have

$$(K^{-1}\mathbf{u}_h, \mathbf{v}_1)_Q = (K^{-1}\mathbf{u}_h, \mathbf{v}_1)_{Q,E_1} + (K^{-1}\mathbf{u}_h, \mathbf{v}_1)_{Q,E_2}. \tag{2.33}$$

The first term on the right above gives

$$\begin{aligned}(K^{-1}\mathbf{u}_h, \mathbf{v}_1)_{Q,E_1} &= (\mathcal{K}^{-1}\hat{\mathbf{u}}_h, \hat{\mathbf{v}}_1)_{\hat{Q},\hat{E}} \\ &= \frac{1}{4}(\mathcal{K}^{-1}_{11,E_1}\hat{u}_1\hat{v}_{1,1} + \mathcal{K}^{-1}_{12,E_1}\hat{u}_4\hat{v}_{1,1}) \\ &= \frac{1}{4}(\mathcal{K}^{-1}_{11,E_1}J_{n_1}u_1 + \mathcal{K}^{-1}_{12,E_1}J_{n_4}u_4)J_{n_1},\end{aligned} \tag{2.34}$$

where we have used (2.12) for the last equality. Here $\mathcal{K}^{-1}_{ij,E_1}$ denotes a component of $\mathcal{K}^{-1}$ in E_1 and all functions are evaluated at the vertex $\hat{\mathbf{r}}_3$ of $\hat{E}$, the vertex corresponding to vertex $\mathbf{r}$ in the mapping F_{E_1}. Similarly,

$$(K^{-1}\mathbf{u}_h, \mathbf{v}_1)_{Q,E_1} = \frac{1}{4}(\mathcal{K}^{-1}_{11,E_2}J_{n_1}u_1 + \mathcal{K}^{-1}_{12,E_2}J_{n_2}u_2)J_{n_1}. \tag{2.35}$$

For the right hand side of (2.29) we write

$$\begin{aligned}(p_h, \nabla\cdot\mathbf{v}_1) &= (p_h, \nabla\cdot\mathbf{v}_1)_{E_1} + (p_h, \nabla\cdot\mathbf{v}_1)_{E_2} \\ &= \langle p_h, \mathbf{v}_1\cdot\mathbf{n}_{E_1}\rangle_{e_1} + \langle p_h, \mathbf{v}_1\cdot\mathbf{n}_{E_2}\rangle_{e_1} \\ &= \langle \hat{p}_h, \hat{\mathbf{v}}_1\cdot\hat{\mathbf{n}}_{E_1}\rangle_{\hat{e}_1} + \langle \hat{p}_h, \hat{\mathbf{v}}_1\cdot\hat{\mathbf{n}}_{E_2}\rangle_{\hat{e}_1} \\ &= \frac{1}{2}(p_1 - p_2)J_{n_1},\end{aligned} \tag{2.36}$$

where we have used the trapezoidal rule for the integrals on $\hat{e}_1$, which is exact since $\hat{p}_h$ is constant and $\hat{\mathbf{v}}_1\cdot\hat{\mathbf{n}}$ is linear. A combination of (2.33)–(2.36) gives the equation

$$\frac{1}{2}((\mathcal{K}^{-1}_{11,E_1} + \mathcal{K}^{-1}_{11,E_2})J_{n_1}u_1 + \mathcal{K}^{-1}_{12,E_1}J_{n_4}u_4 + \mathcal{K}^{-1}_{12,E_2}J_{n_2}u_2) = p_1 - p_2.$$

The other three equations of the local system for $u_1, \ldots, u_4$ are obtained similarly.

REMARK 2.5. The above construction is also valid for vertices on the boundary $\partial\Omega$. In the case of Dirichlet boundary conditions, a 3×3 system allows to express the velocities in terms of cell and boundary pressures. In the case of Neumann boundary conditions, the one unknown vertex velocity is expressed in terms of the two cell pressures and two boundary fluxes.

3. Error analysis. We will make use of the L^2-orthogonal projection onto W_h. For any $\phi \in L^2(\Omega)$, let $\mathcal{Q}_h\phi \in W_h$ be its $L^2(\Omega)$ projection satisfying

$$(\phi - \mathcal{Q}_h\phi, w) = 0 \qquad \forall w \in W_h.$$

It is well known [10] that the L^2-projection has the approximation property

$$\|\phi - \mathcal{Q}_h\phi\| < C\|\phi\|_r h^r, \quad 0 \le r \le 1. \tag{3.1}$$

The convergence analysis of the method (2.29)–(2.30) is similar to the analysis in the case of straight edge quadrilaterals presented in [22]. The following estimates hold.

THEOREM 3.1. *If $K^{-1} \in W^{1,\infty}(E)$ for all elements E, then there exists a constant C independent of h such that*

$$\|\mathbf{u} - \mathbf{u}_h\| \le Ch\|\mathbf{u}\|_1, \tag{3.2}$$

$$\|\nabla \cdot (\mathbf{u} - \mathbf{u}_h)\| \le Ch\|\nabla \cdot \mathbf{u}\|_1, \tag{3.3}$$

$$\|p - p_h\| \le Ch(\|\mathbf{u}\|_1 + \|p\|_1). \tag{3.4}$$

Moreover, if the problem (1.1)–(1.4) has H^2-elliptic regularity, and if $K \in W^{1,\infty}(E)$ and $K^{-1} \in W^{2,\infty}(E)$ for all elements E, then

$$\|\mathcal{Q}_h p - p_h\| \le Ch^2(\|\mathbf{u}\|_1 + \|\nabla \cdot \mathbf{u}\|_1). \tag{3.5}$$

The proof of the theorem uses the following bounds on the quadrature error.

LEMMA 3.1. *If $K^{-1} \in W^{1,\infty}(E)$ for all elements E, then there exists a constant C independent of h such that for all $\mathbf{v} \in \mathbf{V}_h$*

$$|\sigma(K^{-1}\Pi\mathbf{u}, \mathbf{v})| \le Ch\|\mathbf{u}\|_1\|\mathbf{v}\|. \tag{3.6}$$

If $K^{-1} \in W^{2,\infty}(E)$ for all elements E, then, for all $\mathbf{v}, \mathbf{q} \in \mathbf{V}_h$,

$$|\sigma(K^{-1}\mathbf{q}, \mathbf{v})| \le C \sum_{E \in \mathcal{T}_h} h^2\|\mathbf{q}\|_{1,E}\|\mathbf{v}\|_{1,E}. \tag{3.7}$$

The proof of the above lemma follows closely the argument presented in [22] for straight-edge quadrilaterals. The error on any element E is bounded through mapping to the reference element $\hat{E}$, employing bounds on the trapezoidal quadrature error, and mapping back to E. We refer the reader to [22] for details.

Proof of Theorem 3.1. Subtracting the numerical scheme (2.29)–(2.30) from the variational formulation (1.6)–(1.7) gives the error equations

$$\begin{aligned}(K^{-1}(\Pi\mathbf{u} - \mathbf{u}_h), \mathbf{v})_Q = (\mathcal{Q}_h p - p_h, \nabla \cdot \mathbf{v}) + (K^{-1}(\Pi\mathbf{u} - \mathbf{u}), \mathbf{v})\\ -\sigma(K^{-1}\Pi\mathbf{u}, \mathbf{v}), \quad \mathbf{v} \in \mathbf{V}_h,\end{aligned} \tag{3.8}$$

$$(\nabla \cdot (\Pi\mathbf{u} - \mathbf{u}_h), w) = 0, \quad w \in W_h. \tag{3.9}$$

First note that (A.6) implies that on any element E we can choose $w = J_E \nabla \cdot (\Pi\mathbf{u} - \mathbf{u}_h) \in W_h$ in (3.9). Since J_E is uniformly positive, this implies that

$$\nabla \cdot (\Pi\mathbf{u} - \mathbf{u}_h) = 0. \tag{3.10}$$

Bound (3.3) follows from (3.10) and (A.3).

To show (3.2), take $\mathbf{v} = \Pi\mathbf{u} - \mathbf{u}_h$ in (3.9) to obtain

$$\begin{aligned}(K^{-1}(\Pi\mathbf{u} - \mathbf{u}_h), \Pi\mathbf{u} - \mathbf{u}_h)_Q =& (K^{-1}(\Pi\mathbf{u} - \mathbf{u}), \Pi\mathbf{u} - \mathbf{u}_h) \\ &- \sigma(K^{-1}\Pi\mathbf{u}, \Pi\mathbf{u} - \mathbf{u}_h).\end{aligned} \tag{3.11}$$

Using (A.1), the first term on the right above is bounded by

$$|(K^{-1}(\Pi\mathbf{u} - \mathbf{u}), \Pi\mathbf{u} - \mathbf{u}_h)| \leq Ch\|\mathbf{u}\|_1\|\Pi\mathbf{u} - \mathbf{u}_h\|. \tag{3.12}$$

The second term on the right in (3.11) can be bounded using Lemma 3.1,

$$|\sigma(K^{-1}\Pi\mathbf{u}, \Pi\mathbf{u} - \mathbf{u}_h)| \leq Ch\|\mathbf{u}\|_1\|\Pi\mathbf{u} - \mathbf{u}_h\|. \tag{3.13}$$

A combination of (3.11), (3.12), (3.13), (2.31), and (A.1) completes the proof of (3.2).

Using the inf-sup condition (2.18) and (3.9), we obtain

$$\begin{aligned}&\|\mathcal{Q}_h p - p_h\| \\ &\leq \frac{1}{\beta} \sup_{\substack{\mathbf{v} \in \mathbf{V}_h \\ \mathbf{v} \neq 0}} \frac{(\nabla \cdot \mathbf{v}, \mathcal{Q}_h p - p_h)}{\|\mathbf{v}\|_{\text{div}}} \\ &= \frac{1}{\beta} \sup_{\substack{\mathbf{v} \in \mathbf{V}_h \\ \mathbf{v} \neq 0}} \frac{(K^{-1}(\Pi\mathbf{u} - \mathbf{u}_h), \mathbf{v})_Q - (K^{-1}(\Pi\mathbf{u} - \mathbf{u}), \mathbf{v}) + \sigma(K^{-1}\Pi\mathbf{u}, \mathbf{v})}{\|\mathbf{v}\|_{\text{div}}} \\ &\leq \frac{C}{\beta} h\|\mathbf{u}\|_1,\end{aligned}$$

where we have used the Cauchy-Schwarz inequality, (3.2), (A.1), and (3.6) in the last inequality. The proof of (3.4) is completed by an application of the triangle inequality and (3.1).

The proof of (3.5) is based on a duality argument and employs the quadrature error bound (3.7); see [22] for details. □

4. Numerical experiments. In this section we present several numerical experiments that confirm the theoretical results of the previous section. In the first example we take $K = 2 * I$ and solve a problem with Neumann boundary conditions and a known solution

$$p(x, y) = \cos(2\pi(x + 1/2))\cos(2\pi(y + 1/2)).$$

The domain has an irregular shape, see Figure 4. It is partitioned by curved quadrilaterals. Note that the numerical grid is smooth, except across the vertical line that cuts through the middle. Due to (2.22), the non-smoothness of the grid translates into a discontinuous computational permeability $\mathcal{K}$. We test the convergence of our method on a sequence of

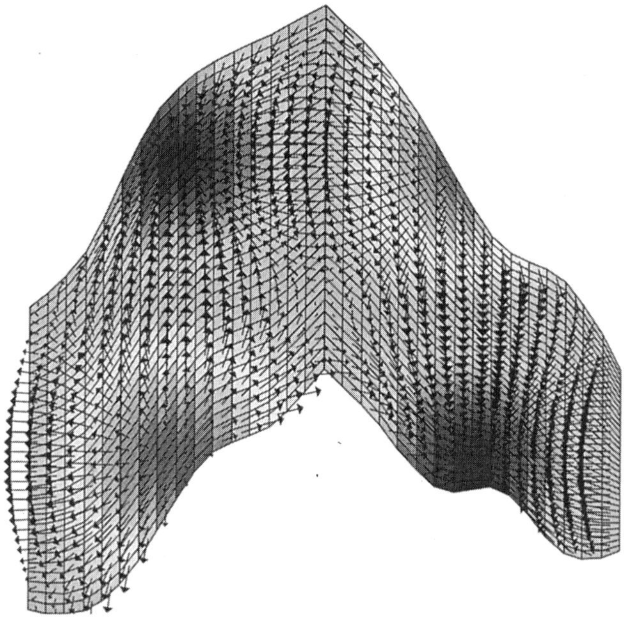

FIG. 4. *Computed pressure (color) and velocity (arrows) with the MFMFE method in Example 1.*

six meshes, from 8×8 to 256×256. The computed solution on the 32×32 mesh is shown in Figure 4. The MFMFE method is compared to the EMFE method of [4, 3]. The two methods have comparable computational costs, as each one reduces to CCFD for the pressure. The discretization errors and asymptotic convergence rates are presented in Table 1. Here $|||p-p_h|||$ denotes a discrete pressure L^2-norm that involves only the function values at the cell-centers and $|||\mathbf{u}-\mathbf{u}_h|||$ denotes a discrete velocity L^2-norm that involves only the normal vector components at the midpoints of the edges. We note that for the MFMFE method the obtained convergence rates of $O(h^2)$ for $|||p-p_h|||$ and $O(h)$ for $\|\mathbf{u}-\mathbf{u}_h\|$ confirm the theoretical results. The $O(h^2)$ accuracy for $|||\mathbf{u}-\mathbf{u}_h|||$ indicates superconvergence for the normal velocities at the midpoints of the edges. At the same time, the EMFE method exhibits only $O(h)$ convergence for the pressure and $O(h^{1/2})$ for the velocity. The slower convergence is due to reduced accuracy along the discontinuity, as it can be seen in Figure 5.

In the second example we test our method on a sequence of meshes obtained by a uniform refinement of an initial rough quadrilateral mesh. It is easy to see that the resulting partitions consist of h^2-parallelograms. We take $K = 2 * I$, Dirichlet boundary conditions, and a true solution

$$p(x, y) = x^3 y + y^4 + \sin(x)\cos(y).$$

TABLE 1
Discretization errors and convergence rates for Example 1.

	MFMFE method			EMFE method	
$1/h$	$\|\|\|p - p_h\|\|\|$	$\|\|\mathbf{u} - \mathbf{u}_h\|\|$	$\|\|\|\mathbf{u} - \mathbf{u}_h\|\|\|$	$\|\|\|p - p_h\|\|\|$	$\|\|\|\mathbf{u} - \mathbf{u}_h\|\|\|$
8	0.17E0	0.37E0	0.17E0	0.21E+1	0.37E0
16	0.60E-1	0.21E0	0.46E-1	0.26E0	0.24E0
32	0.97E-2	0.11E0	0.12E-1	0.74E-1	0.16E0
64	0.25E-2	0.58E-1	0.29E-2	0.31E-1	0.12E0
128	0.67E-3	0.29E-1	0.72E-3	0.14E-1	0.84E-1
256	0.17E-3	0.15E-1	0.18E-3	0.70E-2	0.60E-1
Rate	1.99	0.99	2.00	1.04	0.48

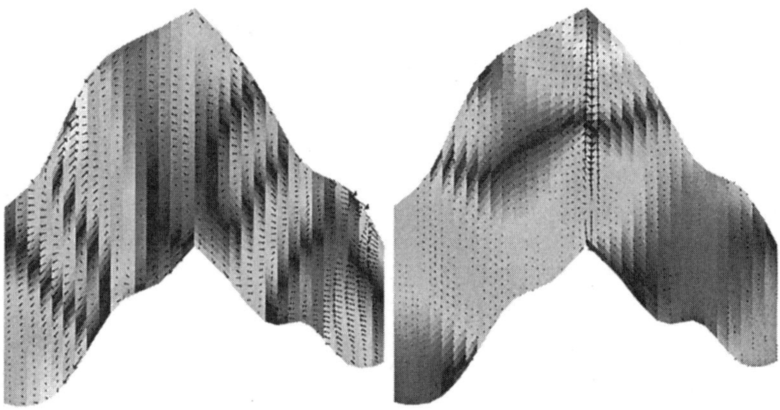

FIG. 5. *Error in the pressure (color) and the velocity (arrows) the MFMFE method (left) and the EMFE method (right) in Example 1. The two graphs are scaled differently. On the left, maximum pressure error (red) is 0.02 and maximum vector length is 0.21. On the right, maximum pressure error is 0.13 and maximum vector length is 9.35.*

The initial 8×8 mesh is generated from a square mesh by randomly perturbing the location of each vertex within a disk centered at the vertex with a radius $h\sqrt{2}/4$. The computed solution on the first level of refinement is shown in Figure 6. The numerical errors and convergence rates are obtained on a sequence of six mesh refinements and are reported in Table 2. As in the first example, the computationally obtained convergence rates for the MFMFE method confirm the theoretical results, while the EMFE method suffers a deterioration of accuracy along the non-smooth interfaces.

REMARK 4.1. We recently learned of the concurrent and related work of Klausen and Winther [14]. They formulate the MPFA method from [1] as a mixed finite element method using an enhanced Raviart-Thomas space and obtain convergence results on quadrilateral grids.

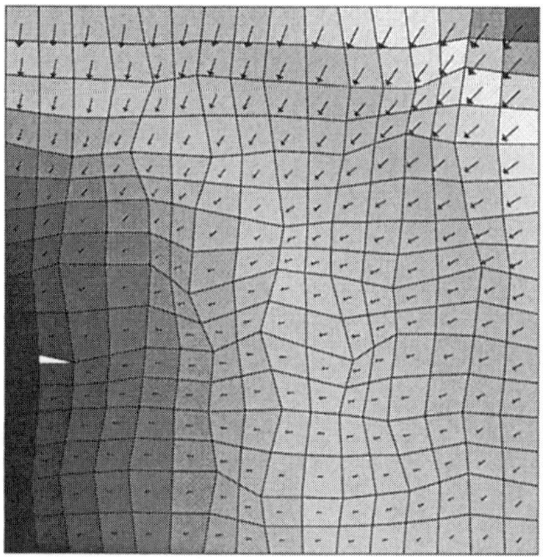

FIG. 6. *Computed solution on the first level of refinement in Example 2.*

TABLE 2
Discretization errors and convergence rates for Example 2.

	MFMFEM			EMFE method	
$1/h$	$\|\|\|p - p_h\|\|\|$	$\|\|\mathbf{u} - \mathbf{u}_h\|\|$	$\|\|\|\mathbf{u} - \mathbf{u}_h\|\|\|$	$\|\|\|p - p_h\|\|\|$	$\|\|\|\mathbf{u} - \mathbf{u}_h\|\|\|$
8	0.10E-1	0.85E-1	0.24E-1	0.19E-1	0.17E0
16	0.27E-2	0.55E-1	0.87E-2	0.88E-2	0.13E0
32	0.70E-3	0.30E-1	0.27E-2	0.45E-2	0.96E-1
64	0.18E-3	0.16E-1	0.73E-3	0.23E-2	0.69E-1
128	0.45E-4	0.81E-2	0.19E-3	0.12E-2	0.50E-1
256	0.11E-4	0.41E-2	0.50E-4	0.59E-3	0.35E-1
Rate	1.99	0.98	1.95	0.99	0.49

APPENDIX

A. Approximation properties of Π.

LEMMA A.1. *If E is a quadrilateral, then*

$$\|\mathbf{q} - \Pi\mathbf{q}\|_E \le C\|\mathbf{q}\|_{2,E}h^2. \tag{A.1}$$

If E is an h^2-quadrilateral, then

$$\|\mathbf{q} - \Pi\mathbf{q}\|_E \le C\|\mathbf{q}\|_{2,E}h. \tag{A.2}$$

If E is an h^2-parallelogram, then

$$\|\nabla \cdot (\mathbf{q} - \Pi\mathbf{q})\|_E \le C\|\nabla \cdot \mathbf{q}\|_{1,E}h. \tag{A.3}$$

Proof. Bound (A.1) has been shown in [20]. The proof of (A.2) is a modification of the argument in [20]. Using Lemma 2.1, the definition of $\hat{\mathbf{V}}(\hat{E})$, and the Bramble-Hilbert lemma,

$$\|\mathbf{q} - \Pi\mathbf{q}\|_E \le C\|\hat{\mathbf{q}} - \hat{\Pi}\hat{\mathbf{q}}\|_{\hat{E}} \le C([\hat{q}_1]_{2,\hat{E},\hat{y}} + [\hat{q}_2]_{2,\hat{E},\hat{x}}), \tag{A.4}$$

where $[\hat{q}_1]_{2,\hat{E},\hat{y}} = \|\partial^2\hat{q}_1/\partial\hat{y}^2\|_{\hat{E}}$ and $[\hat{q}_2]_{2,\hat{E},\hat{x}} = \|\partial^2\hat{q}_2/\partial\hat{x}^2\|_{\hat{E}}$. Letting $\mathbf{g} = \mathbf{r}_{21} - \mathbf{r}_{34}$, it is easy to see from (2.3) that

$$J_E DF_E^{-1} = A + \begin{bmatrix} g_2\hat{x} & -g_1\hat{x} \\ -g_2\hat{y} & g_1\hat{y} \end{bmatrix} + \bar{R}, \tag{A.5}$$

where A is a constant matrix and $\|\bar{R}\|_{j,\infty,\hat{E}} \le Ch^2$, $j = 0, 1, 2$. Using that $\hat{\mathbf{q}} = J_E DF_E^{-1}\tilde{\mathbf{q}}$, where $\tilde{\mathbf{q}}(\hat{x}) = \mathbf{q} \circ F_E(\hat{x})$, (A.4) and (A.5) imply

$$\|\mathbf{q} - \Pi\mathbf{q}\|_E \le C(h([\tilde{q}_1]_{2,\hat{E},\hat{y}} + [\tilde{q}_2]_{2,\hat{E},\hat{x}}) + h^2\|\tilde{\mathbf{q}}\|_{2,\hat{E}}),$$

where we have also used (2.6). Bound (A.2) now follows from a change of variables back to E.

To show (A.3) we first note that (2.11) and $(\nabla\cdot\mathbf{v}, w)_E = (\widehat{\nabla\cdot\mathbf{v}}, \hat{w}J_E)_{\hat{E}}$ imply

$$\nabla\cdot\mathbf{v} = \left(\frac{1}{J_E}\hat{\nabla}\cdot\hat{\mathbf{v}}\right) \circ F_E^{-1}(\mathbf{x}). \tag{A.6}$$

The above relation gives

$$\int_E (\nabla\cdot(\mathbf{q}-\Pi\mathbf{q}))^2\,d\mathbf{x} = \int_{\hat{E}} \frac{1}{J_E^2}(\hat{\nabla}\cdot(\hat{\mathbf{q}}-\hat{\Pi}\hat{\mathbf{q}}))^2 J_E\,d\hat{\mathbf{x}} \le Ch^{-2}|\hat{\nabla}\cdot\hat{\mathbf{q}}|^2_{1,\hat{E}}, \tag{A.7}$$

where we have used (2.6), (2.14), and the Bramble-Hilbert lemma for the inequality. On the other hand,

$$\begin{aligned} |\hat{\nabla}\cdot\hat{\mathbf{q}}|_{1,\hat{E}} &= |J_E\widehat{\nabla\cdot\mathbf{q}}|_{1,\hat{E}} \\ &\le C(\|J_E\|_{\infty,\hat{E}}|\widehat{\nabla\cdot\mathbf{q}}|_{1,\hat{E}} + |J_E|_{1,\infty,\hat{E}}\|\widehat{\nabla\cdot\mathbf{q}}\|_{\hat{E}}) \\ &\le C(h^2|\widehat{\nabla\cdot\mathbf{q}}|_{1,\hat{E}} + h^3\|\widehat{\nabla\cdot\mathbf{q}}\|_{\hat{E}}), \end{aligned} \tag{A.8}$$

using (2.8) for the last inequality. Combining (A.7) – (A.8) and changing variables back to E implies (A.3). □

REFERENCES

[1] I. AAVATSMARK, *An introduction to multipoint flux approximations for quadrilateral grids*, Comput. Geosci., 6 (2002), pp. 405–432.

[2] I. AAVATSMARK, T. BARKVE, Ø. BØE, AND T. MANNSETH, *Discretization on unstructured grids for inhomogeneous, anisotropic media. I. Derivation of the methods*, SIAM J. Sci. Comput., 19 (1998), pp. 1700–1716 (electronic).

[3] T. Arbogast, C.N. Dawson, P.T. Keenan, M.F. Wheeler, and I. Yotov, *Enhanced cell-centered finite differences for elliptic equations on general geometry*, SIAM J. Sci. Comp., 19 (1998), pp. 404–425.

[4] T. Arbogast, M.F. Wheeler, and I. Yotov, *Mixed finite elements for elliptic problems with tensor coefficients as cell-centered finite differences*, SIAM J. Numer. Anal., 34 (1997), pp. 828–852.

[5] K. Aziz and A. Settari, *Petroleum reservoir simulation*, Applied Science Publishers, London and New York, 1979.

[6] J. Baranger, J.-F. Maitre, and F. Oudin, *Connection between finite volume and mixed finite element methods*, RAIRO Modél. Math. Anal. Numér., 30 (1996), pp. 445–465.

[7] F. Brezzi, J. Douglas, Jr., and L.D. Marini, *Two families of mixed elements for second order elliptic problems*, Numer. Math., 88 (1985), pp. 217–235.

[8] F. Brezzi and M. Fortin, *Mixed and Hybrid Finite Element Methods*, vol. 15 of Springer Series in Computational Mathematics, Springer Verlag, Berlin, 1991.

[9] Z. Cai, J.E. Jones, S.F. McCormick, and T.F. Russell, *Control-volume mixed finite element methods*, Comput. Geosci., 1 (1997), pp. 289–315 (1998).

[10] P.G. Ciarlet, *The finite element method for elliptic problems*, North-Holland, New York, 1978.

[11] R.E. Ewing, M. Liu, and J. Wang, *Superconvergence of mixed finite element approximations over quadrilaterals*, SIAM J. Numer. Anal., 36 (1999), pp. 772–787.

[12] J.M. Hyman, M. Shashkov, and S. Steinberg, *The numerical solution of diffusion problems in strongly heterogeneous non-isotropic materials*, J. Comput. Phys., 132 (1997), pp. 130–148.

[13] R.A. Klausen and T.F. Russell, *Relationships among some locally conservative discretization methods which handle discontinuous coefficients.* To appear in Computational Geosciences.

[14] R.A. Klausen and R. Winther, *Convergence of multi point flux approximations on quadrilateral grids.* Preprint.

[15] S. Micheletti, R. Sacco, and F. Saleri, *On some mixed finite element methods with numerical integration*, SIAM J. Sci. Comput., 23 (2001), pp. 245–270 (electronic).

[16] R.A. Raviart and J.M. Thomas, *A mixed finite element method for 2nd order elliptic problems*, in Mathematical Aspects of the Finite Element Method, Lecture Notes in Mathematics, vol. 606, Springer-Verlag, New York, 1977, pp. 292–315.

[17] T.F. Russell and M.F. Wheeler, *Finite element and finite difference methods for continuous flows in porous media*, in The Mathematics of Reservoir Simulation, R. E. Ewing, ed., vol. 1 of Frontiers in Applied Mathematics, SIAM, Philadelphia, PA, 1983, pp. 35–106.

[18] J.M. Thomas, *Sur l'analyse numérique des méthods d'éléments finis hybrides et mixtes*, PhD thesis, Université Pierre et Marie Curie, Paris, 1977.

[19] P.S. Vassilevski, S.I. Petrova, and R.D. Lazarov, *Finite difference schemes on triangular cell-centered grids with local refinement*, SIAM J. Sci. Statist. Comput., 13 (1992), pp. 1287–1313.

[20] J. Wang and T.P. Mathew, *Mixed finite element method over quadrilaterals*, in Conference on Advances in Numerical Methods and Applications, I. T. Dimov, B. Sendov, and P. Vassilevski, eds., World Scientific, River Edge, NJ, 1994, pp. 203–214.

[21] A. Weiser and M.F. Wheeler, *On convergence of block-centered finite-differences for elliptic problems*, SIAM J. Numer. Anal., 25 (1988), pp. 351–375.

[22] M.F. Wheeler and I. Yotov, *A multipoint flux mixed finite element method*, Tech. Rep. TR Math 05-06, Dept. Math., University of Pittsburgh, 2005.

DEVELOPMENT AND APPLICATION OF COMPATIBLE DISCRETIZATIONS OF MAXWELL'S EQUATIONS*

DANIEL A. WHITE†, JOSEPH M. KONING†‡, AND ROBERT N. RIEBEN†§

Abstract. We present the development and application of compatible finite element discretizations of electromagnetics problems derived from the time dependent, full wave Maxwell equations. We review the $H(curl)$-conforming finite element method, using the concepts and notations of differential forms as a theoretical framework. We chose this approach because it can handle complex geometries, it is free of spurious modes, it is numerically stable without the need for filtering or artificial diffusion, it correctly models the discontinuity of fields across material boundaries, and it can be very high order. Higher-order $H(curl)$ and $H(div)$ conforming basis functions are not unique and we have designed an extensible C++ framework that supports a variety of specific instantiations of these such as standard interpolatory bases, spectral bases, hierarchical bases, and semi-orthogonal bases. Virtually any electromagnetics problem that can be cast in the language of differential forms can be solved using our framework. For time dependent problems a method-of-lines scheme is used where the Galerkin method reduces the PDE to a semi-discrete system of ODE's, which are then integrated in time using finite difference methods. For time integration of wave equations we employ the unconditionally stable implicit Newmark-Beta method, as well as the high order energy conserving explicit Maxwell Symplectic method; for diffusion equations, we employ a generalized Crank-Nicholson method. We conclude with computational examples from resonant cavity problems, time-dependent wave propagation problems, and transient eddy current problems, all obtained using the authors massively parallel computational electromagnetics code *EMSolve* .

Key words. Computational electromagnetics, Maxwell's equations, vector finite elements, high order methods, $H(curl)$ and $H(div)$ - conforming methods, discrete differential forms, spurious modes, numerical dispersion, wave propagation, transient eddy currents, electromagnetic diffusion.

1. Introduction.

The equations of electromagnetics can be simply and elegantly cast in the language of differential geometry, more precisely in terms of differential forms or p-forms [1–3]. In this geometrical setting, the fundamental conservation laws are not obscured by the details of coordinate system dependent notation; and, the governing equations can be reformulated in a more compact and clear way using well known differential operators of the exterior algebra such as the exterior derivative, the wedge product, and the Hodge star operator, see [4] for an introduction to differential forms. In this context, a natural framework for the modeling of electromagnetics is provided. For example, the electric potentials can

*This work was performed under the auspices of the U.S. Department of Energy by the University of California, Lawrence Livermore National Laboratory under contract No. W-7405-Eng-48.

†Defense Sciences Engineering Division, Lawrence Livermore National Laboratory, Livermore, CA 94551 (white37@llnl.gov).

‡koning1@llnl.gov.

§rieben1@llnl.gov.

be represented by 0-forms; electric and magnetic fields by 1-forms; electric and magnetic fluxes by 2-forms; and scalar charge density by 3-forms.

In the context of Galerkin approximations, the choice of the finite element space plays a crucial role in the stability and convergence of the discretization. For instance, in numerical approximations of the magnetic and electric field intensities, $H(curl)$ -conforming finite element spaces (or edge elements) are preferred over traditional nodal vector spaces since they eliminate spurious modes in eigenvalue computations and they prevent fictitious charge build-up in time-dependent computations. The lowest order $H(curl)$ -conforming basis functions were developed by Whitney [5] before the advent of finite element programs. Arbitrary order versions were introduced by Nédélec [6, 7] as a generalization of the mixed finite element spaces introduced by P.A. Raviart and J.M. Thomas in [8] for $H(div)$ -conforming methods. For an extensive analysis of several $H(div)$ -conforming methods see [9].

Recently, Hiptmair, motivated by the theory of exterior algebra of differential forms, presented a unified framework for the construction of conforming finite element spaces. Remarkably, both $H(curl)$ and $H(div)$ conforming finite element spaces and the definition of their degrees of freedom and interpolation operators can be derived within this framework, see [10] for more details. In simple terms the finite element basis functions satisfy discrete counterparts of the De Rham exact sequence and related commuting diagrams. The architecture of our *EMSolve* software closely mimics the structure of differential forms. In our software terminology, a discrete differential p-form is a finite element basis used to discretize a p-form field. In *EMSolve* the global discrete exterior derivative and Hodge operators are sparse matrices, and the rules of differential forms define how these matrices can be combined to represent discretizations of PDE's. Given a physical law expressed in the language of differential forms, it is therefore quite straightforward to discretize the problem using *EMSolve* .

One unique feature of *EMSolve* is the emphasis on high-order discretization which can reduce the mesh size, memory usage, and CPU time required to achieve a prescribed error tolerance. This is particularly true for electrically large problems. For these problems it is known that the Galerkin discretization error is larger than the best approximation error of the finite element space. This is sometimes referred to as the pollution effect, and has been more precisely explained in [11, 12]. In the engineering community this is referred to as numerical dispersion, as the computed phase velocity differs from the physical phase velocity and phase error builds up linearly with respect to distance and time. For the popular lowest order edge elements, it is known that the numerical dispersion relation is second order accurate [13–15]. Second order accuracy may seem adequate, but for an electrically large problem the phase error may be such that the global error is 100 percent even though the local truncation error is quite small. A detailed analysis of dispersion for higher order $H(curl)$

finite elements on orthogonal Cartesian meshes is given in [16], with the result that the dispersion error is asymptotically $O(h^{2k})$ where k is the order of the basis functions.

It should be noted that there are numerous numerical schemes for electromagnetics that are based in part on differential forms and related geometrical concepts, such as the cell method [17], finite integration theory [18, 19], and mimetic discretizations [20]. Even the most popular method for time-domain computational electromagnetics, Yee's FDTD method [21], has been reinterpreted from a geometric perspective by numerous authors [22, 23].

2. Numerical formulation. We begin with the generic boundary value problem stated in the language of differential forms from [24]. This problem statement is generic in that the degree of the forms are not specified. By specifying the degree p we have equations involving the divergence, gradient, or curl operators. We assume a 3-dimensional domain Ω with piecewise smooth boundary $\partial\Omega$ partitioned into Γ_D, Γ_N, and Γ_M. The problem statement is

$$du = (-1)^p \sigma \qquad dj = -\Psi + \Phi \quad in \ \Omega \tag{2.1}$$

$$T_D u = f \quad on \ \Gamma_D \qquad T_N j = g \quad on \ \Gamma_N \tag{2.2}$$

$$j = \star_\alpha \sigma \qquad \Psi = \star_\gamma u \quad in \ \Omega \tag{2.3}$$

$$T_M j = (-1)^p \star_\beta T_M u \quad on \ \Gamma_M. \tag{2.4}$$

Here u is a $(p-1)$-form, σ is a p-form, j is a $(3-p)$-form, and both Ψ and Φ are $(3-p+1)$-forms, where $1 \leq p \leq 3$. The variable Φ is a source term. The symbols α, β and γ denote generic material constitutive relations (e.g. electric permittivity or conductivity). In (2.1) the operator d is the exterior derivative which maps p-forms to $(p+1)$-forms. In the boundary conditions (2.2) and (2.4) the symbol T denotes the trace operator, where the trace of a p-form is an integral over a $p-1$-dimensional manifold. In (2.3) and (2.4) the $\star$ symbol denotes the Hodge-star operator, which converts p-forms to $(3-p)$-forms and typically involves material constitutive properties. Equations (2.1) and (2.3) can be combined to yield the general second-order elliptic equation

$$(-1)^p d \star_\alpha du = -\star_\gamma u + \Phi. \tag{2.5}$$

The wedge product of differential forms is used in the definition of bilinear forms. The wedge product of a p-form ω and a q-form η is a $(p+q)$-form ζ

$$\omega^p \wedge \eta^q = \zeta^{(p+q)}, \qquad p+q \leq 3.$$

If $p+q=3$ then $\omega^p \wedge \eta^q$ is an volumetric energy density like quantity and can be integrated over a volume to yield energy. If $p+q=2$ then $\omega^p \wedge \eta^q$

is a flux density like quantity and can be integrated over a surface to yield net flux.

A Galerkin finite element solution of the generic second-order equation (2.5) will require bilinear forms. Using the exterior algebra, the bilinear forms required in the Galerkin finite element method can be easily formulated from the general second-order equation (2.5) by taking the wedge product with a $(p-1)$-form v and integrating over the volume Ω,

$$\int_\Omega (-1)^p d \star_\alpha \, du \wedge v = -\int_\Omega \star_\gamma \, u \wedge v + \int_\Omega \Phi \wedge v.$$

Using the integration-by-parts formula

$$\int_\Omega d\omega \wedge \eta + (-1)^p \int_\Omega \omega \wedge d\eta = \int_{\partial\Omega} \omega \wedge \eta$$

yields the two key symmetric bilinear forms

$$a(u,v) = \int_\Omega \star_\alpha \, (du) \wedge dv, \tag{2.6}$$

$$b(u,v) = \int_\Omega \star_\gamma \, u \wedge v. \tag{2.7}$$

and the additional bilinear forms for source terms and boundary conditions

$$c(u,\Phi) = \int_\Omega u \wedge \Phi$$

$$d(u,g) = \int_{\partial\Omega} \star_\alpha \, du \wedge g.$$

Let $\mathcal{H}^p = \{u \in L_2(\Omega) : \|u\|_p^2 < \infty\}$ and $\mathcal{H}_0^p = \{u \in \mathcal{H}^p : T_D(u) = 0\}$ be generic Hilbert spaces, where $\|u\|_p^2 = \int_\Omega u \wedge \star u + \int_\Omega du \wedge \star du$. Then the Galerkin form of the generic second-order equation (2.5) can now be expressed as follows:

Given the source function Φ and the boundary condition g, find $u \in \mathcal{H}^p$ such that

$$T_D(u) = g \text{ and } a(u,v) = b(u,v) + c(u,\Phi) + d(u,g), \quad \forall v \in \mathcal{H}_0^p.$$

It is not necessary to combine the two 1st-order equations into a single 2nd order equation. If it is desired to formulate the problem as a coupled pair of 1st order equations, as in a mixed method [8, 25, 26, 9], then an additional bilinear form is required, namely

$$e(u,v) = \int_\Omega \star_\alpha \, (du) \wedge v, \tag{2.8}$$

with the requirement that u is a p-form and v is a $(p+1)$ form. With the generic bilinear forms (2.6), (2.7), and (2.8), source terms, and boundary conditions we can construct a wide variety of model equations that can be solved via the finite element method.

We are primarily concerned with time dependent phenomena. The time derivative does not effect the degree of a form. For the generic wave equation we simply add time derivatives to (2.1) which yields

$$du = (-1)^p \frac{\partial \sigma}{\partial t}, \qquad dj = -\frac{\partial \psi}{\partial t} + \Phi \ in \ \Omega \tag{2.9}$$

In the Tonti diagram below we show the time-dependent Maxwell's equations, where d denotes the spatial derivative, $\frac{\partial}{\partial t}$ denotes the time derivative and converging arrows denote summation. In these diagrams ϕ is the 0-form scalar potential; the 1-forms A, E, and H are the magnetic vector potential, the electric field, and the magnetic field, respectively; the 2-forms B, D, and J are the magnetic flux density, the electric flux density, and the electric current density, respectively; and ρ is the 3-form scalar charge density. The left diagram encompasses Faraday's law $dE - \frac{\partial}{\partial t}B = 0$, Coulomb's law for the magnetic field $dB = 0$, and the fact that the electric field E can be written in terms of potentials as $E = d\phi - \frac{\partial}{\partial t}A$. The right diagram encompasses Ampere's law $dH - \frac{\partial}{\partial t}D = J$, Coulomb's law for the electric field $dD = \rho$, and the continuity equation $dJ - \frac{\partial}{\partial t}\rho = 0$. The two diagrams are connected by the constitutive relations $D = \star_\epsilon E$ and $B = \star_\mu H$.

$$\begin{array}{lccccccc}
\text{0-forms :} & & & \phi & & & & \\
 & & & \downarrow d & & & & \\
\text{1-forms :} & A & \xrightarrow{-\frac{\partial}{\partial t}} & E & & & & H \\
 & \downarrow d & & \downarrow d & & & & \downarrow d \\
\text{2-forms :} & B & \xrightarrow{-\frac{\partial}{\partial t}} & 0 & & D & \xrightarrow{-\frac{\partial}{\partial t}} & J \\
 & \downarrow d & & & & \downarrow d & & \downarrow d \\
\text{3-forms :} & 0 & & & & \rho & \xrightarrow{-\frac{\partial}{\partial t}} & 0
\end{array}$$

A wave equation can be derived by combining the two diagrams and solving for E,

$$\frac{\partial^2}{\partial t^2}(\star_\epsilon E) = d \star_{\mu^{-1}} dE - \frac{\partial}{\partial t} J. \tag{2.10}$$

This wave equation resembles the generic second order equation (2.5) with the addition of temporal derivatives. Clearly, if it is determined that one of the time derivatives is small and can be neglected, the result is an electromagnetic diffusion equation (parabolic PDE). Therefore the same bilinear forms (2.6)) and (2.7) are required for spatial discretization of either the elliptic, hyperbolic (wave equation) or parabolic (diffusion equation) problem.

2.1. Local finite element operations. We follow the work of Ciarlet [27] and adhere to the definition of a finite element as a set of three distinct objects $(\Sigma, \mathcal{P}, \mathcal{A})$ such that:

- Σ is the polyhedral domain over which the element is defined
- $\mathcal{P}$ is a finite dimensional polynomial space from which basis functions are constructed
- $\mathcal{A}$ is a set of linear functionals (*Degrees of Freedom*) dual to $\mathcal{P}$

Let Σ_h be a discretization of the problem domain Ω using tetrahedral, hexahedral, or prismatic elements. By using a local change of variables given by the iso-parametric mapping $\Phi(\hat{\Sigma}) = \Sigma$, we re-write the bilinear of (2.6) as follows

$$\begin{aligned} a(u,v) &= \int_\Omega \star_\alpha \, du \wedge dv \\ &= \sum_{\Sigma \in \Sigma_h} \int_{\Sigma = \Phi(\hat{\Sigma})} \star_\alpha \, du \wedge dv \\ &= \sum_{\Sigma \in \Sigma_h} \int_{\hat{\Sigma}} (\star_\alpha \circ \Phi) \; \Phi^* \, (du \wedge dv) \; |D\Phi| \\ &= \sum_{\Sigma \in \Sigma_h} \int_{\hat{\Sigma}} (\star_\alpha \circ \Phi) \; \Phi^*(du) \wedge \Phi^*(dv) \; |D\Phi|. \end{aligned} \tag{2.11}$$

Similarly, the bilinear form of (2.7) can be rewritten as

$$b(u,v) = \sum_{\Sigma \in \Sigma_h} \int_{\hat{\Sigma}} (\star_\beta \circ \Phi) \; \Phi^*(u) \wedge \Phi^*(v) \; |D\Phi|. \tag{2.12}$$

Equations (2.11) and (2.12) show that all calculations for the various bilinear forms can be performed on a standard reference element $\hat{\Sigma}$ (i.e. the unit cube, tetrahedron, or prism). Results are then transformed to physical mesh elements (of arbitrary curvature) via a set of well defined transformation rules based on the properties of differential forms. These rules are summarized in Table 1. Given these transformations the bases need only be evaluated on the reference element and transformed accordingly. In *EMSolve* the bilinear form requires that the reference element, the quadrature rule, and the p-form basis functions be specified just once. The basis functions are then sampled at the quadrature points on the reference element, and this information is cached for latter use. This gives rise to a very computationally efficient algorithm for computing finite element approximations. For a given element topology and basis order, the basis functions only need to be computed once. Then, for every element of the same topology in the mesh, the results from the reference element can simply be mapped according to the transformation rules. This can significantly reduce computational time for a typical finite element computation. In addition, integration over the reference element is much simpler and can often be done exactly using Gaussian quadrature of the appropriate order.

When implementing a finite element space $\mathcal{P}$ in the context of differential forms, the explicit formulation of the space depends on the p-form and the topology of the reference element. The construction of the finite element space $\mathcal{P}$ is not unique, we choose a construction that leads to a simple and efficient implementation. We use uniformly spaced interpolatory polynomials similar to those described in [28] and [29] as a *primitive basis* on a reference element. The actual bases used in the finite element procedure are constructed on this reference element $\hat{\Sigma}$ rather than in the physical coordinate system and are written as a linear combination of the primitive basis. For example, non-uniform interpolatory functions, moment-based functions, orthogonal functions, etc. can all be expressed as a linear combination of the primitive basis

The construction of a p-form basis of order k is as follows. We begin by generating a primitive basis $W = \{w_j\}$. We can then construct a new basis (non-uniform interpolation, hierarchical, etc.) in terms of the primitive bases by imposing a set of constraints of the form

$$\alpha_i(w_j) = \delta_{ij}, \tag{2.13}$$

where $\alpha_i \in \mathcal{A}$ are the known degrees of freedom of the new basis. The degrees of freedom are in general integral moments, but this is not necessary. What is necessary is that the degrees of freedom satisfy the following:

- *Unisolvence*: $\{\alpha_i\}$ is dual to the finite element space $\mathcal{P}$; i.e. there exists a set $\{w_j\} \subset \mathcal{P}$ such that $\alpha_i(w_j) = \delta_{i,j}$.
- *Invariance*: degrees of freedom remain unisolvent upon a change of variables; this implies they are not affected by the pullback operation; i.e. $\Phi^*(\alpha_i) = \hat{\alpha}_i$.
- *Locality*: the trace of a basis function on a sub-simplex is determined by degrees of freedom associated *only* with that sub-simplex.

The procedure requires the formation of a linear system

$$V_{ij} = \alpha_i(w_j); \quad w_j \in W$$

This system, which is similar to a Vandermonde matrix, is a linear mapping which expresses the new basis in terms of the primitive basis and will have a rank equal to the dimension of the primitive basis. The newly defined basis, which we will denote as W' is given by:

$$W' = V^{-1}W$$

In *EMSolve* , the construction and solution of the Vandermonde system is done once and only once on the reference element. For the evaluation of the basis functions on an actual (or global) element, they are first evaluated on the reference element then transformed according to the transformation rules of Table 1, where the "hat" symbol denotes objects defined with respect to the reference element coordinate system. It is important to note that this process implies that the basis functions have units as shown in Table 2. In standard nodal-based finite element methods the basis functions are dimensionless and the unknowns (the unknown coefficients of the basis function expansion of the field) are simply the value of a field at a point, but here the unknowns are integrals of the field. This seems to be a common theme of all compatible discretization schemes of Maxwell's equations whether they are based upon the finite element method given here, or mimetic finite volume [20] and finite difference [22] methods.

TABLE 1
Transformation rules Φ^*.

	$\Phi^*(u)$	$\Phi^*(du)$
0-forms	$\hat{u}$	$D\Phi^{-1}(d\hat{u})$
1-forms	$D\Phi^{-1}\hat{u}$	$\frac{1}{\lvert D\Phi\rvert}D\Phi^T(d\hat{u})$
2-forms	$\frac{1}{\lvert D\Phi\rvert}D\Phi^T\hat{u}$	$\frac{1}{\lvert D\Phi\rvert}(d\hat{u})$

TABLE 2
Units of Electromagnetic Quantities, Basis Functions, and Degrees-of-Freedom.

Form	Basis Function	Electromagnetic Quantity	DOF
0-forms	1	ϕ $(Volts)$	$Volts$
1-forms	m^{-1}	E $(Volts/m)$	$Volts$
1-forms	m^{-1}	H $(Amps/m)$	$Amps$
2-forms	m^{-2}	D $(Coulombs/m^2)$	$Coulombs$
2-forms	m^{-2}	B $(Webers/m^2)$	$Webers$
2-forms	m^{-2}	J $(Amps/m^2)$	$Amps$
2-forms	m^{-2}	$E \times H$ $(Watts/m^2)$	$Watts$
3-forms	m^{-3}	$E \cdot D$ $(Joules/m^3)$	$Joules$
3-forms	m^{-3}	ρ $(Coulombs/m^3)$	$Coulombs$

We have a class hierarchy for each of the p-form bases, the partial hierarchy is shown in Figure 1. Concrete classes are presented in the lowest level of the tree. The other p-forms have a similar inheritance diagram. The complete class library is documented in [30–32]. Our Silvester-Lagrange (SL) bases are similar to the bases defined in [28] which use equidistant and shifted equidistant interpolation points. The difference between our SL bases and the bases proposed in [28] is that ours satisfy the properties in Table 1. The uniformly spaced interpolatory bases are suitable for low

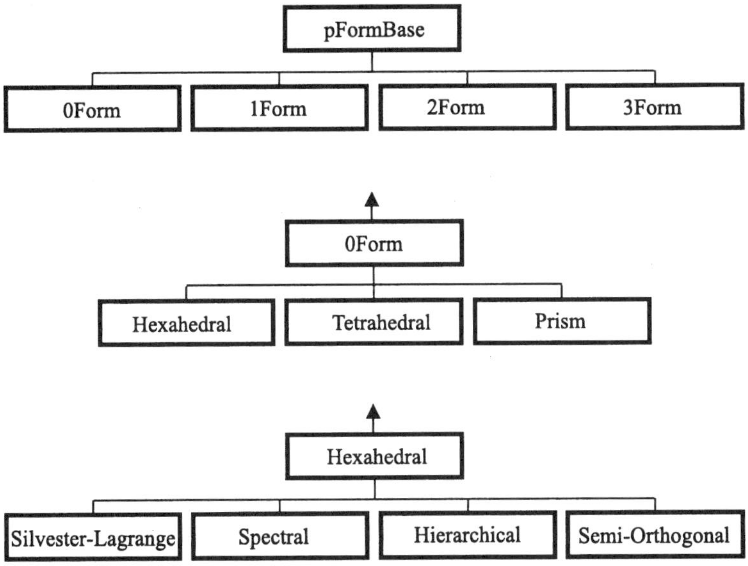

FIG. 1. *p-Form basis function class hierarchy. Only part of the hierarchy is shown. The general idea is that the Application Program Interface is defined at the higher levels, and the unique details of each type of basis function are implemented at the lower levels. Users can easily add new basis functions to the class hierarchy, and the client program need not be modified at all..*

order approximations, *i.e.*, $k = 1$ to 4. It is well known that this particular choice of interpolation points produce badly conditioned mass and stiffness matrices when high order approximations are used. For this reason we have implemented spectral classes that use arbitrary sets of interpolation points, typically Gauss-Lobatto or Tchebyschev points. A study of the conditioning of finite element matrices using various higher-order $H(curl)$ discretizations is given in [33]. An additional class of semi-orthogonal basis functions was developed by the authors in order to increase the efficiency of the method. These basis functions are paired with a custom quadrature rule to minimize the number of non-zeroes in the mass matrices. This is an extension of the standard mass-lumping procedure widely used in computational mechanics. For the special case of an orthogonal Cartesian mesh the mass matrix is made diagonal, resulting in a tremendous increase in efficiency, particularly for higher-order bases applied to time dependent problems. For unstructured meshes the basis functions are not completely orthogonal but the number of non-zeros is decreased by a factor of 5 or more. While these inexact quadratures are often considered a "variational crime", in practice there is no loss of accuracy in the computed solution [34].

2.2. Global finite element operations. The *EMSolve* framework computes sparse matrices which are global versions of the previously described bilinear forms. The basic matrices are

$$\mathbf{M}^p(\alpha)_{ij} = \int_\Omega \alpha W_i^p\ W_j^p\ d\Omega \tag{2.14}$$

$$\mathbf{S}^p(\alpha)_{ij} = \int_\Omega \alpha dW_i^p \cdot dW_j^p\ d\Omega \tag{2.15}$$

$$\mathbf{D}^{p(p+1)}(\alpha)_{ij} = \int_\Omega \alpha dW_i^p\ \cdot W_j^{p+1}\ d\Omega \tag{2.16}$$

which we refer to as the "mass", "stiffness", and "derivative" matrices, respectively. The "mass" matrices $\mathbf{M}$ are square symmetric positive definite, and the "stiffness" matrices $\mathbf{S}$ are square symmetric positive semi-definite. These two matrices map p-forms to p-forms. The "derivative" matrices $\mathbf{D}$ are rectangular and map p-forms to $(p+1)$-forms. It can be shown that

$$\mathbf{D}^{p(p+1)} = \mathbf{M}^{p+1}\mathbf{K}^{p(p+1)} \tag{2.17}$$

$$\mathbf{S}^p = \left(\mathbf{K}^{p(p+1)}\right)^T \mathbf{M}^{(p+1)}\mathbf{K}^{p(p+1)} \tag{2.18}$$

where $\mathbf{K}^{p(p+1)}$ is a "topological derivative" matrix. This matrix is the discretization of the exterior derivative operator d from differential geometry, $dW^p = W^{(p+1)}$. This matrix depends upon the mesh connectivity, but is independent of the nodal coordinates. It does not involve an integral over the element, and it does not involve any material properties. For the special case of first-order basis functions, the topological derivative matrix is a mesh incidence matrix consisting of 0's, +1's, and −1's. While seemingly abstract, the topological derivative matrix is enormously valuable in practice. Given a p-form quantity X with basis function expansion

$$X = \sum_{i=1}^{n} x_i W_i^p, \tag{2.19}$$

and a $(p+1)$-form quantity Y with basis function expansion

$$Y = \sum_{i=1}^{n} y_i W_i^{(p+1)}, \tag{2.20}$$

the exterior derivative (gradient, curl, divergence for $p = 0$, $p = 1$, and $p = 2$, respectively) is given by

$$\mathbf{y} = \mathbf{K}^{p(p+1)}\mathbf{x}. \tag{2.21}$$

It can be shown that

$$\mathbf{K}^{12}\mathbf{K}^{01} = 0 \tag{2.22}$$

$$\mathbf{K}^{23}\mathbf{K}^{12} = 0 \tag{2.23}$$

which are the discrete versions of the identities $\nabla \times \nabla F = 0$ and $\nabla \cdot \nabla \times F = 0$, respectively. These identities are satisfied in the discrete sense, to machine precision, for any mesh and any order basis function. This is a key feature (perhaps the definition of) a compatible discretization. It is these identities that ensure computed solutions of Maxwell's equations are solenoidal, whether in eigenmode computations or time-dependent computations.

EMSolve contains some additional miscellaneous functionality. In some circumstances it is necessary to convert a p-form to a $(3-p)$–form, i.e. a Hodge-star operation. A classic example is converting a "cell-center" quantity to a "nodal" quantity. In our finite element setting the Galerkin procedure prescribes rectangular matrices of the form

$$\mathbf{H}_{ij}^{p(3-p)} = \int_{\Omega} W_i^p \wedge W_j^{(3-p)} \, d\Omega \tag{2.24}$$

which produces optimal (in the least-square error sense) Hodge-star operators for arbitrary order basis functions.

To summarize the overall numerical procedure employed in *EMSolve* , the first step is to identify the correct p-form for the physical quantities. This then dictates the particular basis function expansion of the physical quantity. A generic field variable X is then approximated over each element $\Sigma \in \Sigma_h$ by a basis function expansion of the form

$$X^p(r,t) = \sum_i \alpha_i(t)\, w_i^p(r), \quad w_i^p \in W_h^p \tag{2.25}$$

where $\alpha_i(t)$ are the time-dependent p-form degrees of freedom, $w_i^p(r)$ are the spatially dependent p-form basis functions. The semi-discrete system is formed by applying the Galerkin procedure resulting in combinations of the mass matrices $\mathbf{M}$, the stiffness matrices $\mathbf{S}$, the derivative and topological derivative matrices $\mathbf{D}$ and $\mathbf{K}$. The result is a systematic procedure for discretizing a wide variety of electromagnetics equations.

3. Frequency domain resonant cavity examples. The *EMSolve* framework is well suited for simulations in the frequency domain. Here we focus on the resonant cavity problem, where the goal is to compute the electromagnetic fields within closed perfectly conducting cavities that may contain dielectric and/or magnetic materials. The starting point is the vector Helmholtz equation for the electric field

$$d \star_{\mu^{-1}} dE = -\omega^2 \star_{\epsilon} E \tag{3.1}$$

The electric field is chosen instead of the magnetic field because the perfect electrical conductor conductor boundary condition $n \times E = 0$ is trivial to implement when using a 1-form basis function expansion for E.

Using the Galerkin procedure described in Section 2, the linear system of equations for the discretized eigenvalue problem is

$$\mathbf{S}^{11}_{\mu^{-1}}\mathbf{e} = -\omega^2 \mathbf{M}^{11}_{\epsilon}\mathbf{e}, \tag{3.2}$$

where $\mathbf{e}$ is the vector of 1-form degrees-of-freedom. The exact solution of (3.1) has irrotational eigenmodes corresponding to $\omega = 0$ and solenoidal eigenmodes corresponding to $\omega \neq 0$, with all these modes being orthogonal. This 1-form based discretization preserves the Helmholtz decomposition exactly, with no additional constraints. The irrotational solutions of (3.2) satisfy

$$\mathbf{e}_{irrotational} = \mathbf{K}^{01}\mathbf{f},$$

where $\mathbf{f}$ is an arbitrary discrete scalar potential, and the solenoidal solutions of (3.2) satisfy

$$\left(\mathbf{e}_{solenoidal}\right)^T \mathbf{M}^{11}\mathbf{K}^{01}\mathbf{f} = 0,$$

i.e. they are orthogonal to the gradients of the scalar potentials. Alternative finite element discretizations using vector nodal basis functions discretizations introduced spurious modes, i.e. modes that are not solenoidal. This was first analyzed by Bossavit [35] and Cendes [36], and was historically the primary impetus for using edge-based $H(curl)$ -conforming basis functions in electromagnetics.

Applying general purpose iterative eigenvalue solvers to the $H(curl)$ discretized Helmholtz equation is often problematic due to the large null space of the system. The large degeneracy of zero-eigenvalues can cause iterative methods to fail to converge on the desired smallest non-zero eigenvalues. The authors have developed a method to shift the zero eigenvalues corresponding to the irrotational solutions of the Helmholtz equation arbitrarily to the middle of the spectrum [37]. The implicitly restarted Arnoldi method package (ARPACK) [38] is then used to solve for the smallest extremal eigenvalues which are now non-zero.

3.1. Lowest resonant mode for the Trispal induction cell. The parallel version of the ARPACK code, PARPACK, was used to determine the lowest eigenvalue and eigenmode for the Trispal induction cell. This induction cell is a key component of a proposed proton linear accelerator [39] and is used as a metric for various Helmholtz equation solvers. The Trispal geometry was decomposed into a refined and optimized tetrahedral mesh and a hexahedral mesh as shown in Figure 2. The tetrahedral mesh contained 61,566 zones and 76,838 edges while the hexahedral mesh contained 26,568 zones and 84,807 edges. Results for the computed lowest eigenvalue with $k = 1$ basis functions compared with the measured eigenvalue, frequency=1064.415 MHz, for each mesh are shown in Table 3. The

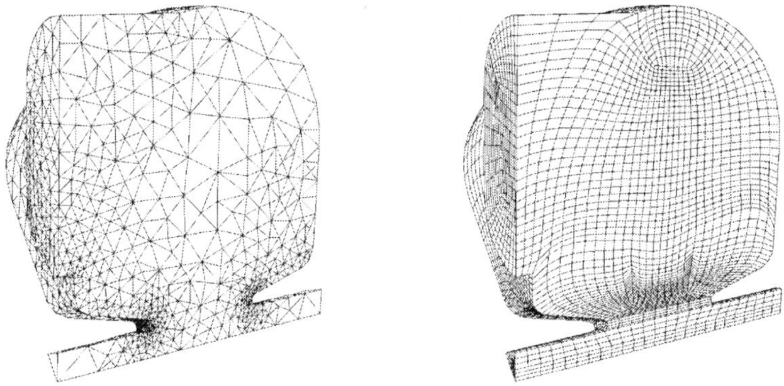

FIG. 2. *The optimized tetrahedral mesh and unoptimized hexahedral mesh for the Trispal geometry. The mesh is of 1/8 of the geometry.*

results were not any better for a higher-order $k = 2$ discretization, indicating that the accuracy for this problem is limited by the discretization of the geometry. The goal for this type of resonant cavity problem is agreement to measurement to within 0.01%, achieving this level of validation requires precise agreement of the CAD geometry and the test article. The z-component of the resulting eigenmode for the hexahedral mesh is shown in Figure 3.

TABLE 3
Trispal eigenvalues.

Mesh	Calculated Frequency (MHz)	Relative Error (%)
Tetrahedral	1066.45	0.19
Hexahedral	1084.12	1.85

4. Electromagnetic wave equations. Consider Maxwell's equations in a charge free region, written in the language of differential forms

$$\star_\epsilon \frac{\partial}{\partial t} E = d(\star_{\mu^{-1}} B) - \star_\sigma E - J \tag{4.1}$$

$$\frac{\partial}{\partial t} B = -dE \tag{4.2}$$

$$d \star_\epsilon E = 0 \tag{4.3}$$

$$dB = 0 \tag{4.4}$$

where the electric field E is a 1-form, the magnetic flux density B is a 2-form, J is an independent 2-form current source, and each of the material property functions are represented by a specific Hodge function. For simplicity the required boundary conditions and initial conditions are not

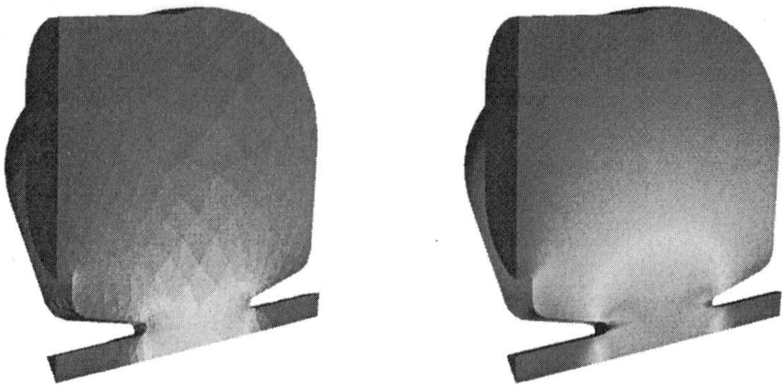

FIG. 3. *The computed Z component of the lowest electric field eigenmode for the Trispal geometry.*

shown here. By applying the exterior derivative operator to both (4.1) and (4.2), it is clear that the divergence constraints of (4.3) and (4.4) are constraints on the initial conditions of the fields. While this is not universally accepted, it is our opinion that a compatible discretization of Maxwell's equations will not require any penalty method or Lagrange multiplier method to satisfy the divergence constraints, they will be intrinsically satisfied by the discretization of equations (4.1) and (4.2). Of course the divergence constraints of (4.3) and (4.4) are satisfied according to a particular, but consistent, discrete metric. For the electric field the divergence is measured in the variational sense

$$\int_\Omega d \star_\epsilon E \wedge \Phi = \int_\Omega \star_\epsilon E \wedge d\Phi, \tag{4.5}$$

for all test functions Φ not on the boundary, since this measure allows for the jump discontinuity in E. For the magnetic flux density the divergence is computed directly.

Using the procedures described in Section 2, the semi-discrete Ampere-Faraday system of equations is

$$\begin{aligned} \mathbf{M}_\epsilon^{11} \frac{\partial}{\partial t}\mathbf{e}(t) &= \left(\mathbf{K}^{12}\right)^T \mathbf{M}_\mu^{22}\, \mathbf{b}(t) - \mathbf{M}_\sigma^{11}\, \mathbf{e}(t) - \mathbf{M}_\epsilon^{11}\, \mathbf{j}(t) \\ \frac{\partial}{\partial t}\mathbf{b}(t) &= -\mathbf{K}^{12}\, \mathbf{e}(t) \end{aligned} \tag{4.6}$$

where $\mathbf{e}(t)$ and $\mathbf{b}(t)$ are the vectors of unknown degrees-of-freedom. The divergence equations are discretized as

$$\begin{aligned} \left(\mathbf{K}^{01}\right)^T \mathbf{M}_\epsilon^{11} e &= 0 \\ \mathbf{K}^{12} b &= 0 \end{aligned}$$

and by the compatibility properties (2.22)–(2.23) these conditions will be satisfied automatically, to the tolerance used in the solution of the mass matrices.

There are several methods for integrating (4.6) in time, the staggered 2nd-order central difference or "leapfrog" method being quite popular. The leapfrog method is conditionally stable and energy conserving. The leapfrog method is an example of a class of methods known as symplectic methods, which were originally developed for Hamiltonian systems. A high order and energy conserving time-integration of (4.6) is given by a generalized symplectic update [40]

$$\begin{bmatrix} \mathbf{e}_{n+1} \\ \mathbf{b}_{n+1} \end{bmatrix} = \left(\prod_{i=1}^{m} Q_i \right) \begin{bmatrix} \mathbf{e}_n \\ \mathbf{b}_n \end{bmatrix} \tag{4.7}$$

where m is the order of the symplectic integration method and the matrices Q_i are of the form

$$Q_i = \begin{bmatrix} I & \beta_i \, \Delta t \, (\mathbf{M}_\epsilon^{11})^{-1} (\mathbf{K}^{12})^T \mathbf{M}_\mu^{22} \\ -\alpha_i \, \Delta t \mathbf{K}^{12} & I - \alpha_i \, \beta_i \, \Delta t^2 \mathbf{K}^{12} \, (\mathbf{M}_\epsilon^{11})^{-1} (\mathbf{K}^{12})^T \mathbf{M}_\mu^{22} \end{bmatrix} \tag{4.8}$$

and Δt is the discrete time step. The specific integration coefficients α_i and β_i of (4.8) can be found in [41]. Note that the standard definition of a symplectic integrator requires that the length of the vectors $\mathbf{e}_n$ and $\mathbf{b}_n$ be the same, and they are not in our case, hence we use the term symplectic loosely. A straightforward but tedious calculation shows that for suitably small Δt the eigenvalues of the Q_i lie on the unit circle, and the eigenvectors of Q_i are linearly independent, hence the time integration method is neutrally stable. The stability condition is given by

$$\Delta t \leq \frac{2}{\sqrt{\rho \left(\alpha_i \, \beta_i \, \mathbf{K}^{12} \, (\mathbf{M}_\epsilon^{11})^{-1} (\mathbf{K}^{12})^T \mathbf{M}_\mu^{22} \right)}}; \quad i = 1, \ldots, m \tag{4.9}$$

where ρ denotes the spectral radius of the matrix, and in practice this is estimated by performing a few power-method iterations to estimate the largest eigenvalue. Stated another way, (4.9) requires that the sampling frequency (determined by Δt) must be less than half the highest resonant frequency of the spatial discretization. The stability condition of (4.9) is valid for all values of k, the order of the polynomial basis functions. However, as k is increased, the value of $\rho \left(\mathbf{K}^{12} \, (\mathbf{M}_\epsilon^{11})^{-1} (\mathbf{K}^{12})^T \mathbf{M}_\mu^{22} \right)$ (and hence the highest resonant frequency of the spatial discretization) will grow, thus requiring a smaller time step Δt. For the special case of lossless materials and no energy entering/exiting the volume through the bounding surface the total electromagnetic energy should be constant. With this class of symplectic time integration the instantaneous energy stored in the electric and magnetic fields is

$$\mathbf{e}^T \mathbf{M}_\epsilon^{11} \, \mathbf{e} + \mathbf{b}^T \mathbf{M}_\mu^{22} \, \mathbf{b} = \tilde{\mathcal{E}} + O(\Delta t^{k+1}) sin(\omega t), \tag{4.10}$$

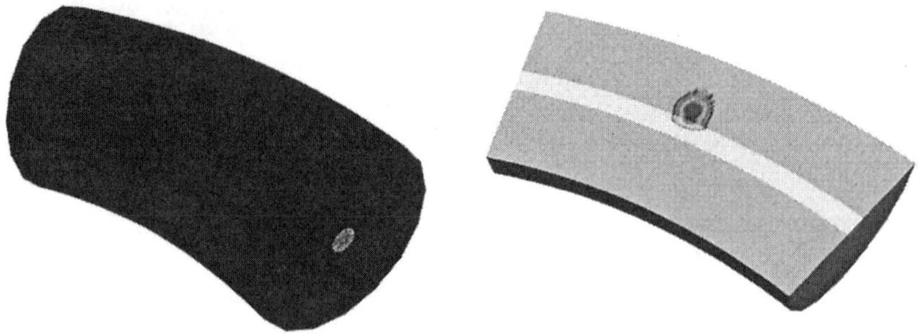

FIG. 4. *Example of bent optical fiber mesh (*left*) and snapshot of the electromagnetic energy in the 30 degree bent fiber at time* $t = 0.4ps$.

the energy oscillates about the constant value $\tilde{\mathcal{E}}$. This is in contrast to non-symplectic methods such as Runge-Kutta, in which the energy is a monotonically decaying function. In [42] it is shown that the symplectic update method can be extended to include electric and magnetic conductivity, for example artificial Perfectly Matched Layers.

4.1. Transmission in a bent optical fiber. There is great interest in analyzing the performance of bent optical fibers [43, 44]. A weak bend can be analyzed using efficient paraxial beam propagation methods, here we demonstrate an accurate full wave simulation of a fiber with an extreme bend. We visualize the propagation of an optical pulse along a $155\mu m$ section of a step index optical fiber. The core of the fiber has a radius of $5\mu m$ and an index of refraction of 1.471 while the cladding has a radius of $40\mu m$ and an index of refraction of 1.456. With these properties, the fiber is capable of propagating a $\lambda = 1.55\mu m$ optical wave. The ratio of problem domain size to wavelength is therefore $\Omega/\lambda = 100$ making this an "electrically large" problem. The problem is excited with a space and time dependent Dirichlet boundary condition applied to the input cap of the fiber representing a TE01 polarized pulse.

We perform the simulation using a straight fiber as reference and four bent fibers with different bend angles. Because the problem is electrically large, it will be subject to the cumulative errors of numerical dispersion. To mitigate this effect, we use high order polynomial basis functions of degree $k = 2$ in conjunction with a high order symplectic (energy conserving) integrator of order $m = 3$ which has been shown to excel at reducing the effects of numerical dispersion for electrically large time domain problems [40]. The computational mesh for each of the five simulations consists of 147,200 hexahedral elements with 4 transverse elements per wavelength, an example of which is shown in Figure 4. Using high order $k = 2$ basis

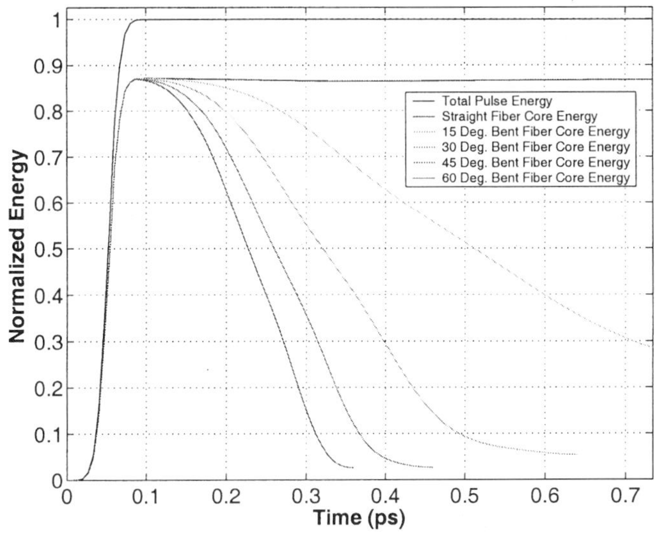

FIG. 5. *Normalized core energy as a function of time for five fiber simulations.*

functions on this mesh results in a semi-discrete linear system (4.6) consisting of 3,562,160 electric field unknowns and 3,547,072 magnetic flux density unknowns. This relatively large problem must therefore be solved in a parallel computational environment. In Figure 5 we plot the normalized energy in the fiber core, computed by (4.10), as a function of time for each of the five fiber simulations. The energy is normalized to the total energy of the optical pulse. As expected, the total pulse energy is conserved. As the fiber is bent, the energy in the core is lost due to radiation in the cladding as the pulse traverses the bend. This effect becomes more drastic as the bend angle increases and as time increases.

4.2. 3D photonic crystal waveguide. Here we simulate a 3D PBG waveguide with a complete photonic band-gap designed to operate in the RF regime. The PBG crystal is based on the "woodpile" structure as investigated by [45]. In particular, we utilize the unit cell originally proposed by [46] which consists of a series of aluminum rods (index of refraction = 3.1) arranged in an alternating, stacked configuration. The lattice constant for this crystal is $1.123cm$ and the unit cell has dimensions of $1.123cm$ by $1.123cm$ by $1.272cm$ making it suitable for operation in the radio frequency regime. We construct a 3D crystal by arranging the unit cell in a 9 by 13 by 7 layer configuration as shown in Figure 6. Our goal is to exploit the complete photonic band gap of this crystal and create a "multi-bend" wave guide where we can make the radio signal traverse two separate 90 degree bends in three dimensional space. Because of the 3D nature of the multi-bend, this type of simulation cannot be performed using standard 2D

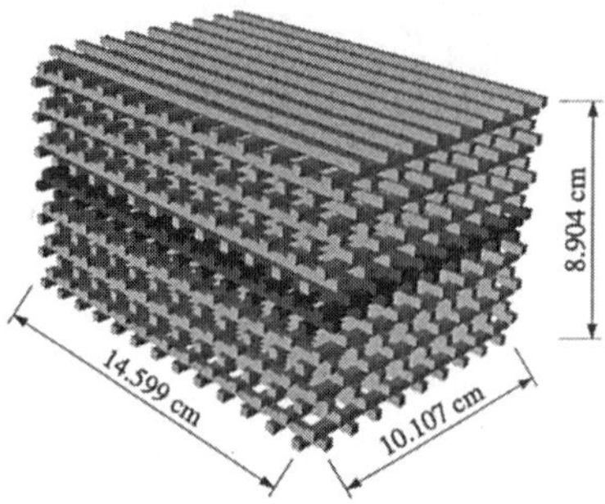

FIG. 6. *3D PBG "woodpile" structure for RF signals. The portion of the mesh representing the air has been removed for visual clarity.*

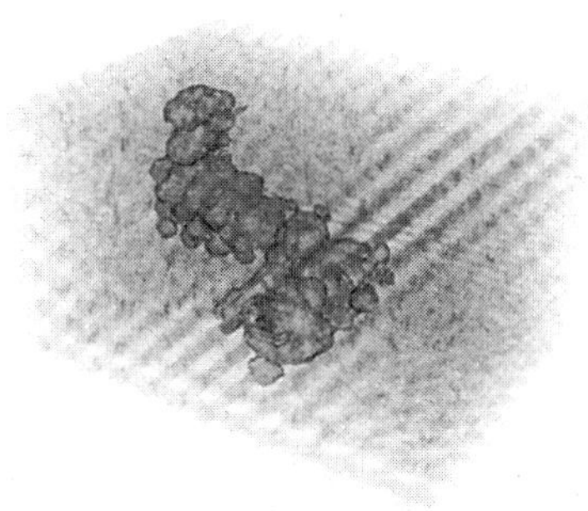

FIG. 7. *Three dimensional iso-surface plot of electric field magnitude for the 3D PBG simulation.*

codes which are extensively used in the study of PBG devices. In addition, trustworthy simulations of PBG waveguides require that phase velocities of propagating waves be computed as accurately as possible. A high order method is therefore highly desirable for an electrically large waveguide such as this.

The computational mesh of Figure 6 consists of 419, 328 hexahedral elements. We excite the problem with a time dependent Dirichlet boundary condition applied at the x-z input plane with an operating frequency of $11GHz$. The rest of the mesh is terminated with a PEC boundary condition. We use high order $k = 2$ basis functions to represent the electric and magnetic fields resulting in a linear system with approximately 10.5 million unknowns. This large linear systems requires that the problem be distributed in parallel across 150 processors. We let the simulation run for a total of 6, 500 time steps. In Figure 7 we show a three dimensional iso-surface plot of the electric field magnitude in the wave guide at the end of the simulation. Note how the wave has made two complete 90 degree bends with a negligible loss due to radiation.

4.3. Simulation of magnetic resonance imaging. When a magnetic field penetrates a conducting object, eddy currents are produced. These eddy currents modify the magnetic field resulting in a non-uniform magnetic field in the conductor. This fact is of significant relevance in the field of magnetic resonance imaging (MRI) where the characterization of field non-uniformities inside of a human head are of great research interest. In a typical MRI experiment, a very large (1-8 Tesla) and static magnetic field (called the $B0$ field) is used to align the magnetic moments of atomic nuclei inside of a human tissue sample. A secondary pulsed RF field (called the $B1$ field) is used to tip the magnetic moments when turned on. When turned off, the magnetic moments relax back to their original state, emitting radiation that is detected by an RF receiver. The $B1$ field determines image intensity and imaging algorithms assume a spatially uniform $B1$ field; non-uniform $B1$ fields lead to artificial variation in image intensity. If the actual non-uniform magnetic field were known, it might be possible to correct for this in the image processing.

Unlike the previous examples which involved propagation in a loss-less region, this application requires the introduction of a lossy term due to finite electrical conductivity. Also, the goal here is to reach a sinusoidal steady-state solution, and due to the fine mesh a very large number of time steps would be required if conditionally stable time integration method were used. We therefore employ an implicit time integration method. In particular, we use the implicit Newmark-Beta method given by

$$\left(\mathbf{M}_\epsilon^{11} + \beta\Delta t^2 \mathbf{S}_{\mu^{-1}}^{11} + \frac{dt}{2}\mathbf{M}_\sigma^{11}\right)\mathbf{e}_{n+1} = \left(2\mathbf{M}_\epsilon^{11} - (1-2\beta)\Delta t^2 \mathbf{S}_{\mu^{-1}}^{11}\right)\mathbf{e}_n$$
$$- \left(\mathbf{M}_\epsilon^{11} + \beta\Delta t^2 \mathbf{S}_{\mu^{-1}}^{11} - \frac{dt}{2}\mathbf{M}_\sigma^{11}\right)\mathbf{e}_{n-1} - dt^2 \mathbf{M}^{11}\mathbf{j}_n'. \qquad (4.11)$$

Note that this is a fully discrete version of the second order electric field wave equation (2.10) with the addition of a lossy conductive term.

In this example we use *EMSolve* to compute the eddy currents and non-uniform magnetic field inside a $10cm$ conducting, dielectric sphere immersed in a spatially uniform and time varying 200 MHz $B1$ magnetic field. The external $B1$ field is created by a pair of Helmholtz coils, driven by a 200 MHz sinusoidal current source represented by the $\mathbf{j}_n'$ term in (4.11). A human head is modeled by a conducting dielectric sphere of conductivity $\sigma = 0.5S/m$ and dielectric constant $85\epsilon_0$ as shown in Figure 8. The fully discrete (4.11) is solved for a net physical time equal to 10 periods of $B1$ field oscillation, enough time for the induced eddy currents to reach a steady state. In Figure 9 we show the induced eddy currents and the resulting non-uniform magnetic field inside of the sphere. Note the appearance of the so called "central-brightening" effect in the magnetic field magnitude, a result in agreement with the theoretical calculations of [47, 48]. Results such as these can be used to calibrate MRI images to account for the non-uniformity of the $B1$ field.

5. Electromagnetic diffusion equations. Solution of the Ampere-Faraday system of equations (4.1)-(4.2) are electromagnetic waves that propagate at the speed of light in the medium. However in many applications the time scales are such that it is not desired to resolve the wave nature of the fields. In some problems the electric field satisfies

$$\left|\star_\epsilon \frac{\partial}{\partial t} E\right| \ll |\star_\sigma E| \qquad (5.1)$$

and the $\star_\epsilon \frac{\partial}{\partial t} E$ term can be neglected without serious consequence. This is a definition of a good conductor, for example copper with $\epsilon = 8.854 \cdot 10^{-12}$

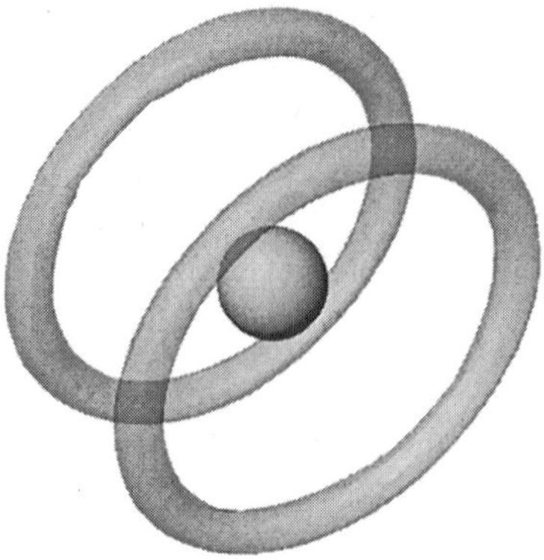

FIG. 8. *Conducting dielectric sphere representing a human head placed between two Helmholtz coils.*

FIG. 9. *Computed eddy currents (*vertical plane*) and non-uniform magnetic field (*horizontal plane*) inside of conducting dielectric sphere.*

and $\sigma = 5.76 \cdot 10^7$ is a good conductor for frequencies up to the MHz range. The electric field is not zero in a good conductor, rather the correct statement is that in a good conductor the displacement current is negligible compared to the conduction current. To continue with the copper example, a dimensional analysis indicates a characteristic field diffusion time of $\tau = \sigma\mu L^2$ where L is the characteristic dimension of the block. Using $\mu = 4\pi \cdot 10^{-7}$ and $L = 1m$, this diffusion time is several orders of magnitude longer than the time it takes the EM wave to traverse the conductor. By not resolving the EM wave, we do not have stability conditions or accuracy conditions that involve the speed of light. This is the motivation for ignoring the displacement current term.

Neglecting the displacement current, Maxwell's equations become

$$d(\star_{\mu^{-1}}B) - \star_\sigma E - J = 0 \tag{5.2}$$

$$\frac{\partial}{\partial t}B = -dE \tag{5.3}$$

$$d \star_\sigma E = 0 \tag{5.4}$$

$$dB = 0 \tag{5.5}$$

Note that (5.4) is not independent from (5.2), it is a consequence of the identity $dd\psi = 0$ (we assume the independent source satisfies $dJ = 0$ also.) Likewise, (5.5) is not completely independent of (5.3), as clearly we will have $dB = 0$ for all time if it is satisfied initially. While not necessary, we will employ potentials to solve this problem. Equation (5.5) implies that $B = dA$ for some magnetic vector potential A. Replacing B with dA in (5.3) gives $E = -\frac{\partial}{\partial t}A - d\phi$ where ϕ is some scalar electric potential. At present ϕ is somewhat arbitrary, and ϕ can be made to agree with the standard electrostatic potential by enforcing the Coulomb gauge condition on A, resulting in

$$d \star_\sigma A = 0 \tag{5.6}$$

$$d \star_\sigma d\phi = 0 \tag{5.7}$$

Combining these equations gives a diffusion equation for A,

$$\star_\sigma \frac{\partial A}{\partial t} = d \star_{\mu^{-1}} dA - \star_\sigma d\phi - J, \tag{5.8}$$

which along with the the constraints (5.6) and (5.7) and appropriate boundary conditions provides a well-posed PDE. Note that as in the discretization of the full-wave Maxwell's equations in Section 4, the divergence constraint on the 1-form field, in this case (5.6), will be implicitly satisfied for all time if it is initially satisfied. The advantage of this formulation compared to an H-based method or an E-based method is that the electrostatic potential ϕ appears explicitly in the PDE, this is useful in solving engineering problems in which the voltage across a conductor is the known boundary condition. The disadvantage of the A-ϕ approach is of course the required elliptic solve for (5.7), but with the advent of scalable multi-grid solvers this is less of an issue than it used to be.

Again using the definitions in Section 2, the semi-discrete equations are given by

$$\mathbf{M}_\sigma^{11} \frac{\partial}{\partial t}\mathbf{a} = -\mathbf{S}_\mu^{11}\mathbf{a} - \mathbf{D}_\sigma^{01}\mathbf{v} + \mathbf{j}^1 \tag{5.9}$$

$$\mathbf{S}_\sigma^{00}\mathbf{v} = \mathbf{f}^0 \tag{5.10}$$

where $\mathbf{a}$ is the vector of degrees-of-freedom of A, $\mathbf{v}$ is the vector of degrees-of-freedom of ϕ, and $\mathbf{j}$ and $\mathbf{f}$ are the discrete volume and surface source terms, respectively.

Given A, it is possible to compute the magnetic flux density B, the electric field E, and the electric current density J. Since A is a 1-form and B is a 2-form and $B = \nabla \times A$ we have

$$\mathbf{b} = \mathbf{K}^{12}\mathbf{a}. \tag{5.11}$$

This is a purely topological operation, no integration or material properties are involved. The computation of E is also trivial, using $\mathbf{e}$ to denote the degrees-of-freedom for the electric field, the semi-discrete electric field equation is

$$\mathbf{e} = -\frac{\partial}{\partial t}\mathbf{a} - \mathbf{K}^{01}\mathbf{v}. \tag{5.12}$$

If required, a 2-form electric current J can be computed from $J = \sigma E$, this is an example of a Hodge-star operation and requires the inversion of a "mass" matrix.

For the numerical time integration, we apply a generalized Crank-Nicholson method by averaging a first-order forward difference at time n with a first-order backward difference at time $(n+1)$. The averaging is performed with a weighting parameter α, where $0 \leq \alpha \leq 1$, such that

$$\alpha = \begin{cases} 0 & \text{Explicit, 1st Order Accurate Forward Euler} \\ 1/2 & \text{Implicit, 2nd Order Accurate Crank Nicholson} \\ 1 & \text{Implicit, 1st Order Accurate Backward Euler.} \end{cases}$$

The fully discrete equations are given by

$$\mathbf{S}^{00}\mathbf{v}_{n+\alpha} = \mathbf{f}^0 \tag{5.13}$$

$$\left(\mathbf{M}^{11}_{\sigma} + \alpha\Delta t\mathbf{S}^{11}_{\mu^{-1}}\right)\mathbf{a}_{n+1} = \left(\mathbf{M}^{11}_{\sigma} - (1-\alpha)\Delta t\mathbf{S}^{11}_{\mu^{-1}}\right)\mathbf{a}_n - \Delta t\mathbf{D}^{01}\mathbf{v}_{n+\alpha} + \Delta t\mathbf{j}^1 \tag{5.14}$$

$$\mathbf{b}_{n+1} = \mathbf{K}^{12}\mathbf{a}_{n+1} \tag{5.15}$$

$$\mathbf{e}_{n+\alpha} = -1/\Delta t\left(\mathbf{a}_{n+1} - \mathbf{a}_n\right) - \mathbf{K}^{01}\mathbf{v}_{n+\alpha} \tag{5.16}$$

$$\mathbf{M}^{22}_{\sigma^{-1}}\mathbf{j}_{n+\alpha} = \mathbf{H}^{12}\mathbf{e}_{n+\alpha} \tag{5.17}$$

where it is assumed that the boundary conditions and current sources can be evaluated at $t = n+\alpha$. Note that to maintain second order accuracy for all variables, the magnetic potential A and the magnetic flux B are known at whole times n, whereas the electric potential Φ and the electric field E are known at intermediate times $(n+\alpha)$. For some problems, striving for accuracy by using $\alpha = 1/2$ will lead to oscillations in the computed solution, and in such cases it is necessary to use standard Backward Euler $(\alpha = 1)$.

5.1. Electromagnetic heating and forces in a simple rail gun model. In this example we use the vector potential formulation of the electromagnetic diffusion equations to compute the $J \times B$ forces and $J \cdot E$ joule heating in a simple rail-gun model. A rail-gun is a device used to launch projectiles using only electromagnetic energy and accurate characterization of the electromagnetic forces and heating is required for trustworthy modeling. The rail gun model consists of two conducting rails and a sliding armature placed between them. Note that in this simple simulation, the motion of the armature is not taken into account, we are simply computing the transient eddy currents and magnetic fields for the case of a fixed armature. The motion of the armature will effect the fields when the velocity is comparable to the diffusion time. The rails and armature are placed in a cubic mesh representing the air. In reality, the conductivity of the air is essentially zero, however due to the nature of the FEM discretization, we cannot simply set this term to zero in the air region, so we make it significantly smaller (7 orders of magnitude) than the conductivity of the rails and armature. While not an elegant solution, this great disparity in conductivity is a good test of the proposed formulation. The problem

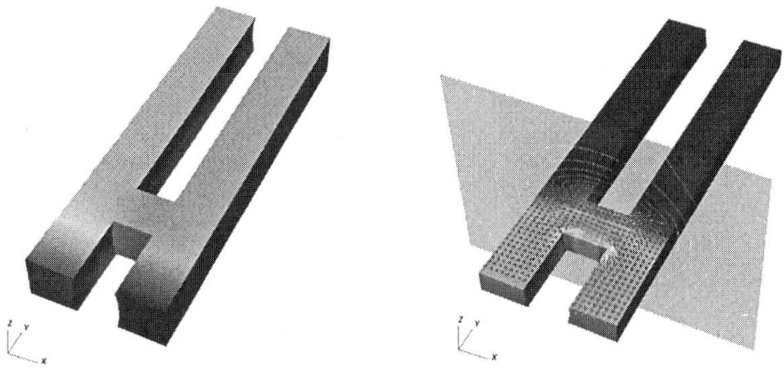

FIG. 10. *Computed scalar potential (*left*) and steady state eddy currents and magnetic fields (*right*) in a simple rail gun model.*

is driven by applying a constant voltage difference across the rail inputs. At time $t = 0$, the scalar Poisson equation is solved via a fast multi-grid method to compute the static scalar potential everywhere in space. This in turn induces transient eddy currents and magnetic fields which gradually build over time to steady state value as shown in Figure 10. The fully discrete vector potential equation (5.15) using $\alpha = 1/2$ is solved at every time step using a diagonally scaled Pre-conditioned Conjugate Gradient (PCG) method with a relative residual error tolerance of 10^{-10}. This linear solver required an average of 300 PCG iterations, in spite of the large contrast in conductivity values. PCG worked well for this relatively modest problem with 161,280 elements and $\sim 500,000$ unknowns, but for larger problems PCG is impractical; a scalable multigrid solver tuned for the $\nabla \times \nabla \times$ operator should be used [49, 50]. In Figure 11 we plot the computed vector force field and scalar Joule heat field for two different armature positions. Note that a net outward force is generated and the Joule heating is strongest at the corners of the armature contact position. Note also that as the armature is moved further along the rails, the net inductance and resistance of the rail gun circuit increases, causing the induced force and heat to decrease.

6. Conclusions. When the Galerkin finite element method is applied to electromagnetics problems using the standard nodal shape functions the results are quite disappointing, and fail to converge for even trivial problems. While adding penalty terms or Lagrange multipliers involving the divergence of the fields improves the situation, these methods cannot be considered a fundamental cure. The problem is not with the Galerkin procedure *per se*, but with the choice of finite element basis functions. Numerous researchers have proposed various $H(curl)$ -conforming and $H(div)$ -conforming based finite element basis functions that result in provably sta-

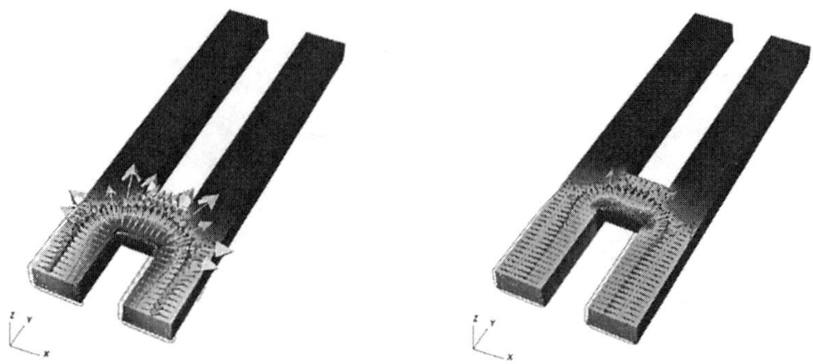

FIG. 11. *Computed $J \times B$ force field and $J \cdot E$ joule heating for two different armature positions.*

ble discrete variational formulations of electromagnetics problems. Aspects of differential forms such as exact sequences have had a significant impact on the development of these basis functions. In addition, we believe that differential forms provide a unified way for organizing and implementing a sophisticated simulation code so that it can be used to solve a wide class of problems, in fact virtually any problem that can be expressed in the language of differential forms. This has been demonstrated in the context of electromagnetics with the *EMSolve* code. However not all PDE's can be simply cast in the language of differential forms. Developing compatible discretizations for multi-physics problems, that involve not just curl and divergence equations but also advection of materials and fields, is likely to be an important area of future research.

REFERENCES

[1] G. Deschamps. Electromagnetics and differential forms. *IEEE Proceedings.*, 69(6):676–687, 1981.

[2] D. Baldomir. Differential forms and electromagnetism in 3-dimensional Euclidean space R3. *IEEE Proceedings.*, 133(3):139–143, 1986.

[3] A. Bossavit. *Computational Electromagnetism: Variational Formulation, Complementarity, Edge Elements.* Academic Press, 1998.

[4] R. Abraham, J.E. Marsden, and T. Ratiu. *Manifolds, Tensor Analysis, and Applications.* Applied Mathematical Sciences. Springer Verlag, second edition edition, 1996.

[5] H. Whitney. *Geometric Integration Theory.* Princeton University Press, 1957.

[6] J.C. Nédélec. Mixed finite elements in R3. *Numer. Math.*, 35:315–341, 1980.

[7] J.C. Nédélec. A new family of mixed finite elements in R3. *Numer. Math.*, 50:57–81, 1986.

[8] P.A. Raviart and J.M. Thomas. A Mixed Finite Element Method for 2nd Order Elliptic Problems. In I. Galligani and E. Mayera, editors, *Mathematical Aspects of the Finite Element Method*, Vol. 606 of *Lect. Notes. on Mathematics*, pp. 293–315. Springer Verlag, 1977.

[9] F. Brezzi and M. Fortin. *Mixed and Hybrid Finite Element Methods.* Springer Series in Computational Mathematics. Springer Verlag, 1991.

[10] R. Hiptmair. Canonical construction of finite elements. *Math. Comp.*, 68(228):1325–1346, 1999.

[11] I. Babuska, T. Strouboulis, C.S. Upadhyay, and S.K. Gangaraj. A posteriori estimation and adaptive control of the pollution error in the h-version of the finite element method for finite element solution of Helmholtz. *Internat. J. Numer. Methods Engrg.*, 38(24):4207–4235, 1995.

[12] I. Babuska, F. Ihlenburg, T. Strouboulis, and S.K. Gangaraj. A posteriori error estimation for finite element solution of Helmholtz. *Internat. J. Numer. Methods Engrg.*, 40(21):3883–3900, 1997.

[13] S. Warren and W. Scott. An investigation of numerical dispersion in the vector finite element method using quadrilateral elements. *IEEE Trans. Ant. Prop.*, 42(11):1502–1508, 1994.

[14] S. Warren and W. Scott. Numerical dispersion in the finite element method using triangular edge elements. *Opt. Tech. Lett.*, 9(6):315–319, 1995.

[15] D.A. White. Numerical dispersion of a vector finite element method on skewed hexahedral grids. *Commun. Numer. Meth. Engng.*, 16:47–55, 2000.

[16] M. Ainsworth. Dispersive properties of high-order Nedelec/edge element approximation of the time-harmonic Maxwell equations. *Philisophical Transactions of the Royal Society of London*, 362(1816):471–491, 2004.

[17] E. Tonti. A direct formulation of field laws: The cell method. *CMES*, 2(2):237–258, 2001.

[18] T. Weiland. Time domain electromagnetic field computation with finite difference methods. *Int. J. Numer. Modelling*, 9:295–319, 1996.

[19] M. Clemens and T. Weiland. Discrete electromagnetism with the finite integration technique. In F. Texeira, editor, *Geometric Methods for Computational Electromagnetics*, Vol. 32 of PIER, pp. 189–206. EMW Publishing, Cambridge, MA, 2001.

[20] J.M. Hyman and M.J. Shashkov. Mimetic discretizations for maxwell's equations. *J. Comput. Phys.*, 151(2):881–909, 1999.

[21] K.S. Yee. Numerical solution of initial boundary value problems involving Maxwell's equations in isotropic media. *IEEE Trans. Ant. Prop.*, 14(3):302–307, 1966.

[22] F.L. Teixeira and W.C. Chew. Lattice electromagnetic theory from a topological viewpoint. *J. Math. Phys.*, 40(1):169–187, 1999.

[23] F.L. Teixeira. *Geometric Methods in Computational Electromagnetics*, Vol. PIER 32. EMW Publishing, Cambridge, Mass., 2001.

[24] R. Hiptmair. Discrete Hodge operators: An algebraic perspective. *J. Electromagnteic Waves Appl.*, 15(3):343–344, 2001.

[25] D.N. Arnold and F. Brezzi. Mixed andnonconforming finite elememnt methods: implementation, postprocessing, and error estimates. *Math. Modelling and Numer. Anal.*, 19:7–32, 1985.

[26] D.N. Arnold. Mixed finite element methods for elliptic problems. *Comput. Methods Appl. Mech. Engrg.*, 82(1-3):281–300, 1990.

[27] P.G. Ciarlet. *The Finite Element Method for Elliptic Problems.* North-Holland, 1978.

[28] R. Graglia, D. Wilton, and A. Peterson. Higher order interpolatory vector bases for computational electromagnetics. *IEEE Trans. Ant. Prop.*, 45(3):329–342, 1997.

[29] R. Graglia, P. Wilton, A. Peterson, and I.-L. Gheorma. Higher order interpolatory vector bases on prism elements. *IEEE Trans. Ant. Prop.*, 46(3):442–450, 1998.

[30] P. Castillo, R. Rieben, and D. White. FEMSTER: An object oriented class library of discrete differential forms. *ACM Trans. Math. Soft.* in press.

[31] P. Castillo, R. Rieben, and D. White. FEMSTER: An object oriented class library of discrete differential forms. In *Proceedings of the 2003 IEEE International*

Antennas and Propagation Symposium, Vol. 2, pp. 181–184, Columbus, Ohio, June 2003.

[32] P. Castillo, J. Koning, R. Rieben, M. Stowell, and D. White. Discrete differential forms: A novel methodology for robust computational electromagnetics. Technical Report UCRL-ID-151522, Lawrence Livermore National Laboratory, Center for Applied Scientific Computing, January 2003.

[33] R. Rieben, D. White, and G. Rodrigue. Improved conditioning of finite element matrices using new high order interpolatory bases. *IEEE Trans. Ant. Prop.*, 52(10):2675–2683, October 2004.

[34] A. Fisher, R. Rieben, G. Rodrigue, and D. White. A generalized mass lumping technique for vector finite element solutions of the time dependent maxwell equations. *IEEE Trans. Ant. Prop.*, December 2004. accepted for publication.

[35] A. Bossavit. Solving Maxwell equations in a closed cavity, and the question of spurious modes. *IEEE Trans. Mag.*, 26(2):702–705, 1990.

[36] Z.J. Cendes. Vector finite elements for electromagnetic field computation. *IEEE Trans. Mag.*, 27(5):3958–3966, 1991.

[37] D.A. White and J.M. Koning. Computing solenoidal eigenmodes of the vector Helmholtz equation: a novel approach. *IEEE Trans. Mag.*, 38(5):3420–3425, 2002.

[38] R. Lehoucq, D. Sorenson, and C. Yang. *ARPACK User's Guide: Solution of Large Scale Eigenvalue Problems with Implicitly Restarted Arnoldi Methods.* SIAM, 1998.

[39] P. Balleyguier. Coupling slots measurements against simulation for Trispal accelerating cavities. In *Linac '98*, p. 130. Chicago, 1998.

[40] R. Rieben, G. Rodrigue, and D. White. A high order mixed vector finite element method for solving the time dependent Maxwell equations on unstructured grids. *J. Comput. Phys.*, 204(2):490–519, April 2005.

[41] R. Rieben, D. White, and G. Rodrigue. High order symplectic integration methods for finite element solutions to time dependent maxwell equations. *IEEE Trans. Ant. Prop.*, 52(8):2190–2195, August 2004.

[42] R. Rieben. *A Novel High Order Time Domain Vector Finite Element Method for the Simulation of Electromagnetic Devices.* PhD thesis, University of California at Davis, Livermore, California, 2004.

[43] D. Marcuse. Curvature loss formula for optical fibers. *J. Opt. Soc. Am.*, 66(3):216–220, 1976.

[44] J. Koning, R. Rieben, and G. Rodrigue. Vector finite element modeling of the full-wave Maxwell equations to evaluate power loss in bent optical fibers. *IEEE/OSA J. Lightwave Tech.*, May 2005. article in press.

[45] H.S. Sözüer and J.P. Dowling. Photonic band calculations for woodpile structures. *J. Mod. Opt.*, 41(2):231–239, 1994.

[46] E Özbay, A. Abeyta, G. Tuttle, M. Trinigides, R. Biswas, C.T. Chan, C.M Soukoulis, and K.M. Ho. Measurement of a three-dimensional photonic band gap in a crystal structure made of dielectric rods. *Phys. Rev. B*, 50(3):1945–1948, 1994.

[47] D.I. Hoult. Sensitivity and power deposition in high-field imaging experiment. *J. Magn. Reson. Imag.*, 12:46–67, 200.

[48] J.S. Tropp. Image brightening in samples of high dielectric constant. *J. Magnetic Resonance*, 167(1):12–24, 2004.

[49] P. Bochev, C. Garasi, J. Hu, A. Robinson, and R. Tuminaro. An improved algebraic multigrid method for solving maxwell's equations. *SIAM J. Sci. Comp.*, 25(2):623–642, 2003.

[50] S. Reitzinger and J. Schoberl. An algebraic multigrid method for finite element discretization with edge elements. *Numer. Linear Algebra Appl.*, 9:223–238, 2002.

LIST OF WORKSHOP PARTICIPANTS

- Ivar Aavatsmark, Center for Integrated Petroleum Research, University of Bergen
- Scot Adams, Institute for Mathematics and its Applications, University of Minnesota
- Peter Arbenz, Institut für Wissenschaftliches Rechnen, ETH Zentrum
- Douglas N. Arnold, Institute for Mathematics and its Applications, University of Minnesota
- Donald G. Aronson, Institute for Mathematics and its Applications, University of Minnesota
- Gerard Awanou, Institute for Mathematics and its Applications, University of Minnesota
- Randolph E. Bank, Department of Mathematics, University of California - San Diego
- Timothy J. Barth, NASA Ames Research Center
- Martin Berggren, Department of Scientific Computing, Uppsala University
- Pavel B. Bochev, Computational Mathematics and Algorithms, Sandia National Laboratories
- Daniele Boffi, Dipartimento di Matematica, Università di Pavia
- Alain Bossavit, Laboratoire de Génie Electrique de Paris
- Olga Brezhneva, Mathematics & Statistics Department, Miami University
- Franco Brezzi, I. A. N. del C. N. R.
- Zhiqiang Cai, Department of Mathematics, Purdue University
- Jose E. Castillo, Department of Mathematics and Statistics, San Diego State University
- Panagiotis Chatzipantelidis, Department of Mathematics, Texas A&M University
- Snorre H. Christiansen, Department of Mathematics, Centre of Mathematics for Applications
- Mark Christon, Sandia National Laboratories
- Bernardo Cockburn, School of Mathematics, University of Minnesota
- Bob Crone, Mechanical R&D, Seagate Technology
- Leszek F. Demkowicz, Department of Aerospace Engineering and Engineering Mechanics,University of Texas - Austin
- Yalchin Efendiev, Department of Mathematics, Texas A&M University
- Richard S. Falk, Department of Mathematics, Rutgers University

- Juergen Geiser, Department of Mathematics, Texas A&M University
- Anne Gundel, Institu fuer Mathematik, Humboldt University Berlin
- Hazem Hamdan, Department of Mathematics, University of Minnesota
- Bo He, Department of Electrical Engineering, Ohio State University
- Jan S. Hesthaven, Division of Applied Mathematics, Brown University
- Ulrich Hetmaniuk, Sandia National Laboratories
- Ralf Hiptmair, Seminar of Applied Mathematics, ETH Zentrum
- Anil N. Hirani, Department of Control and Dynamical Systems, California Institute of Technology
- Michael J. Holst, Department of Mathematics University of California - San Diego
- Ronald Hoppe, Department of Mathematics, University of Houston
- Paul Houston, Department of Mathematics and Computer Science, University of Leicester
- Thomas J.R. Hughes, Institute for Computational Engineering and Sciences, University of Texas - Austin
- E. Mckay Hyde, School of Mathematics, University of Minnesota
- Lili Ju, Department of Mathematics, University of South Carolina
- Ruben Juanes, Department of Petroleum Engineering, Stanford University
- Hye-Ryoung Kim, Seoul National University (BK21)
- Joseph M. Koning, Defense Sciences Engineering Division, Lawrence Livermore National Laboratory
- Robert P. Kotiuga, Department of Elecrical and Computer Engineering, Boston University
- Thomas G. Kurtz, Department of Mathematics and Statistics, University of Wisconsin - Madison
- Yuri Kuznetsov, Department of Mathematics, University of Houston
- Raytcho Lazarov, Department of Mathematics, Texas A&M University
- Richard B. Lehoucq, Computational Mathematics and Algorithms, Sandia National Laboratories,
- Melvin Leok, Department of Control and Dynamical Systems, California Institute of Technology
- Konstantin Lipnikov, Los Alamos National Laboratory
- Mitchell Luskin, School of Mathematics, University of Minnesota
- Antoinette Maniatty, Department of Mechanical, Aerospace, & Nuclear Engineering, Rensselaer Polytechnic Institute

- Elizabeth L. Mansfield, Institute of Mathmematics and Statistics, University of Kent at Canterbury
- Donatella Marini, Dipartimento di Matematica, Università di Pavia
- Ilya D. Mishev ExxonMobil
- Julie C. Mitchell, Departments of Mathematics and Biochemistry, University of Wisconsin - Madison
- Jim Morel, Los Alamos National Laboratory
- J. David Moulton, Los Alamos National Laboratory
- Jean-Claude Nedelec, Centre de Mathematiques Appliquees Ecole Polytechnique
- Roy A. Nicolaides, Department of Mathematical Sciences, Carnegie Mellon University
- Philippe P. Pébay, Reacting Flow Research, Sandia National Laboratories
- Blair Perot, Department of Mechanical and Industrial Engineering, University of Massachusetts
- Ilaria Perugia, Dipartimento di Matematica, Universita di Pavia
- Edward Ratner, Department of Optical CD Metrology, KLA-Tencor
- Fernando Reitich, School of Mathematics, University of Minnesota
- Jean-Francois Remacle, Scientific Computation Research Center, Rensselaer Polytechnic Institute
- Robert N. Rieben, Defense Sciences Engineering Division, Lawrence Livermore National Laboratory
- Beatrice M. Riviere, Department of Mathematics, University of Pittsburgh
- Allen Robinson, Computational Physics R&D, Sandia National Laboratories
- Thomas F. Russell, Division of Mathematical Sciences, National Science Foundation
- Fadil Santosa, Minnesota Center for Industrial Mathematics, University of Minnesota
- Rolf Schuhmann, TEMF Laboratory, Darmstadt University of Technology
- Mikhail Shashkov, Theoretical Division, Los Alamos National Laboratory
- Shagi-Di Shih, Department of Mathematics, University of Wyoming
- Rajen Kumar Sinha, Department of Mathematics, Texas A&M University
- Stanly Steinberg, Department of Mathematics and Statistics, University of New Mexico
- Srdjan Stojanovic, Department of Mathematical Sciences, University of Cincinnati

- Eitan Tadmor, CSCAMM, University of Maryland
- Fernando Lisboa Teixeira, Department of Electrical Engineering, Ohio State University
- Jean-Marie Thomas, Laboratory of Applied Mathematics, University of Pau and the Countries of Adour
- Kathryn A. Trapp, Department of Mathematics, University of Richmond
- Jukka Tuomela, Department of Mathematics, University of Joensuu
- Panayot Vassilevski, Center for Applied Scientific Computing, Lawrence Livermore National Laboratories
- Jing Wang, University of Minnesota
- Tim Warburton, Department of Mathematics and Statistics, University of New Mexico
- Mary Fanett Wheeler, Institute for Computational Engineering and Sciences, University of Texas - Austin
- Daniel A. White, Defense Sciences Engineering Division, Lawrence Livermore National Laboratories
- Ragnar Winther, Centre of Mathematics for Applications, University of Oslo
- Jinchao Xu, Department of Mathematics, Pennsylvania State University
- Ivan Yotov, Department of Mathematics, University of Pittsburgh
- Jun Zhao, University of Minnesota

1999–2000	Reactive Flows and Transport Phenomena
2000–2001	Mathematics in Multimedia
2001–2002	Mathematics in the Geosciences
2002–2003	Optimization
2003–2004	Probability and Statistics in Complex Systems: Genomics, Networks, and Financial Engineering
2004–2005	Mathematics of Materials and Macromolecules: Multiple Scales, Disorder, and Singularities
2005-2006	Imaging
2006-2007	Applications of Algebraic Geometry
2007-2008	Mathematics of Molecular and Cellular Biology

IMA SUMMER PROGRAMS

1987	Robotics
1988	Signal Processing
1989	Robust Statistics and Diagnostics
1990	Radar and Sonar (June 18–29) New Directions in Time Series Analysis (July 2–27)
1991	Semiconductors
1992	Environmental Studies: Mathematical, Computational, and Statistical Analysis
1993	Modeling, Mesh Generation, and Adaptive Numerical Methods for Partial Differential Equations
1994	Molecular Biology
1995	Large Scale Optimizations with Applications to Inverse Problems, Optimal Control and Design, and Molecular and Structural Optimization
1996	Emerging Applications of Number Theory (July 15–26) Theory of Random Sets (August 22–24)
1997	Statistics in the Health Sciences
1998	Coding and Cryptography (July 6–18) Mathematical Modeling in Industry (July 22–31)
1999	Codes, Systems, and Graphical Models (August 2–13, 1999)
2000	Mathematical Modeling in Industry: A Workshop for Graduate Students (July 19–28)
2001	Geometric Methods in Inverse Problems and PDE Control (July 16–27)
2002	Special Functions in the Digital Age (July 22–August 2)
2003	Probability and Partial Differential Equations in Modern Applied Mathematics (July 21–August 1)
2004	n-Categories: Foundations and Applications (June 7–18)
2005	Wireless Communications (June 22–July 1)
2006	Symmetries and Overdetermined Systems of Partial Differential Equations (July 17 - August 4)

IMA "HOT TOPICS" WORKSHOPS

- Challenges and Opportunities in Genomics: Production, Storage, Mining and Use, April 24–27, 1999
- Decision Making Under Uncertainty: Energy and Environmental Models, July 20–24, 1999
- Analysis and Modeling of Optical Devices, September 9–10, 1999
- Decision Making under Uncertainty: Assessment of the Reliability of Mathematical Models, September 16–17, 1999
- Scaling Phenomena in Communication Networks, October 22–24, 1999
- Text Mining, April 17–18, 2000
- Mathematical Challenges in Global Positioning Systems (GPS), August 16–18, 2000
- Modeling and Analysis of Noise in Integrated Circuits and Systems, August 29–30, 2000
- Mathematics of the Internet: E-Auction and Markets, December 3–5, 2000
- Analysis and Modeling of Industrial Jetting Processes, January 10–13, 2001
- Special Workshop: Mathematical Opportunities in Large-Scale Network Dynamics, August 6–7, 2001
- Wireless Networks, August 8–10 2001
- Numerical Relativity, June 24–29, 2002
- Operational Modeling and Biodefense: Problems, Techniques, and Opportunities, September 28, 2002
- Data-driven Control and Optimization, December 4–6, 2002
- Agent Based Modeling and Simulation, November 3–6, 2003
- Enhancing the Search of Mathematics, April 26-27, 2004
- Compatible Spatial Discretizations for Partial Differential Equations, May 11-15, 2004
- Adaptive Sensing and Multimode Data Inversion, June 27–30, 2004
- Mixed Integer Programming, July 25–29, 2005
- New Directions in Probability Theory, August 5–6, 2005

SPRINGER LECTURE NOTES FROM THE IMA:

The Mathematics and Physics of Disordered Media
Editors: Barry Hughes and Barry Ninham
(Lecture Notes in Math., Volume 1035, 1983)

Orienting Polymers
Editor: J.L. Ericksen
(Lecture Notes in Math., Volume 1063, 1984)

New Perspectives in Thermodynamics
Editor: James Serrin
(Springer-Verlag, 1986)

Models of Economic Dynamics
Editor: Hugo Sonnenschein
(Lecture Notes in Econ., Volume 264, 1986)

The IMA Volumes in Mathematics and its Applications

Current Volumes:

1 **Homogenization and Effective Moduli of Materials and Media**
J. Ericksen, D. Kinderlehrer, R. Kohn, and J.-L. Lions (eds.)
2 **Oscillation Theory, Computation, and Methods of Compensated Compactness** C. Dafermos, J. Ericksen, D. Kinderlehrer, and M. Slemrod (eds.)
3 **Metastability and Incompletely Posed Problems**
S. Antman, J. Ericksen, D. Kinderlehrer, and I. Muller (eds.)
4 **Dynamical Problems in Continuum Physics**
J. Bona, C. Dafermos, J. Ericksen, and D. Kinderlehrer (eds.)
5 **Theory and Applications of Liquid Crystals**
J. Ericksen and D. Kinderlehrer (eds.)
6 **Amorphous Polymers and Non-Newtonian Fluids**
C. Dafermos, J. Ericksen, and D. Kinderlehrer (eds.)
7 **Random Media** G. Papanicolaou (ed.)
8 **Percolation Theory and Ergodic Theory of Infinite Particle Systems** H. Kesten (ed.)
9 **Hydrodynamic Behavior and Interacting Particle Systems**
G. Papanicolaou (ed.)
10 **Stochastic Differential Systems, Stochastic Control Theory, and Applications** W. Fleming and P.-L. Lions (eds.)
11 **Numerical Simulation in Oil Recovery** M.F. Wheeler (ed.)
12 **Computational Fluid Dynamics and Reacting Gas Flows**
B. Engquist, M. Luskin, and A. Majda (eds.)
13 **Numerical Algorithms for Parallel Computer Architectures**
M.H. Schultz (ed.)
14 **Mathematical Aspects of Scientific Software** J.R. Rice (ed.)
15 **Mathematical Frontiers in Computational Chemical Physics**
D. Truhlar (ed.)
16 **Mathematics in Industrial Problems** A. Friedman
17 **Applications of Combinatorics and Graph Theory to the Biological and Social Sciences** F. Roberts (ed.)
18 ***q*-Series and Partitions** D. Stanton (ed.)
19 **Invariant Theory and Tableaux** D. Stanton (ed.)
20 **Coding Theory and Design Theory Part I: Coding Theory**
D. Ray-Chaudhuri (ed.)
21 **Coding Theory and Design Theory Part II: Design Theory**
D. Ray-Chaudhuri (ed.)
22 **Signal Processing Part I: Signal Processing Theory**
L. Auslander, F.A. Grünbaum, J.W. Helton, T. Kailath, P. Khargonekar, and S. Mitter (eds.)
23 **Signal Processing Part II: Control Theory and Applications of Signal Processing** L. Auslander, F.A. Grünbaum, J.W. Helton, T. Kailath, P. Khargonekar, and S. Mitter (eds.)

24 **Mathematics in Industrial Problems, Part 2** A. Friedman
25 **Solitons in Physics, Mathematics, and Nonlinear Optics**
P.J. Olver and D.H. Sattinger (eds.)
26 **Two Phase Flows and Waves**
D.D. Joseph and D.G. Schaeffer (eds.)
27 **Nonlinear Evolution Equations that Change Type**
B.L. Keyfitz and M. Shearer (eds.)
28 **Computer Aided Proofs in Analysis**
K. Meyer and D. Schmidt (eds.)
29 **Multidimensional Hyperbolic Problems and Computations**
A. Majda and J. Glimm (eds.)
30 **Microlocal Analysis and Nonlinear Waves**
M. Beals, R. Melrose, and J. Rauch (eds.)
31 **Mathematics in Industrial Problems, Part 3** A. Friedman
32 **Radar and Sonar, Part I**
R. Blahut, W. Miller, Jr., and C. Wilcox
33 **Directions in Robust Statistics and Diagnostics: Part I**
W.A. Stahel and S. Weisberg (eds.)
34 **Directions in Robust Statistics and Diagnostics: Part II**
W.A. Stahel and S. Weisberg (eds.)
35 **Dynamical Issues in Combustion Theory**
P. Fife, A. Liñán, and F.A. Williams (eds.)
36 **Computing and Graphics in Statistics**
A. Buja and P. Tukey (eds.)
37 **Patterns and Dynamics in Reactive Media**
H. Swinney, G. Aris, and D. Aronson (eds.)
38 **Mathematics in Industrial Problems, Part 4** A. Friedman
39 **Radar and Sonar, Part II**
F.A. Grünbaum, M. Bernfeld, and R.E. Blahut (eds.)
40 **Nonlinear Phenomena in Atmospheric and Oceanic Sciences**
G.F. Carnevale and R.T. Pierrehumbert (eds.)
41 **Chaotic Processes in the Geological Sciences** D.A. Yuen (ed.)
42 **Partial Differential Equations with Minimal Smoothness and Applications** B. Dahlberg, E. Fabes, R. Fefferman, D. Jerison, C. Kenig, and J. Pipher (eds.)
43 **On the Evolution of Phase Boundaries**
M.E. Gurtin and G.B. McFadden
44 **Twist Mappings and Their Applications**
R. McGehee and K.R. Meyer (eds.)
45 **New Directions in Time Series Analysis, Part I**
D. Brillinger, P. Caines, J. Geweke, E. Parzen, M. Rosenblatt, and M.S. Taqqu (eds.)
46 **New Directions in Time Series Analysis, Part II**
D. Brillinger, P. Caines, J. Geweke, E. Parzen, M. Rosenblatt, and M.S. Taqqu (eds.)
47 **Degenerate Diffusions**
W.-M. Ni, L.A. Peletier, and J.-L. Vazquez (eds.)
48 **Linear Algebra, Markov Chains, and Queueing Models**
C.D. Meyer and R.J. Plemmons (eds.)

49 **Mathematics in Industrial Problems, Part 5** A. Friedman
50 **Combinatorial and Graph-Theoretic Problems in Linear Algebra**
R.A. Brualdi, S. Friedland, and V. Klee (eds.)
51 **Statistical Thermodynamics and Differential Geometry of Microstructured Materials**
H.T. Davis and J.C.C. Nitsche (eds.)
52 **Shock Induced Transitions and Phase Structures in General Media** J.E. Dunn, R. Fosdick, and M. Slemrod (eds.)
53 **Variational and Free Boundary Problems**
A. Friedman and J. Spruck (eds.)
54 **Microstructure and Phase Transitions**
D. Kinderlehrer, R. James, M. Luskin, and J.L. Ericksen (eds.)
55 **Turbulence in Fluid Flows: A Dynamical Systems Approach**
G.R. Sell, C. Foias, and R. Temam (eds.)
56 **Graph Theory and Sparse Matrix Computation**
A. George, J.R. Gilbert, and J.W.H. Liu (eds.)
57 **Mathematics in Industrial Problems, Part 6** A. Friedman
58 **Semiconductors, Part I**
W.M. Coughran, Jr., J. Cole, P. Lloyd, and J. White (eds.)
59 **Semiconductors, Part II**
W.M. Coughran, Jr., J. Cole, P. Lloyd, and J. White (eds.)
60 **Recent Advances in Iterative Methods**
G. Golub, A. Greenbaum, and M. Luskin (eds.)
61 **Free Boundaries in Viscous Flows**
R.A. Brown and S.H. Davis (eds.)
62 **Linear Algebra for Control Theory**
P. Van Dooren and B. Wyman (eds.)
63 **Hamiltonian Dynamical Systems: History, Theory, and Applications**
H.S. Dumas, K.R. Meyer, and D.S. Schmidt (eds.)
64 **Systems and Control Theory for Power Systems**
J.H. Chow, P.V. Kokotovic, R.J. Thomas (eds.)
65 **Mathematical Finance**
M.H.A. Davis, D. Duffie, W.H. Fleming, and S.E. Shreve (eds.)
66 **Robust Control Theory** B.A. Francis and P.P. Khargonekar (eds.)
67 **Mathematics in Industrial Problems, Part 7** A. Friedman
68 **Flow Control** M.D. Gunzburger (ed.)
69 **Linear Algebra for Signal Processing**
A. Bojanczyk and G. Cybenko (eds.)
70 **Control and Optimal Design of Distributed Parameter Systems**
J.E. Lagnese, D.L. Russell, and L.W. White (eds.)
71 **Stochastic Networks** F.P. Kelly and R.J. Williams (eds.)
72 **Discrete Probability and Algorithms**
D. Aldous, P. Diaconis, J. Spencer, and J.M. Steele (eds.)
73 **Discrete Event Systems, Manufacturing Systems, and Communication Networks**
P.R. Kumar and P.P. Varaiya (eds.)
74 **Adaptive Control, Filtering, and Signal Processing**
K.J. Åström, G.C. Goodwin, and P.R. Kumar (eds.)

75 **Modeling, Mesh Generation, and Adaptive Numerical Methods for Partial Differential Equations** I. Babuska, J.E. Flaherty, W.D. Henshaw, J.E. Hopcroft, J.E. Oliger, and T. Tezduyar (eds.)

76 **Random Discrete Structures** D. Aldous and R. Pemantle (eds.)

77 **Nonlinear Stochastic PDEs: Hydrodynamic Limit and Burgers' Turbulence** T. Funaki and W.A. Woyczynski (eds.)

78 **Nonsmooth Analysis and Geometric Methods in Deterministic Optimal Control** B.S. Mordukhovich and H.J. Sussmann (eds.)

79 **Environmental Studies: Mathematical, Computational, and Statistical Analysis** M.F. Wheeler (ed.)

80 **Image Models (and their Speech Model Cousins)** S.E. Levinson and L. Shepp (eds.)

81 **Genetic Mapping and DNA Sequencing** T. Speed and M.S. Waterman (eds.)

82 **Mathematical Approaches to Biomolecular Structure and Dynamics** J.P. Mesirov, K. Schulten, and D. Sumners (eds.)

83 **Mathematics in Industrial Problems, Part 8** A. Friedman

84 **Classical and Modern Branching Processes** K.B. Athreya and P. Jagers (eds.)

85 **Stochastic Models in Geosystems** S.A. Molchanov and W.A. Woyczynski (eds.)

86 **Computational Wave Propagation** B. Engquist and G.A. Kriegsmann (eds.)

87 **Progress in Population Genetics and Human Evolution** P. Donnelly and S. Tavaré (eds.)

88 **Mathematics in Industrial Problems, Part 9** A. Friedman

89 **Multiparticle Quantum Scattering With Applications to Nuclear, Atomic and Molecular Physics** D.G. Truhlar and B. Simon (eds.)

90 **Inverse Problems in Wave Propagation** G. Chavent, G. Papanicolau, P. Sacks, and W.W. Symes (eds.)

91 **Singularities and Oscillations** J. Rauch and M. Taylor (eds.)

92 **Large-Scale Optimization with Applications, Part I: Optimization in Inverse Problems and Design** L.T. Biegler, T.F. Coleman, A.R. Conn, and F. Santosa (eds.)

93 **Large-Scale Optimization with Applications, Part II: Optimal Design and Control** L.T. Biegler, T.F. Coleman, A.R. Conn, and F. Santosa (eds.)

94 **Large-Scale Optimization with Applications, Part III: Molecular Structure and Optimization** L.T. Biegler, T.F. Coleman, A.R. Conn, and F. Santosa (eds.)

95 **Quasiclassical Methods** J. Rauch and B. Simon (eds.)

96 **Wave Propagation in Complex Media** G. Papanicolaou (ed.)

97 **Random Sets: Theory and Applications** J. Goutsias, R.P.S. Mahler, and H.T. Nguyen (eds.)

98 **Particulate Flows: Processing and Rheology** D.A. Drew, D.D. Joseph, and S.L. Passman (eds.)

99 **Mathematics of Multiscale Materials** K.M. Golden, G.R. Grimmett, R.D. James, G.W. Milton, and P.N. Sen (eds.)
100 **Mathematics in Industrial Problems, Part 10** A. Friedman
101 **Nonlinear Optical Materials** J.V. Moloney (ed.)
102 **Numerical Methods for Polymeric Systems** S.G. Whittington (ed.)
103 **Topology and Geometry in Polymer Science** S.G. Whittington, D. Sumners, and T. Lodge (eds.)
104 **Essays on Mathematical Robotics** J. Baillieul, S.S. Sastry, and H.J. Sussmann (eds.)
105 **Algorithms For Parallel Processing** M.T. Heath, A. Ranade, and R.S. Schreiber (eds.)
106 **Parallel Processing of Discrete Problems** P.M. Pardalos (ed.)
107 **The Mathematics of Information Coding, Extraction, and Distribution** G. Cybenko, D.P. O'Leary, and J. Rissanen (eds.)
108 **Rational Drug Design** D.G. Truhlar, W. Howe, A.J. Hopfinger, J. Blaney, and R.A. Dammkoehler (eds.)
109 **Emerging Applications of Number Theory** D.A. Hejhal, J. Friedman, M.C. Gutzwiller, and A.M. Odlyzko (eds.)
110 **Computational Radiology and Imaging: Therapy and Diagnostics** C. Börgers and F. Natterer (eds.)
111 **Evolutionary Algorithms** L.D. Davis, K. De Jong, M.D. Vose, and L.D. Whitley (eds.)
112 **Statistics in Genetics** M.E. Halloran and S. Geisser (eds.)
113 **Grid Generation and Adaptive Algorithms** M.W. Bern, J.E. Flaherty, and M. Luskin (eds.)
114 **Diagnosis and Prediction** S. Geisser (ed.)
115 **Pattern Formation in Continuous and Coupled Systems: A Survey Volume** M. Golubitsky, D. Luss, and S.H. Strogatz (eds.)
116 **Statistical Models in Epidemiology, the Environment, and Clinical Trials** M.E. Halloran and D. Berry (eds.)
117 **Structured Adaptive Mesh Refinement (SAMR) Grid Methods** S.B. Baden, N.P. Chrisochoides, D.B. Gannon, and M.L. Norman (eds.)
118 **Dynamics of Algorithms** R. de la Llave, L.R. Petzold, and J. Lorenz (eds.)
119 **Numerical Methods for Bifurcation Problems and Large-Scale Dynamical Systems** E. Doedel and L.S. Tuckerman (eds.)
120 **Parallel Solution of Partial Differential Equations** P. Bjørstad and M. Luskin (eds.)
121 **Mathematical Models for Biological Pattern Formation** P.K. Maini and H.G. Othmer (eds.)
122 **Multiple-Time-Scale Dynamical Systems** C.K.R.T. Jones and A. Khibnik (eds.)
123 **Codes, Systems, and Graphical Models** B. Marcus and J. Rosenthal (eds.)
124 **Computational Modeling in Biological Fluid Dynamics** L.J. Fauci and S. Gueron (eds.)
125 **Mathematical Approaches for Emerging and Reemerging Infectious Diseases: An Introduction** C. Castillo-Chavez with S. Blower, P. van den Driessche, D. Kirschner, and A.A. Yakubu (eds.)

126 **Mathematical Approaches for Emerging and Reemerging Infectious Diseases: Models, Methods, and Theory** C. Castillo-Chavez with S. Blower, P. van den Driessche, D. Kirschner, and A.A. Yakubu (eds.)

127 **Mathematics of the Internet: E-Auction and Markets**
B. Dietrich and R.V. Vohra (eds.)

128 **Decision Making Under Uncertainty: Energy and Power**
C. Greengard and A. Ruszczynski (eds.)

129 **Membrane Transport and Renal Physiology**
H. Layton and A.M. Weinstein (eds.)

130 **Atmospheric Modeling**
D.P. Chock and G.R. Carmichael (eds.)

131 **Resource Recovery, Confinement, and Remediation of Environmental Hazards**
J. Chadam, A. Cunningham, R.E. Ewing, P. Ortoleva, and M.F. Wheeler (eds.)

132 **Fractals in Multimedia**
M.F. Barnsley, D. Saupe, and E.R. Vrscay (eds.)

133 **Mathematical Methods in Computer Vision**
P.J. Olver and A. Tannenbaum (eds.)

134 **Mathematical Systems Theory in Biology, Communications, Computation, and Finance**
J. Rosenthal and D.S. Gilliam (eds.)

135 **Transport in Transition Regimes**
N. Ben Abdallah, A. Arnold, P. Degond, I. Gamba, R. Glassey, C.D. Lawrence, and C. Ringhofer (eds.)

136 **Dispersive Transport Equations and Multiscale Methods**
N. Ben Abdallah, A. Arnold, P. Degond, I. Gamba, R. Glassey, C.D. Lawrence, and C. Ringhofer (eds.)

137 **Geometric Methods in Inverse Problems and PDE Control**
C.B. Cooke, I. Lasiecka, G. Uhlmann, and M.S. Vogelius (eds.)

138 **Mathematical Foundations of Speech and Language Processing**
M. Johnson, S. Khudanpur, M. Ostendorf, and R. Rosenfeld (eds.)

139 **Time Series Analysis and Applications to Geophysical Systems**
D.R. Brillinger, E.A. Robinson, and F.P. Schoenberg (eds.)

140 **Probability and Partial Differential Equations in Modern Applied Mathematics** E.C. Waymire and J. Duan (eds.)

141 **Modeling of Soft Matter**
Maria-Carme T. Calderer and Eugene M. Terentjev (eds.)

142 **Compatible Spatial Discretizations**
Douglas N. Arnold, Pavel B. Bochev, Richard B. Lehoucq, Roy A. Nicolaides, and Mikhail Shashkov (eds.)